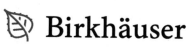

Alexey N. Karapetyants • Vladislav V. Kravchenko

Methods of Mathematical Physics

Classical and Modern

 Birkhäuser

Alexey N. Karapetyants
Institute of Mathematics, Mechanics and
Computer Sciences and Regional
Mathematical Center
Southern Federal University
Rostov-on-Don, Russia

Vladislav V. Kravchenko
Department of Mathematics
Cinvestav-IPN, Campus Queretaro
Santiago de Querétaro, Querétaro, Mexico

ISBN 978-3-031-17847-4 ISBN 978-3-031-17845-0 (eBook)
https://doi.org/10.1007/978-3-031-17845-0

Mathematics Subject Classification: 35-01, 35-02, 35Qxx, 31-01, 35J05

This book is published under the imprint Birkhäuser, www.birkhauser-science.com by the registered
company Springer Nature Switzerland AG
The registered company address is: Gewerbestrasse 11, 6330 Cham, Switzerland

Preface

The book presents a detailed exposition of classical topics of mathematical physics together with recently developed methods. The exposition is kept accessible for undergraduate students of mathematical, physical and engineering sciences.

Together with a standard introduction into main classes of partial differential equations, thorough study of boundary and initial value problems, detailed exposition of integral equations theory, some recent advances in direct and inverse Sturm–Liouville problems, the method of non-orthogonal series, and the Bergman kernel approach for solving boundary value and spectral problems are presented.

The book thus can serve as a textbook for a standard course on equations of mathematical physics as well as a resource for a number of special, more advanced courses and seminars.

The authors, to one extent or another, taught parts of this course in the Mathematics Department of the Southern Federal University (formerly—Rostov State University) and in the Mathematical Department of the Cinvestav (Mexico). Alexey Karapetyants also had the opportunity to partially teach chapters from this course at the University of Padua and at the University of Helsinki.

The authors express their gratitude to Conacyt (Mexico) for partial support of this work via the project 284470 and to the Regional Mathematical Center of the Southern Federal University, Rostov-on-Don, with the support of the Ministry of Science and Higher Education of Russia.

Alexey Karapetyants acknowledges support of the project of the Russian Ministry of Science and Education for the organization and development of scientific and educational mathematical centers, agreement No. 075-02-2022-893.

Rostov-on-Don, Russia
Santiago de Querétaro, Querétaro, Mexico

Alexey N. Karapetyants
Vladislav V. Kravchenko

Contents

Chapter 1
Introduction

Methods of mathematical physics is a vast area of modern applied mathematics which provides mathematical tools to researchers in physics, engineering, chemistry, biology, finances, and of course in mathematics. Both authors taught mathematical physics courses in different universities and research centers engaging a number of well-known existing textbooks and complementing them with selected topics suitable for undergraduate or graduate level courses. From the experience gained the idea of writing this book gradually matured as an attempt to combine standard material with more recent and less known developments which represent powerful and far-reaching techniques and at the same time are simple enough to be taught at the undergraduate level. Thus, the first goal of this book consists in narrowing the existing gap between what is presently taught and what is being developed and used by active researchers.

Classical textbooks in methods and equations of mathematical physics present research developments mostly of XVIII–XIX centuries together with some more recent material from the first half of XX century. Euler (1707–1783), Laplace (1749–1827), Fourier (1768–1830), and Poisson (1781–1840) knew a good deal of presently taught methods of mathematical physics. Hilbert (1862–1943) perhaps knew them all. However, modern mathematical physics is far more developed area than it was a century ago.

The update of mathematical physics textbooks though moved toward inclusion of more sophisticated and abstract mathematical concepts, such as manifolds or distributions, often reducing possible audience to mathematicians and leaving physics and engineering students with classical "battle-tested" textbooks written by the greats in the first half of the past century. Thus, the second goal of this book is to present recent developments in methods of mathematical physics without "raising the bar" and thus keeping the exposition accessible for non-mathematics students and combining the classical style and material with new topics and results.

The ultimate goal of the present book is to share authors' admiration for the work of masters of the past and our own research interests and some results with interested

A. N. Karapetyants, V. V. Kravchenko, *Methods of Mathematical Physics*,
https://doi.org/10.1007/978-3-031-17845-0_1

lectors and readers. We hope that this book will be regarded as an invitation to the journey through selected places in the wonderland of mathematical physics.

The book includes the material of a standard one-year university course of equations of mathematical physics plus several chapters which are usually not encountered in such courses but present powerful and important tools and concepts useful for practical solution of real-world problems. Among such new, non-standard topics we mention the method of non-orthogonal series for numerical solution of boundary value problems (which includes the widely used by specialists method of discrete sources or fundamental solutions) and the method of Bergman reproducing kernels which provides a convenient way for analytical solution of boundary value and spectral problems. Combined with the method of non-orthogonal series the Bergman kernel approach is a very convenient and attractive way for numerical study of elliptic problems.

For solving Sturm-Liouville equations and problems we present the method of spectral parameter power series (SPPS), a simple and universal technique allowing one to solve an initial or a boundary value problem for a general Sturm-Liouville equation, to obtain several eigenvalues and eigenfunctions of a general Sturm-Liouville problem. For Sturm-Liouville equations we introduce the notion of a transmutation (transformation) operator which serves for reducing difficult Sturm-Liouville equations to the elementary ones. Furthermore, we show how with the aid of very recent developments in this field the transmutation operators can be efficiently constructed which leads to interesting series representations for solutions of Sturm-Liouville equations and to powerful and simple methods for solving direct and inverse Sturm-Liouville problems on finite and infinite intervals.

Among the distinctive features of the "standard" part of the book we mention a detailed exposition of the theory of integral equations as well as a profound presentation of the theory of the Helmholtz equation.

The book is aimed at undergraduate students and can serve as a textbook for a regular course in equations of mathematical physics as well as for some special more advanced courses on selected topics.

Besides this introduction, the book contains fifteen chapters.

Chapter 2 presents a quite standard exposition of the classification of partial differential equations and their reduction to respective canonical forms.

In Chap. 3 we derive a considerable number of equations and models of mathematical physics. Usually this part is difficult for mathematics students. We hope that we succeeded in making it less stressful for them.

In Chap. 4 we introduce typical problem statements for linear partial differential equations of mathematical physics, explain in detail the notion of well posedness of a problem.

Chapter 5 is dedicated to Cauchy's problems for hyperbolic (wave) equations. We derive main representations for their solutions, the famous d'Alembert, Poisson, and Kirchhoff formulas.

Chapter 6 presents the Fourier method of separation of variables for solving boundary and initial value problems. Besides a number of examples we present main properties of eigenvalues and eigenfunctions of Sturm-Liouville problems, study Bessel functions, and explain the general scheme of the Fourier method.

Chapter 7 contains recent developments in theory and practice of direct and inverse Sturm-Liouville problems introducing powerful techniques for their study and solution both on finite and infinite intervals with examples and applications. Among the applications we mention the inverse scattering transform method for solving nonlinear evolutionary partial differential equations.

In Chap. 8 we study boundary value problems for the heat equation in finite and infinite domains.

In Chap. 9 properties of harmonic functions are studied.

In Chap. 10 boundary value problems for the Laplace equation are stated and studied with the emphasis on Green's function concept, its properties and meaning as well as some techniques for its construction.

In Chap. 11 we go into details of potential theory, prove theorems on the behavior of volume, single layer and double layer potentials.

Detailed exposition of the theory of integral equations is continued in Chap. 12, where the Fredholm and the Hilbert–Schmidt theories are presented. This chapter is mainly based on the book [71].

In Chap. 13 we apply the Fredholm theory of integral equations to solution of boundary value problems for the Laplace equation.

Chapter 14 is dedicated to the Helmholtz equation and corresponding boundary value problems.

In Chap. 15 we explain the method of non-orthogonal series on the example of the Helmholtz equation and introduce several complete systems of its solutions.

Finally, in Chap. 16 we present the Bergman kernel approach to solution of boundary value problems for elliptic equations with variable coefficients. This chapter is based mainly on the book [12] with important additions explaining how this approach can be used for solving spectral problems and how for constructing Bergman reproducing kernels for equations with variable coefficients the transmutation operators and results from Chaps. 7 and 15 can be used.

When writing this book the authors used many excellent books on mathematical physics, especially [8, 12, 43, 78, 82, 83].

The authors hope that this book will be a useful resource for lectors and students who look for a standard rigorous course of equations of mathematical physics containing modern topics within the reach of undergraduate students.

Chapter 2
Classification of Partial Differential Equations

Mathematical physics mainly relies on the basic theory of partial differential equations. The present chapter contains some first definitions and results of this theory.

A partial differential equation (or briefly a PDE) is a mathematical equation that involves two or more independent variables, an unknown function dependent on those variables, and partial derivatives of the unknown function with respect to the independent variables.

As it is usual, studying an equation we will always keep in mind a certain domain (a non-empty open set) $\Omega \subseteq \mathbb{R}^n$ of variables $(x_1, x_2, \ldots, x_n) \in \mathbb{R}^n$, which may also coincide with the entire space $\mathbb{R}^n$. All objects are *a priori* understood as being properly defined in such a domain. If there is no concern of misunderstanding, we may omit mentioning the domain in definitions or statements.

Definition 1 Let $u = u(x_1, x_2, \ldots, x_n)$ be a function of n independent variables. An equation of the form

$$F\left(x_1, \ldots, x_n, u, \frac{\partial u}{\partial x_1}, \ldots, \frac{\partial u}{\partial x_n}, \frac{\partial^2 u}{\partial x_1^2}, \frac{\partial^2 u}{\partial x_1 \partial x_2}, \ldots, \frac{\partial^k u}{\partial x_n^k}\right) = 0$$

is called **partial differential equation**. The **order** of a partial differential equation is the order of the highest derivative involved.

Definition 2 A function $u = u(x_1, x_2, \ldots, x_n)$ is called a **solution** (or a **particular solution**) to a partial differential equation in some domain $\Omega \subseteq \mathbb{R}^n$ if in this domain it possesses continuous partial derivatives up to the order k of the partial differential equation, and being substituted into the equation turns it into an identity.

Such a solution we will call **classical** or **regular** solution. Later we will briefly discuss solutions which are understood in a weaker sense. However, almost in the

© The Author(s), under exclusive license to Springer Nature Switzerland AG 2022
A. N. Karapetyants, V. V. Kravchenko, *Methods of Mathematical Physics*,
https://doi.org/10.1007/978-3-031-17845-0_2

whole book we will treat the solution in a classical way, without involving theory of distributions.

Example 3 The following equality is an equation of third order

$$\sin^2\left(\frac{\partial^3 u}{\partial x_1^3}\right) - x_1\frac{\partial^2 u}{\partial x_1 \partial x_2} + \frac{\partial u}{\partial x_1} = 0,$$

and the function $u \equiv 1$ is its particular solution.

In this book mainly linear equations of second order will be considered.

Definition 4 A second order partial differential equation is called **linear with respect to the higher order derivative** if it has the form

$$\sum_{1\leqslant i\leqslant j\leqslant n} A_{ij}\frac{\partial^2 u}{\partial x_i \partial x_j} + F\left(x_1, \ldots, x_n, u, \frac{\partial u}{\partial x_1}, \ldots, \frac{\partial u}{\partial x_n}\right) = 0, \qquad (2.1)$$

where the coefficients $A_{ij} = A_{ij}(x_1, \ldots, x_n)$ are functions of independent variables only.

Definition 5 Equation (2.1) is called **quasilinear** if the coefficients A_{ij} also depend on u and on lower order derivatives $\frac{\partial u}{\partial x_1}, \ldots, \frac{\partial u}{\partial x_n}$.

Definition 6 Equation (2.1) is called **linear** if it has the form

$$\sum_{1\leqslant i\leqslant j\leqslant n} A_{ij}\frac{\partial^2 u}{\partial x_i \partial x_j} + \sum_{i=1}^{n} B_i\frac{\partial u}{\partial x_i} + Cu = f,$$

where A_{ij}, B_i, C, and f depend on the independent variables $x_1, \ldots, x_n$ only. If $f \equiv 0$ the equation is called **homogeneous**, otherwise **non-homogeneous**.

2.1   Classification of Linear with Respect to the Highest Derivative Second Order Partial Differential Equations of Two Independent Variables

For $n = 2$ let us write (2.1) in the form

$$A\frac{\partial^2 u}{\partial x^2} + 2B\frac{\partial^2 u}{\partial x \partial y} + C\frac{\partial^2 u}{\partial y^2} + F\left(x, y, u, \frac{\partial u}{\partial x}, \frac{\partial u}{\partial y}\right) = 0, \qquad (2.2)$$

where A, B, and C are functions of the independent variables x and y.

Definition 7 The function $\Delta(x, y) := B^2 - AC$ is called the **discriminant** of Eq. (2.2).

The classification of (2.2) is performed in dependence on the sign of the discriminant $\Delta(x, y)$. It is crucial to observe that the sign of the discriminant does not change upon an arbitrary non-degenerate change of the independent variables. Let us prove it.

Consider a change of variables $\xi = \xi(x, y)$, $\eta = \eta(x, y)$ with a non-degenerate Jacobian

$$\frac{D(\xi, \eta)}{D(x, y)} = \begin{vmatrix} \xi_x & \xi_y \\ \eta_x & \eta_y \end{vmatrix} \neq 0,$$

and suppose that $\xi(x, y)$ and $\eta(x, y)$ are twice continuously differentiable functions. Then using the chain rule for functions of two variables we obtain

$$\frac{\partial u}{\partial x} = \frac{\partial u}{\partial \xi} \cdot \frac{\partial \xi}{\partial x} + \frac{\partial u}{\partial \eta} \cdot \frac{\partial \eta}{\partial x}, \quad \frac{\partial u}{\partial y} = \frac{\partial u}{\partial \xi} \cdot \frac{\partial \xi}{\partial y} + \frac{\partial u}{\partial \eta} \cdot \frac{\partial \eta}{\partial y}$$

and

$$\frac{\partial^2 u}{\partial x^2} = \frac{\partial^2 u}{\partial \xi^2} \cdot \left(\frac{\partial \xi}{\partial x}\right)^2 + 2\frac{\partial^2 u}{\partial \xi \partial \eta} \cdot \frac{\partial \xi}{\partial x} \cdot \frac{\partial \eta}{\partial x} + \frac{\partial^2 u}{\partial \eta^2} \cdot \left(\frac{\partial \eta}{\partial x}\right)^2$$
$$+ \frac{\partial u}{\partial \xi} \cdot \frac{\partial^2 \xi}{\partial x^2} + \frac{\partial u}{\partial \eta} \cdot \frac{\partial^2 \eta}{\partial x^2},$$

$$\frac{\partial^2 u}{\partial x \partial y} = \frac{\partial^2 u}{\partial \xi^2} \cdot \frac{\partial \xi}{\partial x} \cdot \frac{\partial \xi}{\partial y} + \frac{\partial^2 u}{\partial \xi \partial \eta} \cdot \left(\frac{\partial \xi}{\partial x} \cdot \frac{\partial \eta}{\partial y} + \frac{\partial \eta}{\partial x} \cdot \frac{\partial \xi}{\partial y}\right)$$
$$+ \frac{\partial^2 u}{\partial \eta^2} \cdot \frac{\partial \eta}{\partial x} \cdot \frac{\partial \eta}{\partial y} + \frac{\partial u}{\partial \xi} \cdot \frac{\partial^2 \xi}{\partial x \partial y} + \frac{\partial u}{\partial \eta} \cdot \frac{\partial^2 \eta}{\partial x \partial y},$$

$$\frac{\partial^2 u}{\partial y^2} = \frac{\partial^2 u}{\partial \xi^2} \cdot \left(\frac{\partial \xi}{\partial y}\right)^2 + 2\frac{\partial^2 u}{\partial \xi \partial \eta} \cdot \frac{\partial \xi}{\partial y} \cdot \frac{\partial \eta}{\partial y} + \frac{\partial^2 u}{\partial \eta^2} \cdot \left(\frac{\partial \eta}{\partial y}\right)^2$$
$$+ \frac{\partial u}{\partial \xi} \cdot \frac{\partial^2 \xi}{\partial y^2} + \frac{\partial u}{\partial \eta} \cdot \frac{\partial^2 \eta}{\partial y^2}.$$

Substitution of these expressions into (2.2) leads to the equation

$$\widetilde{A}\frac{\partial^2 u}{\partial \xi^2} + 2\widetilde{B}\frac{\partial^2 u}{\partial \xi \partial \eta} + \widetilde{C}\frac{\partial^2 u}{\partial \eta^2} + \widetilde{F}\left(\xi, \eta, u, \frac{\partial u}{\partial \xi}, \frac{\partial u}{\partial \eta}\right) = 0, \qquad (2.3)$$

where

$$\tilde{A}\,(\xi, \eta) = A\xi_x^2 + 2B\xi_x\xi_y + C\xi_y^2,$$

$$\tilde{B}\,(\xi, \eta) = A\xi_x\eta_x + B\left(\xi_x\eta_y + \eta_x\xi_y\right) + C\xi_y\eta_y,$$

$$\tilde{C}\,(\xi, \eta) = A\eta_x^2 + 2B\eta_x\eta_y + C\eta_y^2.$$

Here and often below, for brevity we will denote the partial derivative by the corresponding subindex, so that $\xi_x = \frac{\partial\xi}{\partial x}$, etc. Let us calculate the discriminant $\tilde{\Delta}\,(\xi, \eta)$ of Eq. (2.3). We have

$$\tilde{\Delta} = \tilde{B}^2 - \tilde{A} \cdot \tilde{C}$$

$$= \left(A\xi_x\eta_x + B\left(\xi_x\eta_y + \eta_x\xi_y\right) + C\xi_y\eta_y\right)^2 - \left(A\xi_x^2 + 2B\xi_x\xi_y + C\xi_y^2\right)$$

$$\cdot \left(A\eta_x^2 + 2B\eta_x\eta_y + C\eta_y^2\right)$$

$$= A^2 \cdot 0 + C^2 \cdot 0 + B^2 \cdot \left(\left(\xi_x\eta_y + \eta_x\xi_y\right)^2 - 4\xi_x\xi_y\eta_x\eta_y\right)$$

$$+ AC\left(2\xi_x\xi_y\eta_x\eta_y - \xi_x^2\eta_y^2 - \xi_y^2\eta_x^2\right)$$

$$+ AB \cdot 0 + BC \cdot 0$$

$$= \left(B^2 - AC\right)\left(\xi_x\eta_y - \eta_x\xi_y\right)^2 = \left(B^2 - AC\right)\begin{vmatrix}\xi_x & \xi_y \\ \eta_x & \eta_y\end{vmatrix}^2.$$

Thus, the following equality is valid

$$\tilde{\Delta}\,(\xi, \eta) = \Delta\,(x, y) \cdot \left|\frac{D\,(\xi, \eta)}{D\,(x, y)}\right|^2.$$

Conclusion 8 *The sign of the discriminant is invariant under a non-degenerate change of independent variables. Namely, if $\Delta\,(x_0, y_0) > 0\,(< 0$ or $= 0)$ then $\tilde{\Delta}\,(\xi_0, \eta_0) > 0\,(< 0$ or $= 0)$ where $\xi_0 = \xi\,(x_0, y_0)$, $\eta_0 = \eta\,(x_0, y_0)$.*

Definition 9 Equation (2.2) is said to be of **hyperbolic elliptic**, or **parabolic** type at a point (x_0, y_0) if at that point $\Delta\,(x_0, y_0) > 0$, < 0 or $= 0$, respectively.

Definition 10 Equation (2.2) is said to be of **hyperbolic elliptic**, or **parabolic** type in a domain V of the plane (x, y) if at all points $(x_0, y_0) \in V$ the relation $\Delta\,(x_0, y_0) > 0$, < 0 or $= 0$ holds, respectively.

Note that the above classification is based on the fact that a general second order partial differential equation with two independent variables has the form analogous to the equation for a conic section. Indeed, replacing formally partial derivatives with respect to x and y by the same variables x and y, converts a constant coefficient second order partial differential equation into a quadratic form.

As we shall see below, hyperbolic equations, as an example, describe the propagation process (waves, gas propagation, etc.). Also, an example of parabolic equation is the heat conduction equation. Instead, examples of elliptic equation are, for instance, the equations of elasticity (without inertial terms) and the equation for electromagnetic field.

Assume that $A(x, y)$, $B(x, y)$ and $C(x, y)$ are defined on the whole plane. The discriminant $\Delta(x, y)$ may have different signs at different points of the plane. All points at which the inequality holds $\Delta(x, y) > 0$ are called **points of hyperbolicity** and a totality of them the **domain of hyperbolicity**. Similarly, all points at which the inequality holds $\Delta(x, y) < 0$ are called **points of ellipticity** and a totality of them the **domain of ellipticity**. The equality $\Delta(x, y) = 0$ defines the **points of parabolicity**. In general, the equality $\Delta(x, y) = 0$ may define a **domain of parabolicity** as in the preceding cases or a curve on a plane serving as a boundary separating the domains of hyperbolicity from those of ellipticity. In such case the line $\Delta(x, y) = 0$ is called the **parabolicity line** or the **line of parabolic degeneration**.

Example 11 Consider the equation

$$u_{xx} + 5u_{xy} + 6u_{yy} + 7u_x + 8u_y + u = 1.$$

Then $\Delta(x, y) = \left(\frac{5}{2}\right)^2 - 6 = \frac{1}{4} > 0$ and hence the equation is of hyperbolic type on the whole plane.

Example 12 Consider the Tricomi equation

$$yu_{xx} + u_{yy} = 0.$$

Since $A(x, y) = y$, $B(x, y) = 0$ and $C(x, y) = 1$ we have that $\Delta(x, y) = -y$. Thus, the axis of x ($y = 0$) is a line of parabolic degeneration of the Tricomi equation, the lower half-plane is a domain of hyperbolicity while the upper half-plane is a domain of ellipticity and hence, for example, on any domain contained in the upper half-plane the Tricomi equation is of elliptic type.

Definition 13 If a domain V contains points at which the discriminant $\Delta(x, y)$ has different signs, Eq. (2.2) is said to be of mixed type in the domain V.

Thus, for example, the Tricomi equation is of mixed type in any domain containing points of the axis of x.

2.2   Canonical Forms of the Equations

Definition 14

1. The equation

$$u_{xx} + u_{yy} + F\left(x, y, u, u_x, u_y\right) = 0$$

   is called the **canonical form of the elliptic type equation**.
2. The equation

$$u_{xx} - u_{yy} + F\left(x, y, u, u_x, u_y\right) = 0 \qquad (2.4)$$

   is called the **first canonical form of the hyperbolic type equation** while the
   equation

$$u_{xy} + F\left(x, y, u, u_x, u_y\right) = 0 \qquad (2.5)$$

   is called the **second canonical form of the hyperbolic type equation**.
3. The equation

$$u_{yy} + F\left(x, y, u, u_x, u_y\right) = 0$$

is called the **canonical form of the parabolic type equation**.

Remark 15 The canonical forms (2.4) and (2.5) are equivalent, they reduce to each
other by the non-degenerate change of variables $\xi = x + y, \eta = x - y$.

We will show that any equation of the form (2.2) can be reduced to one of the
canonical forms by a non-degenerate change of variables.

2.2.1   Reduction of Hyperbolic Equations

We consider Eq. (2.2) with $\Delta = B^2 - AC > 0$ in some domain. After the
substitution $\xi = \xi(x, y), \eta = \eta(x, y)$ Eq. (2.2) reduces to the form

$$\widetilde{A}\frac{\partial^2 u}{\partial \xi^2} + 2\widetilde{B}\frac{\partial^2 u}{\partial \xi \partial \eta} + \widetilde{C}\frac{\partial^2 u}{\partial \eta^2} + \widetilde{F} = 0$$

(see Sect. 2.1). Let us show that there exists such a non-degenerate change of
variables ξ, η that $\widetilde{A} \equiv 0$ and $\widetilde{C} \equiv 0$, that is,

$$A\xi_x^2 + 2B\xi_x\xi_y + C\xi_y^2 = 0$$

and

$$An_x^2 + 2Bn_xn_y + Cn_y^2 = 0.$$

Hence the functions ξ and η must be linearly independent solutions of the same equation

$$A\varphi_x^2 + 2B\varphi_x\varphi_y + C\varphi_y^2 = 0. \tag{2.6}$$

Division by φ_y^2 leads to a quadratic equation for the function $\frac{\varphi_x}{\varphi_y}$,

$$A\left(\frac{\varphi_x}{\varphi_y}\right)^2 + 2B\frac{\varphi_x}{\varphi_y} + C = 0$$

from where we obtain that

$$\frac{\varphi_x}{\varphi_y} = \frac{-B \pm \sqrt{B^2 - AC}}{A}$$

supposing that $A(x, y) \neq 0$. Thus, Eq. (2.6) is equivalent to two linear first order equations

$$A\varphi_x + \left(B \pm \sqrt{B^2 - AC}\right)\varphi_y = 0. \tag{2.7}$$

Equation (2.7) can be reduced to an equivalent ordinary differential equation with the aid of the following lemma.

Lemma 16 *The function $\varphi(x_1, x_2, \ldots, x_n)$ is a solution of the equation*

$$X_1\frac{\partial\varphi}{\partial x_1} + X_2\frac{\partial\varphi}{\partial x_2} + \ldots + X_n\frac{\partial\varphi}{\partial x_n} = 0$$

if and only if it is an integral of the following symmetric system

$$\frac{dx_1}{X_1} = \frac{dx_2}{X_2} = \ldots = \frac{dx_n}{X_n}.$$

According to this lemma the following symmetric system corresponds to equations (2.7):

$$\frac{dx}{A} = \frac{dy}{B \pm \sqrt{B^2 - AC}},$$

or in other words to the first order ordinary equations

$$y' = \frac{B \pm \sqrt{B^2 - AC}}{A}. \tag{2.8}$$

Equations (2.8) are a consequence of the equation

$$Ay'^2 - 2By' + C = 0$$

or

$$Ady^2 - 2Bdxdy + Cdx^2 = 0$$

which is called the equation for characteristics of Eq. (2.2). The solutions of (2.8) are called the two families of the characteristics of Eq. (2.2) which can be written in the form of two independent integrals

$$\varphi_1(x, y) = c,$$

$$\varphi_2(x, y) = c.$$

Due to Lemma 16 the functions $\xi = \varphi_1(x, y)$ and $\eta = \varphi_2(x, y)$ satisfy (2.7). Thus, setting $\xi = \varphi_1(x, y)$ and $\eta = \varphi_2(x, y)$ we obtain $\widetilde{A} \equiv 0$ and $\widetilde{C} \equiv 0$, and the transformation $\xi = \varphi_1(x, y), \eta = \varphi_2(x, y)$ is non-degenerate.

Thus, we have

$$2\widetilde{B} \cdot u_{\xi\eta} + \widetilde{F} = 0.$$

Obviously, $\widetilde{B}$ does not vanish because otherwise $\widetilde{\Delta}(x_0, y_0) = \widetilde{B}^2(x_0, y_0) - \widetilde{A}(x_0, y_0) \cdot \widetilde{C}(x_0, y_0) = 0$ that contradicts the condition $\Delta(x, y) > 0$. Hence dividing over $2\widetilde{B}$ we obtain the required canonical form

$$u_{\xi\eta} + \frac{\widetilde{F}}{2\widetilde{B}} = 0.$$

2.2.2 Reduction of Parabolic Equations

We consider Eq. (2.2) with $\Delta = B^2 - AC = 0$ in some domain. Let us look for such transformation $\xi(x, y)$ and $\eta(x, y)$ that $\widetilde{A} \equiv 0$, that is,

$$A\xi_x^2 + 2B\xi_x\xi_y + C\xi_y^2 = 0.$$

Similarly to the previous case (under the supposition $A(x, y) \neq 0$) we again arrive at the equation for characteristics. Since $\Delta = B^2 - AC = 0$, the equation for characteristics (2.8) gives us only one independent integral $\varphi(x, y) = c$. Set

$\xi = \varphi(x, y)$ and choose an arbitrary $\eta = \eta(x, y)$ with the only condition that the transformation ξ, η be non-degenerate. Then $\widetilde{A} \equiv 0$. Obviously, $\widetilde{B} \equiv 0$ as well. Indeed,

$$\widetilde{B}^2 - \widetilde{A} \cdot \widetilde{C} = \left(B^2 - AC\right) \left|\frac{D(\xi, \eta)}{D(x, y)}\right|^2 \equiv 0$$

and $\widetilde{A} \equiv 0$, thus $\widetilde{B} \equiv 0$.

We arrive at the equation

$$\widetilde{C} u_{\eta\eta} + \widetilde{F} = 0.$$

Let us show that $\widetilde{C} \neq 0$. Assume the opposite, that $\widetilde{C} = 0$. Then $A\eta_x^2 + 2B\eta_x\eta_y + C\eta_y^2 = 0$ and η is an integral of the same equation for characteristics which in the case under consideration admits only one linearly independent integral. Hence ξ and η should be linearly dependent that contradicts the non-degeneracy of the transformation. Thus, $\widetilde{C} \neq 0$ and hence we obtain the canonical form

$$u_{\eta\eta} + \frac{\widetilde{F}}{\widetilde{C}} = 0.$$

2.2.3 Reduction of Elliptic Equations

We consider Eq. (2.2) with $\Delta = B^2 - AC < 0$ in some domain. For simplicity we assume additionally that A, B, and C are real analytic functions (the study of a more general case is more difficult). We recall that real analytic functions are usually defined as being locally prescribed by a convergent power series. For instance, a real function which possesses derivatives of all orders and agrees with its Taylor series in a neighborhood of every point, is a real analytic function. Then the equation for characteristics admits a complex valued integral $\varphi(x, y) = \varphi_1(x, y) + i\varphi_2(x, y) = c$ where $\varphi_1(x, y)$ and $\varphi_2(x, y)$ are real valued functions. Substituting $\varphi = \varphi_1 + i\varphi_2 = \xi + i\eta$ into Eq. (2.6) we find that

$$A\left(\xi_x + i\eta_x\right)^2 + 2B\left(\xi_x + i\eta_x\right)\left(\xi_y + i\eta_y\right) + C\left(\xi_y + i\eta_y\right)^2 = 0.$$

Separating the real and the imaginary parts we obtain

$$\widetilde{A} = A\xi_x^2 + 2B\xi_x\xi_y + C\xi_y^2 = A\eta_x^2 + 2B\eta_x\eta_y + C\eta_y^2 = \widetilde{C}$$

and

$$\widetilde{B} = A\xi_x\eta_x + B\left(\xi_x\eta_y + \xi_y\eta_x\right) + C\xi_y\eta_y = 0.$$

Thus, $\tilde{A} = \tilde{C}$ and $\tilde{B} = 0$. Note that $\tilde{A}, \tilde{B}, \tilde{C}$ are real valued since A, B, C, ξ, and η are real valued.

The equation takes the form

$$\tilde{A}\left(u_{\xi\xi} + u_{\eta\eta}\right) + \tilde{F} = 0,$$

where $\tilde{A} \neq 0$ because otherwise $\tilde{C} = 0$ and hence $\tilde{\Delta} = \tilde{B}^2 - \tilde{A} \cdot \tilde{C} = 0$ that contradicts the condition $\Delta < 0$.

Example 17 Consider the equation

$$x^2 \frac{\partial^2 u}{\partial x^2} - y^2 \frac{\partial^2 u}{\partial y^2} = 0.$$

Obviously, $\Delta(x, y) = x^2 y^2$ and thus the equation is of hyperbolic type everywhere except for the coordinate lines. Let the domain V be the first quadrant $x > 0$, $y > 0$. The equation for characteristics takes the form

$$x^2 dy^2 - y^2 dx^2 = 0,$$

from where $xy = c$ and $\frac{y}{x} = c$. Setting $\xi = xy$ and $\eta = \frac{y}{x}$ we obtain the equation

$$\frac{\partial^2 u}{\partial \xi \partial \eta} - \frac{1}{2\xi} \frac{\partial u}{\partial \eta} = 0.$$

Denote $u_\eta =: W$. Then

$$\frac{1}{W} \frac{\partial W}{\partial \xi} = \frac{1}{2\xi}$$

and hence $W = c(\eta) \sqrt{\xi} = u_\eta$. Integration of this equation leads to the solution $u = \sqrt{\xi} c_1(\eta) + c_2(\xi)$ where c_1 and c_2 are arbitrary twice continuously differentiable functions of their arguments, and thus

$$u(x, y) = \sqrt{xy} c_1\left(\frac{y}{x}\right) + c_2(xy).$$

2.3   Classification of Second Order Equations of n Independent Variables

Consider a second order equation, linear with respect to the higher order derivatives (2.1), that is the equation:

$$\sum_{1 \leqslant i \leqslant j \leqslant n} A_{ij} \frac{\partial^2 u}{\partial x_i \partial x_j} + F\left(x_1, \ldots, x_n, u, \frac{\partial u}{\partial x_1}, \ldots, \frac{\partial u}{\partial x_n}\right) = 0.$$

Let us introduce new independent variables

$$\xi_k = \xi_k(x_1, x_2, \ldots, x_n), \quad k = 1, 2, \ldots, n, \tag{2.9}$$

where ξ_k are twice continuously differentiable functions and

$$\frac{D(\xi_1, \xi_2, \ldots, \xi_n)}{D(x_1, x_2, \ldots, x_n)} \neq 0$$

everywhere in a domain of consideration. Then

$$\frac{\partial u}{\partial x_i} = \sum_{k=1}^{n} \frac{\partial u}{\partial \xi_k} \cdot \frac{\partial \xi_k}{\partial x_i},$$

$$\frac{\partial^2 u}{\partial x_i \partial x_j} = \sum_{k=1}^{n}\left(\sum_{l=1}^{n} \frac{\partial^2 u}{\partial \xi_k \partial \xi_l} \cdot \frac{\partial \xi_l}{\partial x_j}\right) \frac{\partial \xi_k}{\partial x_i} + \sum_{k=1}^{n} \frac{\partial u}{\partial \xi_k} \cdot \frac{\partial^2 \xi_k}{\partial x_i \partial x_j}.$$

Substituting these expressions into the initial equation we obtain

$$\sum_{1 \leqslant k \leqslant l \leqslant n} \tilde{A}_{kl} \frac{\partial^2 u}{\partial \xi_k \partial \xi_l} + \tilde{F}\left(\xi_1, \ldots, \xi_n, u, \frac{\partial u}{\partial \xi_1}, \ldots, \frac{\partial u}{\partial \xi_n}\right) = 0,$$

where

$$\tilde{A}_{kl} = \sum_{1 \leqslant i \leqslant j \leqslant n} A_{ij} \frac{\partial \xi_k}{\partial x_i} \cdot \frac{\partial \xi_l}{\partial x_j}.$$

This formula for the transformation of the coefficients is analogous to the formula for the transformation of the coefficients of the quadratic form. Indeed, consider the quadratic form

$$\sum_{1 \leqslant i \leqslant j \leqslant n} A_{ij} p_i p_j \tag{2.10}$$

with the coefficients $A_{ij} = A_{ij}(x_1^0, \ldots, x_n^0)$ calculated at a fixed point $x_1^0, \ldots, x_n^0$.
    A linear transformation

$$q_i = \sum_{k=1}^{n} \alpha_{ki} q_k \tag{2.11}$$

transforms the quadratic form (2.10) into

$$\sum_{1 \leqslant k \leqslant l \leqslant n} \widetilde{A}_{kl} q_k q_l,$$

where the coefficients $\widetilde{A}_{kl}$ have the form

$$\widetilde{A}_{kl} = \sum_{1 \leqslant i \leqslant j \leqslant n} A_{ij} \alpha_{ki} \cdot \alpha_{lj}.$$

Thus, the coefficients of the equation are transformed according to the same formulas as the coefficients of a quadratic form. It is sufficient to set

$$\alpha_{ki} = \left. \frac{\partial \xi_k}{\partial x_i} \right|_{x_1 = x_1^0, \, \dots, x_n = x_n^0}.$$

It is known that by a nonsingular transformation (which always exists!) a quadratic form can be reduced to the form

$$\sum_{i=1}^{m} \left(\pm q_i^2 \right), \quad m \leq n, \tag{2.12}$$

and due to the law of inertia for the quadratic forms the numbers of each 1 and -1 are invariants of the quadratic form, independent of the choice of the transformation. Let (2.11) be a transformation reducing the quadratic form to the form (2.12). Then the transformation of variables (2.9) satisfying the equalities

$$\frac{\partial \xi_k \left(x_1^0, x_2^0, \dots, x_n^0 \right)}{\partial x_i} = \alpha_{ki}$$

transforms equation (2.1) to the form

$$\sum_{i=1}^{m} \left(\pm \frac{\partial^2 u}{\partial \xi_i^2} \right) + \widetilde{F} = 0. \tag{2.13}$$

This equation is called the canonical form of Eq. (2.1) at the point $\left(x_1^0, x_2^0, \dots, x_n^0 \right)$.

Definition 18

1. If $m = n$ and all the coefficients in (2.13) are of a same sign the original equation (2.1) is said to be **elliptic** at the point $\left(x_1^0, x_2^0, \dots, x_n^0 \right)$.
2. If $m = n$ and $n - 1$ coefficients in (2.13) are of a same sign equation (2.1) is said to be **hyperbolic** at the point $\left(x_1^0, x_2^0, \dots, x_n^0 \right)$. If $m = n$ and the number of

coefficients of a same sign is less than $n-1$ equation is said to be **ultrahyperbolic** at the point $(x_1^0, x_2^0, \ldots, x_n^0)$.

3. If $m < n$ Eq. (2.1) is called **parabolic in a wide sense**. If, in particular, $m = n-1$ equation is called **parabolic in a narrow sense** or just **parabolic**.

Remark 19 Unlike in the case $n = 2$ the transformation reducing the equation to the canonical form differs for different points, and in general there is no a single transformation of variables for all points (notice that such a transformation naturally exists in the case of constant coefficients).

An attempt to find a single transformation in the case of arbitrary coefficients leads to the system of $n(n-1)/2$ equalities:

$$\widetilde{A}_{kl} = \sum_{i,j=1}^n A_{ij} \frac{\partial \xi_k}{\partial x_i} \frac{\partial \xi_l}{\partial x_j} = 0, \quad k \neq l.$$

When $n > 3$ this system in general may have no solution $(n(n-1)/2 > n)$. When $n = 3$ the number of equalities coincides with the number of unknowns $(= 3)$ and thus the system is solvable but then there is no possibility to transform appropriately the coefficients at the unmixed partial derivatives $\frac{\partial^2 u}{\partial \xi_i^2}$.

Example 20 Here we present some concrete examples of equations:

$n = 3$, $\quad \frac{\partial^2 u}{\partial x_1^2} + \frac{\partial^2 u}{\partial x_2^2} + \frac{\partial^2 u}{\partial x_3^2} + f\left(x_1, x_2, x_3, u, \frac{\partial u}{\partial x_1}, \frac{\partial u}{\partial x_2}, \frac{\partial u}{\partial x_3}\right) = 0$—elliptic type.

$n = 3$, $\quad \frac{\partial^2 u}{\partial x_1^2} + \frac{\partial^2 u}{\partial x_2^2} - \frac{\partial^2 u}{\partial x_3^2} = 0$—hyperbolic type.

$n = 4$, $\quad \frac{\partial^2 u}{\partial x_1^2} + \frac{\partial^2 u}{\partial x_2^2} - \frac{\partial^2 u}{\partial x_3^2} - \frac{\partial^2 u}{\partial x_4^2} = 0$—ultrahyperbolic type.

$n = 4$, $\quad \frac{\partial^2 u}{\partial x_1^2} + \frac{\partial^2 u}{\partial x_2^2} + \frac{\partial^2 u}{\partial x_3^2} = 0$—equation of parabolic type in a narrow sense.

Chapter 3
Some Models of Mathematical Physics Reduced to Partial Differential Equations

3.1 Equation of Small Vibrations of a String

Here and in further sections of this chapter we present the most common examples of equations of the hyperbolic, parabolic, and elliptic type. The purpose of this chapter is to guide the reader into the process of formulating and describing real model problems in mathematical terms that lead to a partial differential equation.

In order to derive an equation describing a vibrating string we will make the following assumptions.

1. The string is perfectly flexible, that is, does not resist to bending. In particular, this means that the tension force is directed tangentially along the string.
2. The string is elastic and obeys Hooke's law that states that the force needed to extend or compress a string by some distance scales linearly with respect to that distance.
3. The string performs transverse plane vibrations. This means that the string remains in a plane and that each point of the string oscillates orthogonally to a certain fixed axis.
4. The string undergoes small vibrations. That is, the magnitude of $\tan(\alpha(x, t))$ is small, where $\alpha(x, t)$ is the angle of inclination of the tangent to the axis x at the instant t. The square of this magnitude will be neglected, $\tan^2 \alpha(x, t) \sim 0$.
5. The tension force in the string is sufficiently large, such that the gravity force can be neglected.

For instance, it could be a guitar string or a plucked violin string. However, if you take a thin rod its vibration picture differs from that of a string due to the bending stresses. A substantial role is played by internal bending stresses that prevent a change in its shape; therefore, it cannot be regarded as a string.

Let us show that under these assumptions the string is not extensible and hence according to Hooke's law the tension does not depend on time.

© The Author(s), under exclusive license to Springer Nature Switzerland AG 2022
A. N. Karapetyants, V. V. Kravchenko, *Methods of Mathematical Physics*,
https://doi.org/10.1007/978-3-031-17845-0_3

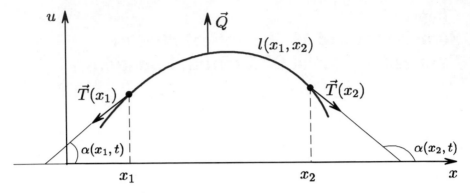

Fig. 3.1 The tension in the string is constant

Indeed, let $u(x, t)$ be the deviation of the string at the instant t. Then for any two points of the string x_1 and x_2 the length of the curvilinear segment of the string $\overset{\smile}{x_1 x_2}$ equals approximately the difference $x_2 - x_1$, because

$$l(x_1, x_2) = \int_{x_1}^{x_2} \sqrt{1 + u_x^2(x, t)}dx = \int_{x_1}^{x_2} \sqrt{1 + \tan^2 \alpha(x, t)}dx \approx x_2 - x_1.$$

By Hooke's law the tension in the string does not change and hence is independent of time: $T(x, t) = T(x)$.

Let us show that the tension is independent of the coordinate x either.

Consider again the curvilinear segment of the string $\overset{\smile}{x_1 x_2}$ replacing the outside pieces by the corresponding values of tension $T(x_1)$ and $T(x_2)$ (see Fig. 3.1). The projection of the selected part of the string onto the axis x is affected only by the forces $T(x_1)$ and $T(x_2)$. Thus, we have

$$T(x_2) \cos(\alpha(x_2, t)) - T(x_1) \cos(\alpha(x_1, t)) = 0.$$

Taking into account that

$$\cos^2 \alpha(x, t) = \frac{1}{1 + \tan^2 \alpha(x, t)} \approx 1,$$

we obtain that $T(x_1) \approx T(x_2)$. That is, under the assumptions made $T(x) = T_0 =$ Const.

In order not to trace the cosine sign we may consider the segment of the string where $\cos \alpha(x, t) > 0$. For the sections of the string where the cosine sign is negative the reasonings are similar.

Now let us derive the equation modeling small vibrations of the string. We will fix the segment $[x_1, x_2]$ and apply the d'Alembert[1] principle (principle of kinetostatics), according to which, if we are given (active) forces acting on points of a mechanical system and if we add the force(s) of inertia, then we get a balanced system of forces. For our needs this means that the sum of projections of all forces to the axis u equals zero.

On the segment $[x_1, x_2]$ the following forces act: the forces $\vec{T}(x_1)$, $\vec{T}(x_2)$ and an external force

$$Q = \int_{x_1}^{x_2} p(x, t)dx,$$

where $p(x, t)$ is the density of the acting forces. The force of inertia is given by

$$I = \int_{x_1}^{x_2} \frac{\partial^2 u}{\partial t^2} \rho(x)dx,$$

where $\rho(x)$ is the linear density of the string. Hence, by the d'Alembert principle we have

$$\text{Pr}_u \vec{T}(x_2) + \text{Pr}_u \vec{T}(x_1) + Q - I = 0,$$

where by Pr_u we denote the projection onto the u-axis. Since $T(x) = T_0 = \text{Const}$, we obtain

$$T_0 \sin \alpha(x_2, t) - T_0 \sin \alpha(x_1, t) + \int_{x_1}^{x_2} \left(p(x, t) - \rho(x) \frac{\partial^2 u}{\partial t^2} \right) dx = 0$$

or

$$T_0 \left(\frac{\partial u}{\partial x} \bigg|_{x=x_2} - \frac{\partial u}{\partial x} \bigg|_{x=x_1} \right) + \int_{x_1}^{x_2} \left(p(x, t) - \rho(x) \frac{\partial^2 u}{\partial t^2} \right) dx = 0$$

since $\sin \alpha(x, t) = \tan \alpha(x, t) \cos \alpha(x, t) \approx \tan \alpha(x, t)$ for small angles $\alpha(x, t)$ and

$$\sin \alpha(x_1, t) \approx \frac{\partial u}{\partial x} \bigg|_{x=x_1} \quad \text{and} \quad \sin \alpha(x_2, t) \approx \frac{\partial u}{\partial x} \bigg|_{x=x_2}.$$

Application of the mean value theorem with $\Delta x = x_2 - x_1$ gives

---

[1]The d'Alembert principle states that the sum of the differences between the forces acting on a system of massive particles and the time derivatives of the momenta of the system itself projected onto any virtual displacement consistent with the constraints of the system is zero.

$$\Delta x \left(T_0 \left.\frac{\partial^2 u}{\partial x^2}\right|_{x=x_1+\theta_1 \Delta x} + \left(p(x,t) - \rho(x)\frac{\partial^2 u}{\partial t^2} \right)\bigg|_{x=x_1+\theta_2 \Delta x} \right) = 0.$$

Dividing over Δx and letting $\Delta x \to 0$ leads to the following equation

$$T_0 \frac{\partial^2 u}{\partial x^2} - \rho(x)\frac{\partial^2 u}{\partial t^2} + p(x,t) = 0,$$

modeling small vibrations of an inhomogeneous string.

If $\rho(x) \equiv \rho_0 = \text{Const}$, denoting

$$a^2 := \frac{T_0}{\rho_0} \quad \text{and} \quad f := \frac{p}{\rho_0}$$

we obtain the equation modeling small vibrations of a homogeneous string

$$\frac{\partial^2 u}{\partial t^2} = a^2 \frac{\partial^2 u}{\partial x^2} + f(x,t). \tag{3.1}$$

This equation is obtained under the assumption of the absence of a resistivity. For a string vibrating in the air the resistivity of the medium is proportional to the speed of the vibrations which means that the equation takes the form

$$\frac{\partial^2 u}{\partial t^2} = a^2 \frac{\partial^2 u}{\partial x^2} - k\frac{\partial u}{\partial t} + f(x,t). \tag{3.2}$$

In a denser medium the resistivity can be proportional to the square of the speed; therefore, as an example one may obtain the following equation

$$\frac{\partial^2 u}{\partial t^2} = a^2 \frac{\partial^2 u}{\partial x^2} - k\left(\frac{\partial u}{\partial t}\right)^2 + f(x,t). \tag{3.3}$$

Equations (3.1)–(3.3) are hyperbolic because $\Delta = a^2 > 0$. Equations (3.1) and (3.2) are linear while Eq. (3.3) is linear with respect to the higher derivatives.

3.2   Energy of Free Vibrations of the String

For simplicity let us assume that the endpoints of the string are fixed, then

$$u(0,t) = u(l,t) = 0$$

and hence $u_t(0,t) = u_t(l,t) = 0$ that corresponds to the zero speed of motion of the ends.

We denote the full energy of the string by E,

$$E = K + W,$$

where K is the kinetic energy and W is the potential energy. For the kinetic energy we have the equalities

$$\Delta K = \frac{\Delta m}{2} \left(\frac{\partial u}{\partial t} \right)^2 = \frac{1}{2} \rho u_t^2 \, \Delta x,$$

where ρ is a (constant) linear density of the string, and hence

$$K = \frac{1}{2} \int_0^l \rho u_t^2 \, dx,$$

while $W = -A$, where A is the amount of work required to translate the string from the equilibrium state

$$u(x, 0) = 0 \tag{3.4}$$

to a given state $u(x, t)$. We have

$$\Delta A = \Delta F \Delta h = \Delta F \frac{\partial u}{\partial t} \Delta t,$$

$$\Delta F = \mathrm{Pr}_u \, \overrightarrow{T}(x_2) + \mathrm{Pr}_u \, \overrightarrow{T}(x_1) = T_0 \left. \frac{\partial^2 u}{\partial x^2} \right|_{x=x_1+\theta \Delta x} \Delta x.$$

Hence

$$\Delta A = \frac{\partial u}{\partial t} T_0 \left. \frac{\partial^2 u}{\partial x^2} \right|_{x=x_1+\theta \Delta x} \Delta x \Delta t$$

and using τ instead of t inside the integral we have

$$A = T_0 \int_0^t \int_0^l u_\tau u_{xx} dx d\tau = T_0 \int_0^t d\tau \int_0^l u_\tau u_{xx} dx.$$

Integration by parts leads to the equality

$$A = T_0 \int_0^t d\tau \left\{ u_\tau u_x \big|_0^l - \int_0^l u_x u_{\tau x} dx \right\}$$

$$= -T_0 \int_0^t d\tau \int_0^l u_x u_{\tau x} dx.$$

Here the boundary conditions were used.

Since

$$u_x u_{\tau x} = \frac{1}{2} \frac{\partial}{\partial \tau} \left(u_x^2 \right),$$

we have

$$A = -\frac{T_0}{2} \int_0^t \frac{\partial}{\partial \tau} \left(\int_0^l u_x^2 dx \right) d\tau$$

$$= -\frac{T_0}{2} \int_0^l u_x^2 dx \Big|_0^t$$

$$= -\frac{T_0}{2} \int_0^l u_x^2(x, t) dx + \frac{T_0}{2} \int_0^l u_x^2(x, 0) dx.$$

The last integral equals zero due to (3.4). Thus,

$$A = -\frac{T_0}{2} \int_0^l u_x^2 dx$$

and

$$W = \frac{T_0}{2} \int_0^l u_x^2 dx.$$

Finally, we obtain the expression for the energy of the string

$$E = \frac{1}{2} \int_0^l \left(\rho u_t^2 + T_0 u_x^2 \right) dx.$$

Let us prove that for the free vibrations of the string the energy conservation law is fulfilled. Consider

$$\frac{dE}{dt} = \int_0^l (\rho u_t u_{tt} + T_0 u_x u_{xt}) \, dx$$

$$= \int_0^l \rho u_t u_{tt} dx + T_0 \int_0^l u_x d\,(u_t)$$

$$= \int_0^l \rho u_t u_{tt} dx + T_0 \left\{ u_x u_t \Big|_0^l - \int_0^l u_{xx} u_t dx \right\}.$$

Using again the boundary conditions we obtain

$$\frac{dE}{dt} = \rho \int_0^l u_t \left(u_{tt} - \frac{T_0}{\rho} u_{xx} \right) dx = \rho \int_0^l u_t \left(u_{tt} - a^2 u_{xx} \right) dx = 0.$$

That is, $E = \text{Const.}$

3.3   Longitudinal Vibrations of a Rod

A rod is a body of cylindrical or prismatic, for instance, shape, for stretching or compression of which a certain force must be applied. Let us agree that the forces causing its elongation obey Hooke's law.

Let us consider an elastic cylindrical body and suppose that all physical forces act along the axis of the rod whose transverse size is much smaller than its length. We assume that during the vibrations the cross sections of the rod remain plane and parallel to each other.

The rod is fixed at the point $x = 0$ and stretched at the moment $t = 0$. If the rod is then released, it starts vibrating, i.e., we have free longitudinal vibrations. The displacement of a transverse section at any moment t is denoted by $u(x, t)$. Then, the displacement of the section whose abscissa is $x + \Delta x$ will be equal to

$$u(x + \Delta x, t) \approx u(x, t) + u_x \Delta x.$$

Thus, the derivative u_x characterizes the relative lengthening of the rod at the cross section whose abscissa is x.

According to Hooke's law,

$$T = E \cdot S \frac{\partial u}{\partial x},$$

where E is the modulus of elasticity of the material of which the rod is composed (Young's modulus), and S is the cross sectional area.

Consider the element of the rod included between two cross sections whose abscissas, when the rod is at rest, are x and $x + \Delta x$. It is acted on by the forces of tension $\overrightarrow{T}_{x+\Delta x}$, $\overrightarrow{T}_x$, the external force $\overrightarrow{Q}$ and the force of inertia $\overrightarrow{I}$. Hence

$$\overrightarrow{T}_{x+\Delta x} + \overrightarrow{T}_x + \overrightarrow{Q} + \overrightarrow{I} = 0$$

or considering their magnitudes only

$$T_{x+\Delta x} - T_x + Q - I = 0.$$

Let $p(x, t)$ be the external force density and $\rho = \rho(x)$ be the volume density of the rod. Then

$$Q = p(x, t)\Delta v = p(x, t)S\Delta x,$$

where $\Delta v = S\Delta x$ is an element of volume, and

$$I = \rho(x)S\Delta x \frac{\partial^2 u}{\partial t^2}.$$

We have

$$T_{x+\Delta x} - T_x = E \cdot S \left(\left. \frac{\partial u}{\partial x} \right|_{x+\Delta x} - \left. \frac{\partial u}{\partial x} \right|_x \right),$$

and thus,

$$E \cdot S \left(\left. \frac{\partial u}{\partial x} \right|_{x+\Delta x} - \left. \frac{\partial u}{\partial x} \right|_x \right) + pS\Delta x - \rho S\Delta x \frac{\partial^2 u}{\partial t^2} = 0.$$

Dividing over $S\Delta x$ and letting $\Delta x \to 0$ we obtain

$$E \frac{\partial^2 u}{\partial x^2} + p = \rho \frac{\partial^2 u}{\partial t^2}$$

or

$$\frac{\partial^2 u}{\partial t^2} = a^2 \frac{\partial^2 u}{\partial x^2} + f(x, t), \tag{3.5}$$

where

$$a^2 = \frac{E}{\rho} \quad \text{and} \quad f = \frac{p}{\rho}.$$

Equation (3.5) describes longitudinal vibrations of a rod and has the same form as the equation of forced string vibrations.

3.4   Equation of Vibrations of a Membrane

A stretched film that can be freely bent is called a membrane. Suppose that a membrane (in its equilibrium position) is situated in the xy-plane and that it occupies some region V bounded by a closed curve L. We will assume the fulfillment of the following conditions.

1. The membrane is flexible. The tensions arising in it are distributed uniformly. This means that if we draw a line on the membrane, say dl, in any arbitrary direction, the force between the two parts of the membrane that are separated by a given element of the line will be proportional to the length of the element and directed perpendicularly to it. The magnitude of the force acting on the element dl will be equal to $T dl$, that is, $\left| \overrightarrow{T}_{dl} \right| = T dl$.

2. The membrane is elastic, and obeys the Hooke's law.

3. The membrane oscillates in the transverse direction, that is, its points move perpendicularly to the plane xy.

4. The vibrations of the membrane are small, that is, the tangent of the angle between the axis u and the normal vector to the membrane surface is small. The square of this tangent will be neglected.

5. The weight of the membrane will be neglected.

Let us prove that under the assumptions made, the tension $T(x, y, t)$ is independent of the time t and of the point (x, y). Let $u(x, y, t)$ denote the position of the point (x, y) of the membrane at the moment t. Let $\overrightarrow{i}$, $\overrightarrow{j}$, $\overrightarrow{k}$ be the unit direction vectors of the axes x, y, u, and $\overrightarrow{n}$ denote the unit normal vector to the surface of the membrane. Denote $\alpha = (\widehat{\overrightarrow{n}, \overrightarrow{k}})$, the angle between the vectors. As it is known,

$$\cos \alpha = \frac{\left(\overrightarrow{n}, \overrightarrow{k} \right)}{\left| \overrightarrow{n} \right| \left| \overrightarrow{k} \right|} = \frac{1}{\sqrt{1 + u_x^2 + u_y^2}} \approx 1.$$

For the area of an arbitrary part of the membrane in our approximation we have

$$\int_\sigma d\sigma = \int_{\sigma'} \frac{d\sigma'}{\cos \alpha} = \int_{\sigma'} \sqrt{1 + u_x^2 + u_y^2} d\sigma' \approx \int_{\sigma'} d\sigma',$$

where σ' is a projection of σ onto the plane xy, and $\int_\sigma d\sigma$ stands for area integral over σ. Thus, the membrane is inextensible. Due to Hooke's law the tension is independent of time. It can be shown by analogy with the string that the tension does not depend on σ either and hence is constant.

Let us apply the d'Alembert principle to an arbitrary area σ. The following physical forces act on it: the force of tension $\overrightarrow{T}$, the external force $\overrightarrow{Q}$, and the force of inertia $\overrightarrow{I}$. The external force is assumed to be parallel to the u-axis, and hence in projection onto the u-axis we have

$$\mathrm{Pr}_u \overrightarrow{T} + Q - I = 0. \tag{3.6}$$

Let $p(x, y, t)$ be the density of the external force, then $\Delta Q = p(x, y, t)\Delta\sigma$ and

$$Q = \int_{\sigma'} p(x, y, t)d\sigma'.$$

For I we have the formula

$$I = \int_{\sigma'} \rho(x, y)\frac{\partial^2 u}{\partial t^2}d\sigma',$$

where $\rho(x, y)$ is the surface density.

Let us find the projection of the force of tension onto the u-axis. The force of tension acts on the boundary l of the area σ (see Fig. 3.2). The force of tension acting on an element dl of the boundary l equals Tdl. Denote by $\overrightarrow{dl}$ the vector directed tangentially with respect to l such that $\left|\overrightarrow{dl}\right| = dl$. The cosine of the angle between this vector and the u-axis is obviously equal to $\frac{\partial u}{\partial \overrightarrow{n}}$ where $\overrightarrow{n}$ is the normal (directed outward) to the curve l. It follows that the projection onto the u-axis of the tension $\overrightarrow{T}$ calculated for an element dl of the curve l will be equal to

$$T\frac{\partial u}{\partial \overrightarrow{n}}dl.$$

Integration of this product over the entire curve l leads to the expression for uniformly acting tension along this curve,

$$T\int_l \frac{\partial u}{\partial \overrightarrow{n}}dl.$$

Since dl is approximately equal to its projection dl' onto the plane xy when the vibrations of the membrane are of small amplitude, we may replace the path of integration l with l'. Hence

$$\mathrm{Pr}_u \overrightarrow{T} = T\int_{l'} \frac{\partial u}{\partial \overrightarrow{n}}dl' = T\int_{l'} \left(-u_y dx + u_x dy\right).$$

Using Green's formula, we obtain

$$\mathrm{Pr}_u \overrightarrow{T} = T\int_{\sigma'} \left(\frac{\partial^2 u}{\partial x^2} + \frac{\partial^2 u}{\partial y^2}\right) d\sigma'.$$

Thus, from (3.6) we have

$$\int_{\sigma'} \left(T\left(u_{xx} + u_{yy}\right) + p - \rho u_{tt}\right) d\sigma' = 0.$$

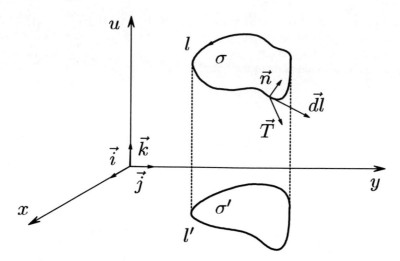

Fig. 3.2 The force of tension acts on the boundary l of the area σ

Since σ' is arbitrary from above formula we obtain that

$$T \left(u_{xx} + u_{yy}\right) + p - \rho u_{tt} = 0$$

and hence the equation modeling small vibrations of a membrane has the form

$$u_{tt} = a^2 \left(u_{xx} + u_{yy}\right) + g(x, y, t),$$

where

$$a^2 = \frac{T}{\rho} \quad \text{and} \quad g = \frac{p}{\rho}.$$

3.5   Equations of Hydrodynamics

We naturally consider the model of the three-dimensional space of variables x, y, and z. The motion of a liquid or gas is characterized by the following five scalar magnitudes.

1. The velocity $\vec{v}$ with three scalar components v_x, v_y, v_z.
2. The pressure $p = p(x, y, z, t)$.
3. The density $\rho = \rho(x, y, z, t)$.

Let us consider an ideal liquid, that is the friction is absent. Our aim is to derive the equations relating the magnitudes $\vec{v}$, p, and ρ.

Consider an arbitrary volume Ω with the boundary S. Let $f_1(x, y, z, t)$ be the power of sources (incoming liquid or gas) while $f_2(x, y, z, t)$ the power of sinks leaking the liquid or gas. Denote $f(x, y, z, t) = f_1 - f_2$.

Let $Q_\Omega(t)$ be the total mass of the liquid in the volume Ω,

$$Q_\Omega(t) = \int_\Omega \rho \, d\Omega.$$

Then the rate of change of the mass is equal to $\frac{d}{dt} Q_\Omega(t)$. This magnitude is composed of the following components

$$\frac{d}{dt} Q_\Omega(t) = \text{(the flux through the boundary)}$$

$$+ \text{(the liquid from the sources)-(the liquid to the sinks)},$$

that is,

$$\frac{d}{dt} \int_\Omega \rho \, d\Omega = - \int_S \rho \vec{v}_n dS + \int_\Omega (f_1 - f_2) \, d\Omega.$$

From the Gauss theorem we have

$$\int_S \rho \vec{v}_n dS = \int_\Omega \text{div} \left(\rho \vec{v} \right) d\Omega.$$

Thus,

$$\frac{d}{dt} \int_\Omega \rho \, d\Omega = - \int_\Omega \text{div} \left(\rho \vec{v} \right) d\Omega + \int_\Omega f(x, y, z, t) \, d\Omega.$$

Since Ω is an arbitrary (small) volume, the integrand is zero. Hence

$$\frac{\partial \rho}{\partial t} + \text{div} \left(\rho \vec{v} \right) = f(x, y, z, t). \tag{3.7}$$

This equation is called the **continuity equation** for fluid or gas. In the case of absence of sources and sinks ($f = 0$) the equation takes the form

$$\frac{\partial \rho}{\partial t} + \text{div} \left(\rho \vec{v} \right) = 0.$$

Now let us obtain an equation for the pressure p. Again we consider an arbitrary volume Ω. Due to the assumption of the ideal liquid, the action of the liquid from the outside of Ω reduces to the pressure directed at each point of the boundary S along the internal normal vector. Let $\vec{p}$ be the resulting force of pressure

$$\vec{p} = - \int_S p \vec{n} \, dS,$$

where $\vec{n}$ denotes the external normal vector. Using one of the corollaries of the Gauss theorem we obtain that

$$\vec{p} = -\int_\Omega \operatorname{grad} p \, d\Omega.$$

Let $\vec{F}$ be the resulting external force (with the density $\vec{f}$ calculated with respect to the unit of mass), and $\vec{I}$ the force of inertia. Let us apply to the volume Ω the d'Alembert principle

$$\vec{p} + \vec{F} + \vec{I} = 0. \tag{3.8}$$

Obviously,

$$\vec{I} = -\int_\Omega \rho \frac{d\vec{v}}{dt} \, d\Omega, \quad \Delta \vec{I} = -\Delta m \frac{d\vec{v}}{dt} = -\rho \frac{d\vec{v}}{dt} \Delta \Omega,$$

where $\frac{d\vec{v}}{dt}$ is the total time derivative of the velocity $\vec{v}$ and

$$\vec{F} = \int_\Omega \rho \vec{f} (x, y, z, t) d\Omega.$$

Thus, from (3.8) we obtain

$$\int_\Omega \left(-\rho \frac{d\vec{v}}{dt} - \operatorname{grad} p + \rho \vec{f} \right) d\Omega = 0$$

and due to the arbitrariness of the volume Ω,

$$\frac{d\vec{v}}{dt} = -\frac{1}{\rho} \operatorname{grad} p + \vec{f}.$$

This equation is called **Euler's equation** of fluid dynamics. Let us transform it to another form. We have

$$\frac{d\vec{v}}{dt} = \frac{\partial \vec{v}}{\partial t} + \frac{\partial \vec{v}}{\partial x} \frac{\partial x}{\partial t} + \frac{\partial \vec{v}}{\partial y} \frac{\partial y}{\partial t} + \frac{\partial \vec{v}}{\partial z} \frac{\partial z}{\partial t}$$

$$= \frac{\partial \vec{v}}{\partial t} + v_x \frac{\partial \vec{v}}{\partial x} + v_y \frac{\partial \vec{v}}{\partial y} + v_z \frac{\partial \vec{v}}{\partial z}.$$

Let us introduce a symbolic vector

$$\vec{\nabla} = \vec{i} \frac{\partial}{\partial x} + \vec{j} \frac{\partial}{\partial y} + \vec{k} \frac{\partial}{\partial z}.$$

The scalar product $\left(\vec{v}, \vec{\nabla}\right)$ is understood as follows

$$\left(\vec{v}, \vec{\nabla}\right) = v_x \frac{\partial}{\partial x} + v_y \frac{\partial}{\partial y} + v_z \frac{\partial}{\partial z},$$

and by $\left(\vec{v}, \vec{\nabla}\right) \vec{v}$ we understand the vector

$$\left(\vec{v}, \vec{\nabla}\right) \vec{v} = \left(v_x \frac{\partial}{\partial x} + v_y \frac{\partial}{\partial y} + v_z \frac{\partial}{\partial z}\right) \vec{v} = v_x \frac{\partial \vec{v}}{\partial x} + v_y \frac{\partial \vec{v}}{\partial y} + v_z \frac{\partial \vec{v}}{\partial z}.$$

Thus,

$$\frac{d\vec{v}}{dt} = \frac{\partial \vec{v}}{\partial t} + \left(\vec{v}, \vec{\nabla}\right) \vec{v}$$

and Euler's equation takes the form

$$\frac{\partial \vec{v}}{\partial t} + \left(\vec{v}, \vec{\nabla}\right) \vec{v} = -\frac{1}{\rho} \operatorname{grad} p + \vec{f}. \tag{3.9}$$

Equation (3.9) together with (3.7) give us a system of four scalar equations for five unknown scalar functions v_x, v_y, v_z, p, ρ. They are usually combined with a so-called thermodynamic equation of state $\Phi(p, \rho) = 0$ relating the pressure with the density. For example, in the case of an incompressible liquid the equation of state is $\rho = \mathrm{Const}$. In the case of adiabatic motion of a compressible liquid or gas the functional relation between p and ρ is expressed by the formula

$$p = p_0 \left(\frac{\rho}{\rho_0}\right)^\gamma, \quad \gamma = \frac{C_p}{C_v},$$

where ρ_0, p_0 are the initial values of the density and pressure, C_p and C_v are the specific heats at constant pressure and constant volume. Thus, we have five equations for five unknowns. Let us apply these equations to the sound propagation process in a gas.

3.6   Sound Propagation and Equations of Acoustics

Let us derive the equation modeling the propagation of acoustic waves under the following assumptions.

1. External forces are absent.
2. There are no gas leaks or sources.

3. The process of wave propagation is adiabatic (there is no heat transfer due to the gas compression and rarefaction).
4. The oscillations of the gas are small, that is, the magnitudes of the form $v_x \frac{\partial v_x}{\partial x}$, ... are negligible.

It follows from the assumptions that the equations describing the motion of the gas have the form

$$\frac{\partial \rho}{\partial t} + \operatorname{div}\left(\rho \vec{v}\right) = 0,$$

$$\frac{\partial \vec{v}}{\partial t} + \frac{1}{\rho} \operatorname{grad} p = 0,$$

$$\frac{p}{p_0} = \left(\frac{\rho}{\rho_0}\right)^\gamma.$$

The relative change of the gas density

$$\frac{\rho - \rho_0}{\rho_0} = s(x, y, z, t)$$

is called the **gas condensation**. Substituting $\rho = \rho_0(1 + s)$ we write the equations of the gas dynamics in the form

$$\rho_0 \frac{\partial s}{\partial t} + \rho_0(1 + s) \operatorname{div} \vec{v} + \rho_0 \left(\vec{v}, \operatorname{grad} s\right) = 0,$$

$$\frac{\partial \vec{v}}{\partial t} = -\frac{1}{\rho_0(1 + s)} \operatorname{grad} p,$$

$$p = p_0 (1 + s)^\gamma.$$

Let us assume the magnitude of condensation to be small, so that the magnitudes of order s^2 and combined magnitudes such as $v_x \frac{\partial s}{\partial x}$, ... are supposed to be negligible. Then $s \operatorname{div} \vec{v} \approx 0$, $\left(\vec{v}, \operatorname{grad} s\right) \approx 0$ and

$$\frac{1}{1 + s} = 1 - s + s^2 - s^3 + \ldots \approx 1 - s,$$

$$(1 + s)^\gamma = 1 + \gamma s + \frac{\gamma(\gamma - 1)}{2} s^2 + \ldots \approx 1 + \gamma s.$$

Thus, we obtain

$$\frac{\partial s}{\partial t} + \operatorname{div} \vec{v} = 0,$$

$$\frac{\partial \vec{v}}{\partial t} = -\frac{p_0\,(1-s)}{\rho_0}\,\operatorname{grad}(1+\gamma s) = -\frac{p_0\gamma}{\rho_0}\,\operatorname{grad} s.$$

Let us eliminate $\vec{v}$ from this couple of equations. Differentiating the first equation with respect to t, and applying div to the second one, we obtain

$$\frac{\partial}{\partial t}\operatorname{div}\vec{v} = -\frac{\partial^2 s}{\partial t^2},$$

and

$$\frac{\partial}{\partial t}\operatorname{div}\vec{v} = -\frac{p_0\gamma}{\rho_0}\operatorname{div}\operatorname{grad} s.$$

Then

$$\frac{\partial^2 s}{\partial t^2} = a^2 \Delta s, \quad a = \sqrt{\frac{p_0\gamma}{\rho_0}}, \tag{3.10}$$

where $\Delta = \operatorname{div}\operatorname{grad}$ is the **Laplace operator**,

$$\Delta u = \frac{\partial^2 u}{\partial x^2} + \frac{\partial^2 u}{\partial y^2} + \frac{\partial^2 u}{\partial z^2}, \quad u = u(x, y, z).$$

Thus, the gas condensation (as well as the pressure, due to the relation $p = p_0\,(1+\gamma s)$) in absence of external forces satisfies Eq. (3.10) which in expanded form can be written as

$$\frac{\partial^2 s}{\partial t^2} = a^2\left(\frac{\partial^2 s}{\partial x^2} + \frac{\partial^2 s}{\partial y^2} + \frac{\partial^2 s}{\partial z^2}\right).$$

Suppose that initially, for $t = 0$ the field of velocities was conservative:

$$\vec{v}\,\big|_{t=0} = -\operatorname{grad} u_0(x, y, z).$$

Then from the equality

$$\frac{\partial \vec{v}}{\partial t} = -a^2\operatorname{grad} s$$

we have

$$\vec{v} = -a^2 \operatorname{grad} \int_0^t s\, dt + \vec{v}\,|_{t=0}$$

or

$$\vec{v} = -\operatorname{grad}\left(u_0 + a^2 \int_0^t s\, dt\right).$$

Denote

$$u(x, y, z, t) = u_0(x, y, z) + a^2 \int_0^t s\, dt.$$

Hence

$$\vec{v} = -\operatorname{grad} u.$$

This means that if the field of velocities was conservative at a moment $t = 0$, it remains conservative at all times $t > 0$.

Let us show that the scalar potential u of $\vec{v}$ also satisfies Eq. (3.10). We have

$$\frac{\partial^2 u}{\partial t^2} = \frac{\partial^2}{\partial t^2}\left(a^2 \int_0^t s\, dt + u_0\right) = a^2 \frac{\partial s}{\partial t} = -a^2 \operatorname{div} \vec{v} = a^2 \operatorname{div} \operatorname{grad} u = a^2 \Delta u$$

and thus

$$\frac{\partial^2 u}{\partial t^2} = a^2 \Delta u.$$

This equation is called the **equation of acoustics**.

Remark 21 Different wave processes, such as the transverse vibrations of a string or membrane, the longitudinal vibrations of a rod, oscillations of liquids and gases under certain simplifications are modeled by the equation

$$\frac{\partial^2 u}{\partial t^2} = a^2 \Delta u + f.$$

This is why the equation is called the **wave equation**.

3.7 Heat Equation

Consider a solid body and let $u = u(x, y, z, t)$ denote the temperature at the point (x, y, z) at the moment t. In the theory of thermal conduction there is a Fourier's law which is analogous to Hooke's law in elasticity.

Fourier's Law *Let s be an arbitrary surface located inside a body and Δs be a small neighborhood of a point (x, y, z) of the surface. The amount of heat ΔQ passing through the area Δs per time Δt is proportional to Δt, Δs and to the change of the temperature along the normal to the area Δs,*

$$\Delta Q = -k(x, y, z)\Delta s \Delta t \frac{\partial u}{\partial n},$$

where n is the normal directed to the side of the decrease of temperature.

Here $k(x, y, z) > 0$ is the coefficient of the thermal conductivity. The body is assumed to be isotropic, that is, the coefficient k is independent of the orientation of Δs in space and depends only on the coordinates of the point (x, y, z).

Let us derive the heat balance equation for an arbitrary volume v located inside a physical body. Let t_1 and t_2 be two instants, $\Delta t = t_2 - t_1$. The temperature at any point of the physical body changes at the magnitude

$$u(x, y, z, t_2) - u(x, y, z, t_1).$$

The following total amount of the heat Q is required for the change of the temperature in the volume v:

$$Q = \int_v (u(x, y, z, t_2) - u(x, y, z, t_1))\, c\rho\, dv = \int_{t_1}^{t_2} dt \int_v \frac{\partial u}{\partial t} c\rho\, dv,$$

where c is the specific heat capacity of the body, ρ is its density. This heat Q is composed of two parts

$$Q = Q_1 + Q_2, \tag{3.11}$$

where Q_1 is the heat entering through the boundary S of the volume v and Q_2 is the heat absorbed or generated in the same volume.

The heat flux q, which is the amount of heat passing through a unit of the surface area per unit of time, due to Fourier's law equals

$$q = -k\frac{\partial u}{\partial n}.$$

Hence

$$Q_1 = -\int_{t_1}^{t_2} dt \int_S k(x, y, z)\frac{\partial u}{\partial n} ds,$$

where n is the interior normal vector.

Let $f(x, y, z, t)$ denote the density of the absorbed or generated heat per unit of time and per unit of volume. Then

$$Q_2 = \int_{t_1}^{t_2} dt \int_v f(x, y, z, t) dv.$$

Substituting the expressions for Q, Q_1, and Q_2 into (3.11) we obtain the equality

$$\int_{t_1}^{t_2} \left(\int_v \left(c\rho\frac{\partial u}{\partial t} - f \right) dv + \int_S k(x, y, z)\frac{\partial u}{\partial n} ds \right) dt = 0.$$

Due to the arbitrariness of the time interval we have

$$\int_v \left(c\rho\frac{\partial u}{\partial t} - f \right) dv + \int_S k\frac{\partial u}{\partial n} ds = 0. \tag{3.12}$$

Applying the Gauss theorem to the second integral we obtain

$$\int_S k\frac{\partial u}{\partial n} ds = \int_S k\,(\operatorname{grad} u)_n\, ds = \int_S (k\operatorname{grad} u)_n\, ds = -\int_v \operatorname{div}(k\operatorname{grad} u)dv,$$

where we took into account that n is the interior normal.

Equation (3.12) takes the form

$$\int_v \left(c\rho\frac{\partial u}{\partial t} - f - \operatorname{div}(k\operatorname{grad} u) \right) dv = 0.$$

Since v is an arbitrary volume, we can write

$$c\rho\frac{\partial u}{\partial t} = \operatorname{div}(k\operatorname{grad} u) + f(x, y, z, t). \tag{3.13}$$

This equation is called the equation of heat conduction or the **heat equation**.

In the case when the body is homogeneous, hence c, ρ, and k are all constants, the heat equation takes the form

$$\frac{\partial u}{\partial t} = a^2 \Delta u + f_1, \quad a = \sqrt{\frac{k}{c\rho}}, \quad f_1 = \frac{f}{c\rho}.$$

In particular, for a plane body when one of the sizes is negligible (e.g., a disk) we have

$$\frac{\partial u}{\partial t} = a^2 \left(\frac{\partial^2 u}{\partial x^2} + \frac{\partial^2 u}{\partial y^2} \right) + f(x, y, t).$$

If the body is linear (a rod), the heat equation takes the form

$$\frac{\partial u}{\partial t} = a^2 \frac{\partial^2 u}{\partial x^2} + f(x, t).$$

Obviously, the heat equation (3.13) is of parabolic type at all points. This is easy to see by considering the terms containing the second derivatives

$$c\rho \frac{\partial u}{\partial t} = k\Delta u + (\operatorname{grad} k, \operatorname{grad} u) + f.$$

Dividing over k we obtain a canonical form in which the term $\frac{\partial^2 u}{\partial t^2}$ is absent, thus, the equation is parabolic.

3.8 Diffusion Equation

Let us consider the diffusion phenomenon in a motionless medium filled nonuniformly with a substance. The convection will be neglected (it inevitably occurs in liquids and gases).

For the diffusing substance **Fick's law** is valid. It postulates that the flux goes from regions of high concentration to regions of low concentration, with a magnitude that is proportional to the concentration gradient

$$q = -D \frac{\partial u}{\partial n}.$$

Here $u = u(x, y, z, t)$ is the concentration of the substance defined as the limit

$$u = \lim_{dv \to 0} \frac{dQ}{dv},$$

where dQ is the amount of the substance in the volume dv. The coefficient D is called the diffusion coefficient.

The amount of the substance which is required for changing the concentration of the substance in the volume v per time Δt and by the quantity Δu equals

$$\Delta Q = C\Delta u \Delta v = C \frac{\partial u}{\partial t} \Delta v \Delta t,$$

where C is called the porosity coefficient. Hence

$$Q = \int_{t_1}^{t_2} dt \int_v C \frac{\partial u}{\partial t} \, dv.$$

We have that $Q = Q_1 + Q_2$ where Q_1 is the amount of substance diffusing into the volume v through its boundary S according to Fick's law, and Q_2 is the amount of substance diffusing in the volume due to the proper sources.

All the reasonings are analogous to those from the previous section. Thus, the diffusion equation has the form

$$C \frac{\partial u}{\partial t} = \operatorname{div}(D \operatorname{grad} u) + f,$$

where f is the intensity of the sources per unit of volume.

If the diffusion coefficient is constant we have

$$\frac{\partial u}{\partial t} = a^2 \Delta u + f_1, \quad a^2 = \frac{D}{C}, \quad f_1 = \frac{f}{C}.$$

As a conclusion, the phenomena revealing an oscillatory behavior are described by wave equations of hyperbolic type, meanwhile the phenomena related to the heat conduction or diffusion of a substance are described by equations of the form (3.13) of parabolic type.

3.9   Problems Reducing to Elliptic Equations

Consider the simplest equations of elliptic type

$$\Delta u := \frac{\partial^2 u}{\partial x^2} + \frac{\partial^2 u}{\partial y^2} + \frac{\partial^2 u}{\partial z^2} = f(x, y, z).$$

The homogeneous equation

$$\Delta u = 0$$

is called the **Laplace equation**, and the non-homogeneous

$$\Delta u = f$$

is called the **Poisson equation**.

The description of the following phenomena leads to these partial differential equations.

The steady-state temperature distribution.
The steady-state conservative flow of a liquid.
The electrostatic field.

1. Consider the heat equation

$$\frac{\partial u}{\partial t} = a^2 \Delta u + f$$

and assume that the temperature u in the body has settled, that is, does not depend on time. Obviously, this is possible if only f is independent of time. Then $u_t \equiv 0$ and

$$\Delta u = -\frac{f}{a^2}.$$

Thus, the steady-state temperature distribution is modeled by the Poisson equation. If, additionally, there are no sources of heat we arrive at the Laplace equation $\Delta u = 0$.

2. Consider a steady-state conservative flow of a liquid

$$\vec{v} = -\operatorname{grad} u.$$

Since the flow is steady-state, $\frac{\partial \rho}{\partial t} \equiv 0$, and from the continuity equation

$$\frac{\partial \rho}{\partial t} + \operatorname{div}(\rho \vec{v}) = 0$$

it follows that $\operatorname{div}(\rho \vec{v}) = 0$ and hence $\operatorname{div}(\rho \operatorname{grad} u) = 0$. Then

$$\rho \Delta u + (\operatorname{grad} \rho, \operatorname{grad} u) = 0.$$

This is an elliptic equation. If the liquid is incompressible, that is, $\rho = \text{Const}$, we obtain the Laplace equation.

3. Similar reasoning applies to the electrostatic field

$$\vec{E} = -\operatorname{grad} \varphi.$$

From electrostatics it is known that $\operatorname{div} \vec{E} = 4\pi\rho$ where ρ is the volume charge density. Then

$$\operatorname{div} \vec{E} = -\operatorname{div} \operatorname{grad} \varphi = -\Delta\varphi = 4\pi\rho.$$

That is, $\Delta\varphi = -4\pi\rho$ which is a Poisson equation. Again, in the absence of the sources (charges) of the electrostatic field, we obtain the Laplace equation.

3.10 Helmholtz Equation

The elliptic partial differential equation of the form

$$\Delta u + k^2 u = f,$$

where k is a constant, is called the **Helmholtz equation**. The heat equation

$$\frac{\partial u}{\partial t} = a^2 \Delta u + f$$

and the wave equation

$$\frac{\partial^2 u}{\partial t^2} = a^2 \Delta u + f$$

both reduce to the Helmholtz equation in the case when the corresponding physical phenomenon is time-harmonic. Namely, let us consider the situation when the density of the external forces is periodic in time (time-harmonic)

$$f = f_0(x, y, z)e^{i\omega t}$$

and let us look for solutions in a time-harmonic form $u = u_0(x, y, z)e^{i\omega t}$. Substitution to the wave equation gives us the equality

$$-\omega^2 u_0 e^{i\omega t} = a^2 e^{i\omega t} \Delta u_0 + f_0 e^{i\omega t}$$

and hence

$$\Delta u_0 + \left(\frac{a}{\omega}\right)^2 u_0 = -\frac{f_0}{\omega^2}.$$

Thus, for the complex amplitude u_0 we obtain a Helmholtz equation.

In a similar way the heat equation reduces to the Helmholtz equation whenever the intensity of the heat sources decays in time as

$$f = f_0(x, y, z)e^{-\alpha t}, \quad \alpha > 0.$$

Then considering $u = u_0(x, y, z)e^{-\alpha t}$ we arrive at the equation

$$\Delta u_0 + \frac{\alpha}{a^2} u_0 = -\frac{f_0}{a^2}.$$

3.11   Equation of Electric and Electromagnetic Oscillations

Let C denote the capacitance, R the resistance, L the self-inductance, and $\mathcal{I}$ the leakage. For the voltage v and the current i the following equations can be derived

$$\frac{\partial v}{\partial x} + L\frac{\partial i}{\partial t} + Ri = 0,$$

$$\frac{\partial i}{\partial x} + C\frac{\partial v}{\partial t} + \mathcal{I}v = 0.$$

Differentiating the first equation with respect to t and the second with respect to x gives us the equations

$$\frac{\partial^2 v}{\partial x \partial t} + L\frac{\partial^2 i}{\partial t^2} + R\frac{\partial i}{\partial t} = 0,$$

$$\frac{\partial^2 i}{\partial x^2} + C\frac{\partial^2 v}{\partial x \partial t} + \mathcal{I}\frac{\partial v}{\partial x} = 0.$$

Eliminating $\frac{\partial^2 v}{\partial x \partial t}$ we obtain

$$L\frac{\partial^2 i}{\partial t^2} + R\frac{\partial i}{\partial t} - \frac{1}{C}\frac{\partial^2 i}{\partial x^2} - \frac{1}{C}\mathcal{I}\frac{\partial v}{\partial x} = 0.$$

Substituting the expression for $\frac{\partial v}{\partial x}$ from the first equation we have

$$L\frac{\partial^2 i}{\partial t^2} - \frac{1}{C}\frac{\partial^2 i}{\partial x^2} + R\frac{\partial i}{\partial t} + \frac{\mathcal{I}L}{C}\frac{\partial i}{\partial t} + \frac{\mathcal{I}R}{C}i = 0$$

or

$$\frac{\partial^2 i}{\partial x^2} = LC\frac{\partial^2 i}{\partial t^2} + (RC + \mathcal{I}L)\frac{\partial i}{\partial t} + \mathcal{I}Ri.$$

Similarly, v satisfies the equation

$$\frac{\partial^2 v}{\partial x^2} = LC\frac{\partial^2 v}{\partial t^2} + (RC + \mathcal{I}L)\frac{\partial v}{\partial t} + \mathcal{I}Rv.$$

Thus, the electric current and voltage satisfy the same partial differential equation

$$\frac{\partial^2 w}{\partial x^2} = a_0\frac{\partial^2 w}{\partial t^2} + 2b_0\frac{\partial w}{\partial t} + c_0 w,$$

where $a_0 = LC$, $2b_0 = RC + \mathcal{I}L$, $c_0 = \mathcal{I}R$. This equation is called the telegraph equation.

Setting $w = e^{-\frac{b_0}{a_0}t}u$ simplifies the equation

$$\frac{\partial^2 u}{\partial t^2} = a^2 \frac{\partial^2 u}{\partial x^2} + b^2 u,$$

where $a = \sqrt{\frac{1}{a_0}}$ and $b = \frac{1}{a_0}\sqrt{b_0^2 + a_0 c_0}$.

3.12   Schrödinger Equation

Let ψ denote the wave function of a particle of mass m subject to a potential V such as that due to an electric field, the modulus of ψ is related to the probability the particle is in some spatial configuration at some instant of time. The wave function is a solution of the parabolic equation called the Schrödinger equation

$$i\hbar \frac{\partial \psi}{\partial t} = -\frac{\hbar^2}{2m}\Delta\psi + V\psi,$$

where $\hbar$ stands for the reduced **Planck constant** ($\hbar = h/(2\pi)$). We recall that the Planck constant h, is a fundamental physical constant. For instance, a photon's energy is equal to its frequency multiplied by the Planck constant.

Chapter 4
Boundary Value Problem Statements for Partial Differential Equations

Partial differential equations similarly to ordinary differential equations admit numerous solutions. For example, a general solution of the equation

$$\frac{\partial^2 u}{\partial x \partial y} = 0$$

has the form

$$u(x, y) = \varphi(x) + \psi(y),$$

where φ and ψ are arbitrary twice continuously differentiable functions of one variable. The equations that we consider describe real physical phenomena which should admit a unique description and hence the corresponding solutions of the equations should be unique. In order to distinguish a solution describing certain physical process additional conditions need to be imposed which would guarantee the uniqueness of the solution. Let us clarify this by the following example.

Example 22 The transverse forced vibrations of a finite string are described by the wave equation

$$\frac{\partial^2 u}{\partial t^2} = a^2 \frac{\partial^2 u}{\partial x^2} + f,$$

(see Sect. 3.1). Let l be the length of the string. If the string is fixed at the end points $x = 0$, $x = l$, we have the additional conditions

$$u(0, t) = u(l, t) = 0. \tag{4.1}$$

However, even these conditions are insufficient for specifying a unique solution. It is necessary to know additionally the position of the string at the initial time as well

© The Author(s), under exclusive license to Springer Nature Switzerland AG 2022
A. N. Karapetyants, V. V. Kravchenko, *Methods of Mathematical Physics*,
https://doi.org/10.1007/978-3-031-17845-0_4

as the initial velocity

$$u(x, 0) = \varphi(x), \quad u_t(x, 0) = \psi(x), \tag{4.2}$$

where φ and ψ are some given functions.

Conditions of the form (4.1) are called boundary conditions, while conditions of the form (4.2) are called initial conditions. Conditions (4.1) and the first condition from (4.2) specify the function on the following sets in the (x, t) plane: $\{0 \leq x \leq l, \, t = 0\}$, $\{x = 0, \, t > 0\}$, and $\{x = l, \, t > 0\}$. Boundary conditions may have another form, different from (4.1). For example, at the end points of the string the following conditions can be imposed

$$u_x(0, t) = u_x(l, t) = 0$$

meaning that the ends of the string are free

4.1   Cauchy's Problem Statement and Boundary Conditions. Equations of Normal Type

Consider the partial differential equation of a general form

$$F\left(x_1, x_2, \ldots, x_n, u, \frac{\partial u}{\partial x_1}, \ldots, \frac{\partial u}{\partial x_n}, \frac{\partial^2 u}{\partial x_1^2}, \ldots\right) = 0.$$

When considering a nonstationary process one of the variables is usually chosen as a time variable. Let us introduce t into the equation. For a technical reason it is convenient to preserve n variables $x_1, \ldots x_n$ in the equation, so we will consider from now on the following equation:

$$F\left(t, x_1, x_2, \ldots, x_n, u, \frac{\partial u}{\partial t}, \frac{\partial^2 u}{\partial t^2}, \ldots, \frac{\partial^r u}{\partial t^r}, \frac{\partial u}{\partial x_1}, \ldots\right) = 0.$$

Suppose that this equation is solvable with respect to the derivative

$$\frac{\partial^r u}{\partial t^r},$$

which we assume to be the highest derivative in t variable which is presented in the equation. Therefore, we have

$$\frac{\partial^r u}{\partial t^r} = \phi\left(t, x_1, x_2, \ldots, x_n, u, \frac{\partial u}{\partial t}, \ldots, \frac{\partial^{(r-1)} u}{\partial t^{(r-1)}}, \frac{\partial u}{\partial x_1}, \ldots\right) = 0. \tag{4.3}$$

Definition 23 Equation (4.3) is called normal with respect to the variable t.

The initial conditions for the equation are imposed with respect to the variable t, i.e., to the variable in which the equation is normal.

Definition 24 The conditions

$$\begin{cases} u|_{t=0} = \varphi_1\,(x_1, x_2, \ldots, x_n)\,, \\ \frac{\partial u}{\partial t}\big|_{t=0} = \varphi_2\,(x_1, x_2, \ldots, x_n)\,, \\ \qquad \ldots \\ \frac{\partial^{(r-1)} u}{\partial t^{(r-1)}}\big|_{t=0} = \varphi_r\,(x_1, x_2, \ldots, x_n) \end{cases} \tag{4.4}$$

are called initial conditions for Eq. (4.3)

Definition 25 The problem of finding a solution of (4.3) satisfying conditions (4.4) is called **Cauchy's problem** for Eq. (4.3).

Let Eq. (4.3) with respect to the variables $x_1, x_2, \ldots, x_n$ be considered in a domain V of an n-dimensional space, and let Γ be a boundary of the domain and $\overrightarrow{n}$ the exterior normal to Γ.

Definition 26 The condition of the form

$$u|_\Gamma = \psi(s, t), \quad s \in \Gamma, \quad t > 0$$

is called boundary condition for Eq. (4.3). It is often called the first kind boundary condition or the Dirichlet condition. The condition

$$\frac{\partial u}{\partial \overrightarrow{n}}\bigg|_\Gamma = \psi(s, t), \quad s \in \Gamma, \quad t > 0$$

is called the second kind boundary condition or the Neumann condition. The condition

$$\left(\alpha u + \beta \frac{\partial u}{\partial \overrightarrow{n}}\right)\bigg|_\Gamma = \psi(s, t), \quad s \in \Gamma, \quad t > 0$$

is called the mixed type (or third kind, or Robin's) condition.

In next sections we specify these conditions in more details in hyperbolic, parabolic and elliptic cases.

4.2   Boundary Value Problems for the Wave Equation

In this section we formulate several typical problems for the wave equation

$$\frac{\partial^2 u}{\partial t^2} = a^2 \Delta u + f.$$

1. Vibrations of an unbounded medium (a string, a membrane of an infinitely large size, gas in an unbounded domain). In this case only initial conditions are imposed, that is the following Cauchy problem is considered

$$\begin{cases} u|_{t=0} = \varphi_1 (x, y, z), \\ \frac{\partial u}{\partial t}\big|_{t=0} = \varphi_2 (x, y, z). \end{cases} \tag{4.5}$$

   Clearly, there are no boundary conditions in that case.

2. Vibrations of a bounded medium. For definiteness, let us consider a finite string. Again the Cauchy problem is considered

$$u|_{t=0} = \varphi_1 (x), \qquad \frac{\partial u}{\partial t}\bigg|_{t=0} = \varphi_2 (x).$$

Additionally, boundary conditions need to be imposed. Let us consider possible boundary conditions.

(a) the string is fixed at end points

$$u|_{x=0} = 0, \qquad u|_{x=l} = 0,$$

(b) the string is not fixed at end points, and the law of their motion is known

$$u|_{x=0} = g_1(t), \qquad u|_{x=l} = g_2(t),$$

(c) the end points of the string are not fixed, and the forces acting on them are known, then due to Hooke's law,

$$u_x|_{x=0} = g_1(t), \qquad u_x|_{x=l} = g_2(t), \tag{4.6}$$

(d) the end points of the string are fixed elastically

$$\begin{cases} u_x(0, t) - h_1 u(0, t) = g_1(t), \\ u_x(l, t) + h_2 u(l, t) = g_2(t), \end{cases}$$

(e) the ends of the string are free

$$u_x|_{x=0} = 0, \quad u_x|_{x=l} = 0. \tag{4.7}$$

The conditions (4.7) represent a special case of the conditions (4.6).

Definition 27 The first boundary value problem for the wave equation

$$\frac{\partial^2 u}{\partial t^2} = a^2 \frac{\partial^2 u}{\partial x^2} + f(x, t)$$

is the problem of finding a solution of this equation satisfying the Cauchy conditions

$$u(x, 0) = \varphi_1(x), \quad u_t(x, 0) = \varphi_2(x)$$

and the boundary conditions

$$\begin{cases} u(0, t) = g_1(t), \\ u(l, t) = g_2(t). \end{cases} \tag{4.8}$$

If instead of the conditions (4.8) the following boundary conditions are imposed

$$\begin{cases} u_x(0, t) = g_1(t), \\ u_x(l, t) = g_2(t), \end{cases}$$

we have **the second boundary value problem**. Finally, under the conditions

$$\begin{cases} \alpha u(0, t) + \beta u_x(0, t) = g_1(t), \\ \gamma u(l, t) + \delta u_x(l, t) = g_2(t) \end{cases} \tag{4.9}$$

we have **the mixed** or **the third boundary value problem**.

Remark 28 The boundary conditions can be of different type at the ends, for example,

$$\begin{cases} u(0, t) = g_1(t), \\ u_x(l, t) = g_2(t). \end{cases}$$

Obviously, this is a special case of the conditions (4.9)

$$\begin{cases} 1 \cdot u(0, t) + 0 \cdot u_x(0, t) = g_1(t), \\ 0 \cdot u(l, t) + 1 \cdot u_x(l, t) = g_2(t). \end{cases}$$

In a similar way three boundary value problems are formulated for the n-dimensional wave equation

$$\frac{\partial^2 u}{\partial t^2} = a^2 \left(\frac{\partial^2 u}{\partial x_1^2} + \frac{\partial^2 u}{\partial x_2^2} + \ldots + \frac{\partial^2 u}{\partial x_n^2} \right) + f(x_1, x_2, \ldots, x_n, t)$$

with the only difference that instead of u_x the normal derivative of u is used with respect to the normal vector of the boundary Γ of a domain Ω in which the variables $(x_1, x_2, \ldots, x_n)$ range.

4.3   Boundary Value Problems for the Heat Equation

Consider the heat equation

$$\frac{\partial u}{\partial t} = a^2 \Delta u + f.$$

When the phenomenon of the heat transfer in a physical body of an infinite size is studied, it is sufficient to impose the initial condition

$$u|_{t=0} = \varphi(x_1, x_2, \ldots, x_n),$$

where φ is a given function. In particular, in the case of a rod (one space dimension) the initial condition becomes

$$u(x, 0) = \varphi(x). \tag{4.10}$$

When the heat transfer phenomenon is considered in a physical body of finite dimensions additional boundary conditions are necessary. For simplicity, let us discuss the propagation of heat in a rod of length l. In addition to the initial condition (4.10) the following boundary conditions can be imposed.

1. At the endpoints of the rod a prescribed temperature is maintained

$$u(0, t) = g_1(t), \quad u(l, t) = g_2(t). \tag{4.11}$$

2. At the endpoints of the rod the heat flux is given

$$q|_0 = g_1(t), \quad q|_l = g_2(t).$$

Since

$$q|_0 = k \frac{\partial u}{\partial x}\bigg|_{x=0}, \quad q|_l = -k \frac{\partial u}{\partial x}\bigg|_{x=l}, \tag{4.12}$$

we obtain the boundary conditions of the form

$$u_x(0, t) = g_1(t), \quad u_x(l, t) = g_2(t).$$

3. At the endpoints of the rod the transfer of the heat takes place. Let $\theta_1(t)$ and $\theta_2(t)$ be the temperature of the surrounding medium near $x = 0$ and $x = l$, respectively. We suppose that the heat transfer follows the Newton law, which states that the rate of change of the temperature of an object is proportional to the difference between its own temperature and the ambient temperature (i.e., the temperature of its surroundings)

$$q|_{\text{across the boundary}} = \lambda\left(u(s, t) - \theta(t)\right),$$

where $\theta(t)$ is the ambient temperature and $u(s, t)$ is the temperature of the object on the boundary $s = (x_1, x_2, \ldots, x_n) \in \Gamma$. In particular, for the rod we have

$$q|_{x=0} = \lambda\left(u(0, t) - \theta_1(t)\right), \quad q|_{x=l} = -\lambda\left(u(l, t) - \theta_2(t)\right),$$

or, according to (4.12),

$$u_x(0, t) - h_1 u(0, t) = g_1(t), \quad u_x(l, t) + h_2 u(l, t) = g_2(t),$$

where $h_1 = h_2 = \lambda/k$.

4. The endpoints of the rod are heat insulated that means that there is no heat transfer at the endpoints and hence

$$u_x(0, t) = u_x(l, t) = 0.$$

Thus, the first, the second, and the mixed boundary value problems for the heat equation are formulated analogously to the corresponding problems for the wave equation. The only difference is the number of the initial conditions.

4.4 Boundary Value Problems for the Poisson and Laplace Equations

Consider the Poisson equation

$$\Delta u \equiv \frac{\partial^2 u}{\partial x_1^2} + \frac{\partial^2 u}{\partial x_2^2} + \ldots + \frac{\partial^2 u}{\partial x_n^2} = f. \tag{4.13}$$

Let $(x_1, x_2, \ldots, x_n) \in V$, and Γ be the boundary of V. There are three basic types of boundary value problems, depending on the form of the boundary condition.

Definition 29 The **first boundary value problem** or the **Dirichlet problem** consists in finding a solution of the Poisson equation (4.13) satisfying the (Dirichlet) boundary condition

$$u|_\Gamma = \psi(s), \quad s \in \Gamma.$$

That is the values of the solution in all points of the boundary are known.

The **second boundary value problem** or the **Neumann problem** consists in finding a solution of the Poisson equation (4.13) satisfying the (Neumann) boundary condition

$$\frac{\partial u}{\partial n}\bigg|_\Gamma = \psi(s), \quad s \in \Gamma.$$

That is the values of the normal derivative of the solution in all points of the boundary are known.

The **mixed boundary value problem** or the **Robin problem** consists in finding a solution of the Poisson equation (4.13) satisfying the boundary condition

$$\left(\alpha u + \beta \frac{\partial u}{\partial n}\right)\bigg|_\Gamma = \psi(s), \quad s \in \Gamma.$$

That is the values of a linear combination of the normal derivative and the solution in all points of the boundary are known. Here α and β can be some given functions.

If V is a bounded domain, then the above problems are called interior. The boundary value problems in unbounded domains are also of a great interest. Let Γ be a closed surface enclosing a simply connected domain V which we denote by V^+. The exterior domain which is the complement of $\overline{V^+}$ we denote by V^-. If the solution is sought in V^-, satisfying one of the above boundary conditions, the corresponding boundary value problem is called exterior. In such case, usually a condition of the boundedness of the solution at infinity is added

$$|u(M)| \leq \text{Const}, \quad M = (x_1, x_2, \ldots, x_n) \to \infty.$$

Sometimes instead of this condition a more restrictive condition is imposed

$$u(M) \to 0, \quad M \to \infty.$$

4.5   Boundary and Initial Conditions for the Telegraph Equation

In Sect. 3.11 the following (telegraph) equation was derived both for the voltage v and the current i

$$\frac{\partial^2 v}{\partial x^2} = LC\frac{\partial^2 v}{\partial t^2} + (RC + \mathcal{I}L)\frac{\partial v}{\partial t} + \mathcal{I}Rv$$

and

$$\frac{\partial^2 i}{\partial x^2} = LC\frac{\partial^2 i}{\partial t^2} + (RC + \mathcal{I}L)\frac{\partial i}{\partial t} + \mathcal{I}Ri.$$

Let us make use of the original first order system

$$\frac{\partial v}{\partial x} + L\frac{\partial i}{\partial t} + Ri = 0,$$

$$\frac{\partial i}{\partial x} + C\frac{\partial v}{\partial t} + \mathcal{I}v = 0.$$

The initial conditions have the form $v|_{t=0} = f(x)$, $i|_{t=0} = g(x)$. Let us transform them in order to have the initial conditions for one of the functions only. We have

$$v|_{t=0} = f(x) \quad \text{and} \quad v_t|_{t=0} = -\frac{1}{C}\left(g'(x) + \mathcal{I}f(x)\right)$$

and

$$i|_{t=0} = g(x) \quad \text{and} \quad i_t|_{t=0} = -\frac{1}{L}\left(f'(x) + Rg(x)\right).$$

The following boundary conditions are usually imposed.

1. At the beginning of the electric circuit, turn on the electromotive force E and the endpoint of the circuit is given,

$$v|_{x=0} = E, \quad v|_{x=l} = 0.$$

2. The beginning of the circuit is under a sinusoidal voltage and the endpoint is isolated,

$$v|_{x=0} = E\sin\omega t, \quad i|_{x=l} = 0.$$

From here we find that

$$i|_{x=l} = 0 \quad \text{and} \quad \frac{\partial i}{\partial x}\bigg|_{x=0} = -CE\omega\cos\omega t - \mathcal{I}E\sin\omega t$$

or for v,

$$v|_{x=0} = E\sin\omega t, \quad \text{and} \quad \frac{\partial v}{\partial x}\bigg|_{x=l} = 0.$$

4.6 Well Posedness of Problems of Mathematical Physics

Since the boundary value problems model real physical phenomena, they should satisfy the following three natural conditions.

1. The solution of the problem **exists** in some class of functions B, that is, there exists a function $\varphi \in B$ which solves the problem.
2. The solution of the problem is **unique** in this class of functions.
3. The solution is **stable**, that is it depends continuously on the initial data (initial and boundary conditions).

The condition of stability means that small changes in the initial data produce small changes in the solution. This is very important when solving practical problems. Here we deal with an operator equation

$$Au = \Phi,$$

where the right hand side can be calculated approximately. Hence the obtained "approximate" solution should be close to the real solution. In order to give a rigorous definition of the stability of the problem we need to learn how to evaluate the closeness of initial data as well as that of solutions. In other words, how to measure a "distance" between functions. This can be done in different ways. Note that here we give only some of the most common examples, without going into details.

1. **The uniform closeness of zero order.** Let $u_1(M)$ and $u_2(M)$ be continuous functions in a closed bounded domain V ($\overline{V} = V$) of an n-dimensional space. The closeness of the functions is measured as follows

$$\sup_{M\in V} |u_1(M) - u_2(M)| < \varepsilon.$$

The difference of a function $u(M)$ from the function zero is called the uniform norm or the maximum norm of the function $u(M)$: $\|u\| := \max_{M\in V} |u(M)|$.

2. **The uniform closeness of m-th order,**

$$\sup_{M \in V} \left\{ |u_1(M) - u_2(M)|, \left| \frac{\partial u_1}{\partial x_1} - \frac{\partial u_2}{\partial x_1} \right|, \ldots, \left| \frac{\partial^m u_1}{\partial x_n^m} - \frac{\partial^m u_2}{\partial x_n^m} \right| \right\} < \varepsilon.$$

That is, the closeness of all partial derivatives up to the m-th order is also measured.

3. **The closeness in the mean (integral norm) of order** p ($1 \leq p < \infty$),

$$\int_V |u_1(M) - u_2(M)|^p \, dV_M < \varepsilon.$$

In particular, the mean squared closeness is measured as

$$\int_V |u_1(M) - u_2(M)|^2 \, dV_M < \varepsilon.$$

There exist other ways to measure the closeness of functions. The general idea consists in associating to any pair of functions from a certain functional class a distance between them, and their closeness to each other is understood as the smallness of this distance.

In order to formulate the "closeness" in the language of functional analysis, using normed space and the distance provided by the norm we recall some elementary facts and definitions.

Definition 30 A set of elements is called a **metric space** if to every pair of its elements u_1, u_2 there corresponds a number $\rho(u_1, u_2)$ called a **distance** (or metric) between u_1 and u_2 which satisfies the following requirements.

1. $\rho(u_1, u_2) \geq 0$, and $\rho(u_1, u_2) = 0$ if only $u_1 = u_2$.
2. $\rho(u_1, u_2) = \rho(u_2, u_1)$.
3. $\rho(u_1, u_2) \leq \rho(u_1, u_3) + \rho(u_3, u_2)$, this inequality is called the **triangle inequality**.

The following metric spaces are among the most frequently used.

1. $\mathbb{R}$—the real line with the distance defined as $\rho(x, y) = |x - y|$.
2. The space $C\left(\overline{V}\right)$ of all continuous functions defined on $\overline{V}$ with the metric

$$\rho(u_1, u_2) = \max_{M \in \overline{V}} |u_1(M) - u_2(M)|.$$

3. The space $C^m\left(\overline{V}\right)$ of all functions defined on $\overline{V}$ which are continuously differentiable up to m-th order with the metric

$$\rho(u_1, u_2) = \max_{M \in \overline{V}} \left\{ |u_1(M) - u_2(M)|, \left| \frac{\partial u_1}{\partial x_1} - \frac{\partial u_2}{\partial x_1} \right|, \ldots, \left| \frac{\partial^m u_1}{\partial x_n^m} - \frac{\partial^m u_2}{\partial x_n^m} \right| \right\}.$$

Obviously, $C\left(\overline{V}\right) = C^0\left(\overline{V}\right)$.

4. The space $L_p(V)$ $(1 \leq p < \infty)$ of all (Lebesgue measurable) functions (equivalence classes) u whose absolute value raised to the p-th power has a finite integral, or equivalently, that

$$\int_V |u(M)|^p \, dV_M < \infty.$$

The distance is defined as

$$\rho(u_1, u_2) = \left(\int_V |u_1(M) - u_2(M)|^p \, dV_M \right)^{\frac{1}{p}}.$$

The following two definitions serve us for distinguishing an important class of complete metric spaces.

Definition 31 A sequence u_n, $n = 1, 2, \ldots$ of elements of a metric space is called a **Cauchy sequence** if for any $\varepsilon > 0$ there exists a number N, such that for all n_1, $n_2 > N$ the inequality holds

$$\rho(u_{n_1}, u_{n_2}) < \varepsilon.$$

Definition 32 A metric space is said to be **complete** if the limit of any Cauchy sequence of its elements belongs to this space.

Example 33 The space $\mathbb{R}$ is complete, while the space of all rational numbers although being metric (with $\rho(x, y) = |x - y|$) is incomplete. For example, the sequence of its elements 1, 1.4, 1.41,... is a Cauchy sequence, but its limit $\sqrt{2}$ does not belong to the space.

In what follows we consider linear spaces, that is, such sets of elements in which the operation of the addition of elements is defined as well as the multiplication of elements by numbers and all other algebraic axioms of a linear space are valid (there exist a zero element, a unit element, etc.).

Definition 34 A linear metric space is said to be **normed** if the distance $\rho(u_1, u_2)$ satisfies the following two axioms:

1. $\rho(u_1, u_2) = \rho(u_1 - u_3, u_2 - u_3)$, that is, it is invariant with respect to the shift.
2. $\rho(\lambda u_1, \lambda u_2) = |\lambda| \, \rho(u_1, u_2)$, that is, it is homogeneous. Here λ is a number.

If additionally such metric space is complete then it is called a **complete normed space** or **Banach space**.

This definition is equivalent to the following definition.

Definition 35 A linear space is said to be **normed** if to every its element u there corresponds a number $\|u\|$, called the norm of u, satisfying the following conditions.

1. $\|u\| \geq 0$ and $\|u\| = 0$ if only $u = 0$ (zero element of space).
2. $\|\lambda u\| = |\lambda| \, \|u\|$.
3. $\|u_1 + u_2\| \leq \|u_1\| + \|u_2\|$—the triangle inequality.

The equivalence of these two definitions follows from the equality $\|u\| = \rho(u, 0)$. Then the first condition from Definition 35 becomes obvious, the second follows from the second condition of Definition 34. Finally, we have

$$\|u_1 + u_2\| =^{\text{def}} \rho(u_1 + u_2, 0) =^{\text{Def 34}} \rho(u_1, -u_2) \leq^{\text{Def 30}} \rho(u_1, 0) + \rho(0, -u_2)$$

$$\leq^{\text{Def 30}} \rho(u_1, 0) + \rho(u_2, 0) =^{\text{def}} \|u_1\| + \|u_2\|.$$

The spaces $\mathbb{R}$, $C\left(\overline{V}\right)$, $C^m\left(\overline{V}\right)$, $L_p(V)$ $(p \geq 1)$ are Banach spaces.

According to Definitions 34 and 35, two functions are said to be close to each other if the norm of their difference $\|u_1 - u_2\|$ is small ($\|u_1 - u_2\| = \rho(u_1 - u_2, 0) = \rho(u_1, u_2)$), that is, the definition of the closeness of functions is the same as before.

Definition 36 (The definition of well posedness). A problem of mathematical physics is said to be well posed if it is possible to indicate such pair of normed spaces B_1 and B_2 that for any input data from B_2 the problem is solvable and the solution is unique in the space B_1. Moreover, this solution is stable, that is, a small change in the input data with respect to the norm of B_2 produces a small change in the solution with respect to the norm of B_1.

Thus, the meaning of the stability of a problem can be formulated as follows. Let u be a solution corresponding to the input data φ, and $\tilde{u}$ be a solution corresponding to the input data $\tilde{\varphi}$. Then for any $\varepsilon > 0$ there exists such $\delta > 0$ that $\|u - \tilde{u}\|_{B_1} < \varepsilon$ when $\|\varphi - \tilde{\varphi}\|_{B_2} < \delta$. The spaces B_1 and B_2 are called then the well posedness classes of the problem.

Not all problems, of course, are well posed. Let us show that the Cauchy problem for the Laplace equation can be ill posed.

Example 37 (Hadamard's Example) In the domain

$$V = \left\{(x, y) \in \mathbb{R}^2 : \ -\infty < x < \infty, \ 0 \leq y \leq a\right\}$$

consider the elliptic equation

$$\frac{\partial^2 u}{\partial x^2} + \frac{\partial^2 u}{\partial y^2} = 0 \tag{4.14}$$

for a function $u(x, y)$ subject to the initial conditions

$$u(x, 0) = 0, \quad u_y(x, 0) = \frac{1}{n} \sin nx. \tag{4.15}$$

The functions $\varphi_1(x) = 0$ and $\varphi_2(x) = \frac{1}{n}\sin nx$ belong to the space $C(-\infty, \infty)$. The problem is considered in the domain V which is a strip, and the solution u is sought in the class $C^2(V)$.

The unique solution of the Laplace equation (4.14) subject to the homogeneous initial conditions $u(x, 0) = 0$, $u_y(x, 0) = 0$ is zero.

It is a matter of substitution to check that the function

$$u(x, y) = \frac{1}{n^2}\sin nx \sinh ny$$

is a solution of the Cauchy problem (4.14), (4.15).

Observe that the initial (input) data are quite close to zero with respect to the norm of $B_2 = C(-\infty, \infty)$. Indeed,

$$\|\varphi - \widetilde{\varphi}\|_{B_2} = \left\|\frac{\sin nx}{n} - 0\right\|_{C(-\infty,\infty)} = \sup_{-\infty < x < \infty}\left|\frac{\sin nx}{n}\right| < \frac{1}{n}$$

and for n being sufficiently large, the difference in norm between $\varphi(x) = \frac{\sin nx}{n}$ and $\widetilde{\varphi} = 0$ is arbitrarily small. At the same time for the corresponding solutions u and $\widetilde{u}$ we have

$$\|u - \widetilde{u}\|_{B_1} = \left\|\frac{1}{n^2}\sin nx \sinh ny - 0\right\|_{C^2(V)}$$

$$= \sup_{(x,y)\in V}\left\{|u(x, y)|, \left|\frac{\partial u}{\partial x}\right|, \left|\frac{\partial u}{\partial y}\right|, \left|\frac{\partial^2 u}{\partial x^2}\right|, \left|\frac{\partial^2 u}{\partial y^2}\right|, \left|\frac{\partial^2 u}{\partial x \partial y}\right|\right\}$$

$$\geq \sup_{\substack{-\infty < x < \infty \\ 0 \leq y \leq a}}\left|\frac{1}{n^2}\sin nx \sinh ny\right|$$

$$= \sup_{\substack{-\infty < x < \infty \\ 0 \leq y \leq a}}\frac{1}{n^2}|\sinh ny| \to \infty \quad \text{when } n \to \infty.$$

We have that for small deviations of the input data from zero, the solution can differ arbitrarily from the trivial solution. Thus, the problem (4.14), (4.15) is unstable and hence ill posed.

Another example of an ill-posed problem is the following Dirichlet problem for a hyperbolic equation.

Example 38 Consider the equation

$$\frac{\partial^2 u}{\partial x \partial y} = 0$$

in the rectangle $V = \{x, y \mid 0 < x < a,\ 0 < y < b\}$ with the boundary Γ and the boundary condition $u|_\Gamma = \varphi$. More explicitly, the boundary condition has the form

$$\left.\begin{array}{l} u(0, y) = \varphi_1(y) \\ u(a, y) = \varphi_2(y) \end{array}\right\} \quad 0 \le y \le b,$$

$$\left.\begin{array}{l} u(x, 0) = \psi_1(x) \\ u(x, b) = \psi_2(x) \end{array}\right\} \quad 0 \le x \le a$$

(see Fig. 4.1).

Let the compatibility conditions be fulfilled,

$$\varphi_1(b) = \psi_2(0), \quad \psi_2(a) = \varphi_2(b),$$

$$\varphi_1(0) = \psi_1(0), \quad \psi_1(a) = \varphi_2(0).$$

As the spaces B_1 and B_2 let us choose $B_1 = C^2\left(\overline{V}\right)$ and $B_2 = C\left(\Gamma\right)$. Let us show that this Dirichlet problem may possess no solution, and hence it is ill posed.

The general solution of the equation has the form

$$u(x, y) = f(x) + g(y),$$

where f and g are arbitrary functions from $C^2\left(\Gamma\right)$. From the boundary conditions we have that

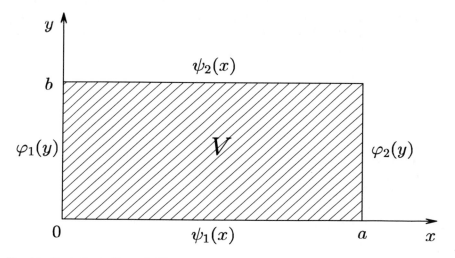

Fig. 4.1 Illustration to Example 38

$$u(0, y) = f(0) + g(y) = \varphi_1(y),$$

$$u(x, 0) = f(x) + g(0) = \psi_1(x),$$

from where we obtain $f(x) = \psi_1(x) - g(0)$ and $g(y) = \varphi_1(y) - f(0)$. Consequently, in general, it is impossible to fulfill the rest of the boundary conditions.

Remark 39 Suppose that the problem is ill posed in a pair of the spaces B_1 and B_2, and its ill-posedness reveals itself in the instability only, but the solution exists and is unique. Often it turns out that the instability is a consequence of an inappropriate choice of the distance. Sometimes it is possible to choose it in another way, and the problem becomes stable. It is said that the problem admits a **regularization** if it is possible to propose such pair of spaces $\widetilde{B}_1$ and $\widetilde{B}_2$ in which it becomes well posed.

4.7   Notion of Generalized Solutions

We have already learned how to use partial differential equations to describe various processes. However, it may happen that some data used to derive the corresponding model are uniform, but may break in some places. For example, a field or medium may not be continuous if there is a phase transition (for example, a transition between a liquid and a solid). In such cases, there will be places where functions describing the process are not differentiable, and hence the partial differential equation does not make sense. We would still like to have mathematically meaningful concept of solving a partial differential equation, even in the presence of singularities. This leads to different ideas about the generalized solution, one of which is described below.

Consider, for example, the wave equation

$$\frac{\partial^2 u}{\partial t^2} = a^2 \Delta u. \tag{4.16}$$

A function u will be its solution if it is twice continuously differentiable with respect to all independent variables and satisfies the equation. Such solutions are called **regular**. Consider a sequence u_n of regular solutions. Suppose this sequence tends to some limit function u_0. It is natural to consider u_0 as a solution of (4.16). However, it can happen that the limit function u_0 does not possess continuous partial second derivatives. It can be non-differentiable at all, and its substitution into (4.16) may have no sense. Hence such a limit function is not a solution of (4.16) in a usual sense.

Definition 40 A function is called a **generalized solution** of Eq. (4.16) if it is a limit of a sequence of regular solutions of the same equation.

The use of generalized solutions is often convenient because they can exist under less restrictive conditions on initial data.

The above approach to defining generalized solution was based on the idea of approximating the rough and even putative solution by smoother objects which were already known to be solutions. One can go further and allow these object to be approximate solutions.

Another approach is simply to accept that the solution contains singularities, and try to make sense of the partial differential equation in a way which does not require the solution to be smooth (in a regular sense). For instance, one might consider distributional solutions, which leads to the well known theory of **generalized functions** or **distributions**. The most famous distribution is the Dirac **delta-function** (δ-function), also known as the unit impulse symbol. It is a generalized function or distribution over the real numbers, whose value is zero everywhere except at zero, and whose integral over the real line is equal to one.

Summarizing the above said, we conclude that in a variety of notions of generalized (distributional, weak) solutions a major task in this field is to identify to what extent these notions are acceptable when considering a particular task, taking into account the real meaning of the process under consideration.

Chapter 5
Cauchy Problem for Hyperbolic Equations

5.1 Cauchy Problem for the One-Dimensional Wave Equation

Let us reduce the wave equation

$$\frac{\partial^2 u}{\partial t^2} = a^2 \frac{\partial^2 u}{\partial x^2}$$

to the canonical form. The equation for characteristics

$$dx^2 - a^2 dt^2 = 0$$

possesses the following integrals $x \pm at = \text{Const}$. Thus, the change of the variables to be considered is $\xi = x + at$, $\eta = x - at$, which reduces the equation to the canonical form

$$\frac{\partial^2 u}{\partial \xi \partial \eta} = 0.$$

The general solution of this equation can be written as $u = \varphi(\xi) + \psi(\eta)$ or, equivalently,

$$u(x, t) = \varphi(x - at) + \psi(x + at), \tag{5.1}$$

where φ and ψ are arbitrary twice continuously differentiable functions.

Now, let us choose the functions φ and ψ such that the Cauchy problem conditions be fulfilled

$$u(x, 0) = \varphi_1(x),$$

© The Author(s), under exclusive license to Springer Nature Switzerland AG 2022
A. N. Karapetyants, V. V. Kravchenko, *Methods of Mathematical Physics*,
https://doi.org/10.1007/978-3-031-17845-0_5

$$u_t(x, 0) = \varphi_2(x)$$

for given functions φ_1 and φ_2.

Substituting (5.1) into the Cauchy conditions leads to the equalities

$$\varphi(x) + \psi(x) = \varphi_1(x)$$

and

$$-a\left(\varphi'(x) - \psi'(x)\right) = \varphi_2(x).$$

Thus,

$$\varphi(x) + \psi(x) = \varphi_1(x)$$

and

$$\varphi(x) - \psi(x) = -\frac{1}{a} \int_{x_0}^x \varphi_2(s) ds + C.$$

From here we find out that

$$\varphi(x) = \frac{1}{2}\varphi_1(x) - \frac{1}{2a} \int_{x_0}^x \varphi_2(s) ds + \frac{C}{2},$$

$$\psi(x) = \frac{1}{2}\varphi_1(x) + \frac{1}{2a} \int_{x_0}^x \varphi_2(s) ds - \frac{C}{2}.$$

Substitution of these expressions into (5.1) gives us the general solution of the one-dimensional wave equation in the form

$$u(x, t) = \frac{1}{2}\left(\varphi_1(x - at) + \varphi_1(x + at)\right) + \frac{1}{2a} \int_{x_0}^{x+at} \varphi_2(s) ds - \frac{1}{2a} \int_{x_0}^{x-at} \varphi_2(s) ds,$$

which can be written as follows

$$u(x, t) = \frac{\varphi_1(x - at) + \varphi_1(x + at)}{2} + \frac{1}{2a} \int_{x-at}^{x+at} \varphi_2(s) ds. \tag{5.2}$$

This representation for the general solution of the one-dimensional wave equation is called the **d'Alembert formula**.

Thus, if the Cauchy problem admits a solution, then it necessarily has the form (5.2). Moreover, due to the way of its construction, this solution is unique. Substitution of (5.2) into the wave equation shows that this function is a regular solution if φ_1 is twice continuously differentiable and φ_2 is once continuously differentiable.

As a result, we obtain that for any $\varphi_1 \in C^2(\mathbb{R})$ and $\varphi_2 \in C^1(\mathbb{R})$ the solution of the Cauchy problem exists and is unique in the space $C^2(\Omega)$, where Ω is the half-plane $\Omega = \{(x, t) \mid -\infty < x < \infty, \quad t > 0\}$. Let us prove that the solution is stable.

Obviously, the identical zero is the only solution corresponding to the trivial initial values. Hence it is sufficient to prove that small deviations from zero of the initial data produce a solution close to zero, and the stability has place in any subdomain

$$\Omega_T = \{(x, t) \mid -\infty < x < \infty, \quad 0 < t < T\}$$

of Ω, that is, for any finite period of time T.

Thus, assume that

$$\|\varphi_1\|_{C^2(\mathbb{R})} \leq \delta \quad \text{and} \quad \|\varphi_2\|_{C^1(\mathbb{R})} \leq \delta.$$

Let us estimate the norm of the corresponding solution,

$$\|u\|_{C^2(\Omega_T)} = \sup_{\Omega_T} \{|u(x, t)|, |u_x(x, t)|, \ldots, |u_{xt}(x, t)|\}.$$

We have

$$|u(x, t)| \leq \frac{|\varphi_1(x - at)| + |\varphi_1(x + at)|}{2} + \frac{2aT}{2a} \sup_{-\infty < x < \infty} |\varphi_2(x)|$$

$$\leq \frac{1}{2} \sup_{-\infty < x < \infty} |\varphi_1(x)| + \frac{1}{2} \sup_{-\infty < x < \infty} |\varphi_1(x)| + T \sup_{-\infty < x < \infty} |\varphi_2(x)|$$

$$\leq \frac{\delta}{2} + \frac{\delta}{2} + T\delta$$

and thus,

$$\sup_{\Omega_T} |u(x, t)| \leq (1 + T)\delta.$$

Similarly,

$$|u_x(x, t)| \leq \frac{|\varphi_1'(x - at)| + |\varphi_1'(x + at)|}{2} + \frac{1}{2a}|\varphi_2(x - at)| + \frac{1}{2a}|\varphi_2(x + at)|$$

$$\leq \frac{\delta}{2} + \frac{\delta}{2} + \frac{\delta}{2a} + \frac{\delta}{2a} = \left(1 + \frac{1}{a}\right)\delta.$$

The other derivatives can be estimated in a similar way. Hence

$$\|u\|_{C^2(\Omega_T)} \leq C\delta,$$

where the constant C is independent of δ. We note, however, that the constant C depends on T. Thus, to obtain the estimate $\|u\|_{C^2(\Omega_T)} \leq \varepsilon$ for a given $\varepsilon > 0$, it is sufficient to choose such initial data that $\|\varphi_1\|_{C^2(\mathbb{R})} \leq \delta$ and $\|\varphi_2\|_{C^1(\mathbb{R})} \leq \delta$, where $\delta = 2/c$. Hence for any finite period of time, i.e., for $0 \leqslant t \leqslant T$, where T is arbitrary fixed, the solution of the Cauchy problem for the one-dimensional wave equation is stable.

Conclusion 41 *The Cauchy problem for the one-dimensional wave equation is well posed provided $B_1 = C^2(\Omega_T)$ and*

$$B_2 = \left\{ C^2(\mathbb{R}) \text{ for } \varphi_1, \text{ and } C^1(\mathbb{R}) \text{ for } \varphi_2 \right\}.$$

5.2 Physical Meaning of d'Alembert Formula

The d'Alembert formula is written often in the form

$$u(x, t) = u_1(x - at) + u_2(x + at),$$

where

$$u_1(x - at) = \frac{\varphi_1(x - at)}{2} - \frac{1}{2a} \int_0^{x-at} \varphi_2(z) dz$$

and

$$u_2(x + at) = \frac{\varphi_1(x + at)}{2} + \frac{1}{2a} \int_0^{x+at} \varphi_2(z) dz.$$

Consider the first term $u = u_1(x - at)$. In the plane (x, t), called the **phase plane**, consider the straight lines $x - at = $ Const (Fig. 5.1). They are characteristics for the wave equation.

Along these lines we have that $u|_{x-at} = u_1(c) = $ Const. That is, the function $u_1(x - at)$ remains constant all along the characteristic lines $x - at = $ Const. In other words, if the observer, who initially, for $t = 0$ was at the point $x = c$, moves at a constant speed $v = a$ along the x-axis, then for her $x = c + at$, and hence she will observe the same value

$$u = u_1(x - at) = u_1(c) = \text{Const}.$$

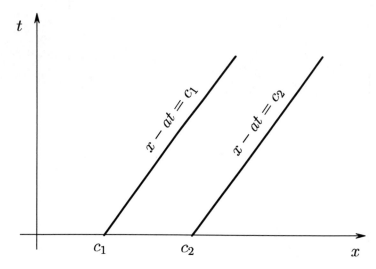

Fig. 5.1 Straight lines $x - at = $ Const

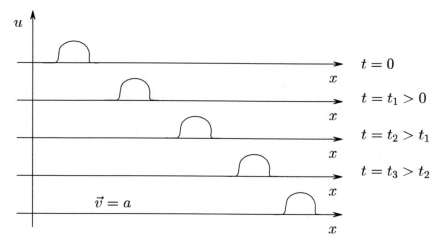

Fig. 5.2 Direct wave

Thus, the deviation of the string $u_1(x - at)$ (see Section 3.1) propagates along the x-axis at the velocity equal to $a = \sqrt{\frac{T}{\rho}}$ (Fig. 5.2).

This propagation of the deviation of the string is called the **direct wave** (or the waveform traveling to the right). Similarly, the function $u = u_2(x + at)$ describes the movement in the opposite direction and is called the **inverse wave** (or the waveform traveling to the left). In general,

$$u(x, t) = u_1(x - at) + u_2(x + at),$$

that is, the movement of the string is a sum of both waves.

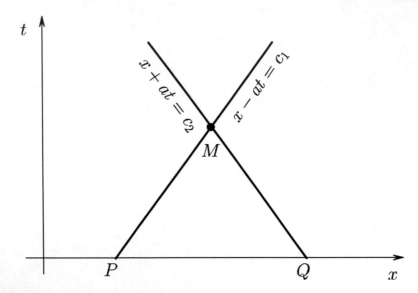

Fig. 5.3 Characteristics crossing the point M

Remark 42 Obviously, under the condition

$$\frac{\varphi_1(x+at)}{2} + \frac{1}{2a}\int_0^{x+at} \varphi_2(z)dz = u_2(x+at) \equiv 0,$$

which can be fulfilled by an appropriate choice of the initial data, we obtain the direct wave alone $u_1(x-at)$, propagating to the right at the velocity a.

Let us take a point $M = (x_0, t_0)$, $t_0 > 0$ of the phase plane and draw the characteristic lines $x \pm at = $ Const crossing it. We will have $c_1 = x_0 - at_0$, $c_2 = x_0 + at_0$, and the points of intersection of the characteristics with the x-axis will be

$$P = (x_0 - at_0, 0) \text{ and } Q = (x_0 + at_0, 0)$$

(see Fig. 5.3).

The d'Alembert formula can be written then as follows

$$u(M) = \frac{1}{2}\left(\varphi_1(P) + \varphi_1(Q)\right) + \frac{1}{2a}\int_P^Q \varphi_2(z)dz. \tag{5.3}$$

The triangle MPQ is called the **characteristic triangle** of the point $M = (x_0, t_0)$.

Conclusion 43 *The deviation of the point x_0 of the string at the instant of time $t_0 > 0$ is fully defined by the initial deviation φ_1 at the vertices P and Q and by*

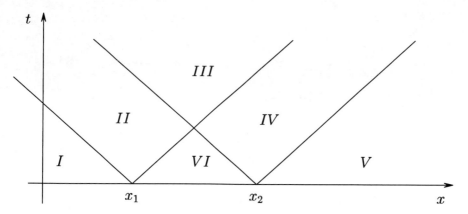

Fig. 5.4 Characteristic lines crossing the points $(x_1, 0)$ and $(x_2, 0)$ divide the upper half-plane into six zones

the initial velocity φ_2 along the base PQ of the characteristic triangle of the point $M = (x_0, t_0)$.

Let us study the wave propagation when the initial data equal zero identically outside a segment $[x_1, x_2]$. For this let us draw the characteristic lines through the points $(x_1, 0)$ and $(x_2, 0)$ (see Fig. 5.4).

They divide the upper half-plane into six zones.

Let us consider the following two situations.

I. Suppose that $\varphi_2 \equiv 0$ for all $x \in (-\infty, \infty)$ and $\varphi_1(x) = 0$ for $x \notin [x_1, x_2]$. Then with the aid of the d'Alembert formula (5.3) it is easy to see that $u \equiv 0$ in zones I, III, V and can be different from zero in zones II, IV, VI. Moreover, zone IV receives the direct wave only, zone II receives the inverse wave only, while both waves arrive at zone VI. It is worth observing that every point of the string vibrates during a finite period of time, after which passes to the equilibrium position and remains in it indefinitely. The direct and inverse waves diverge further along the string at the velocity $v = a = \sqrt{\frac{T}{\rho}}$ (as it is depicted on Fig. 5.5).

II. Let, on the contrary, the initial velocity be not identically zero on $[x_1, x_2]$, while all points of the string be in the equilibrium position initially. Thus, $\varphi_1 \equiv 0$ for all $x \in (-\infty, \infty)$ and $\varphi_2(x) = 0$ for $x \notin [x_1, x_2]$. In this case they say that the string is given an initial impulse. The reasonings are analogous to the previous case, however, now in zone III there appears a nontrivial solution (Fig. 5.6).

From formula (5.3) we obtain

$$u(M) = \frac{1}{2a} \int_P^Q \varphi_2(z)dz = \frac{1}{2a} \int_{x_1}^{x_2} \varphi_2(z)dz = \text{Const}.$$

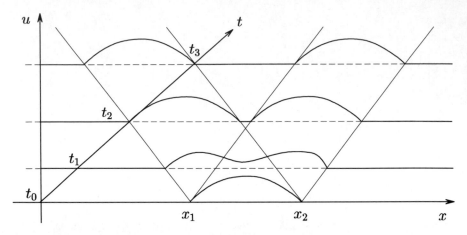

Fig. 5.5 The direct and inverse waves diverge along the string

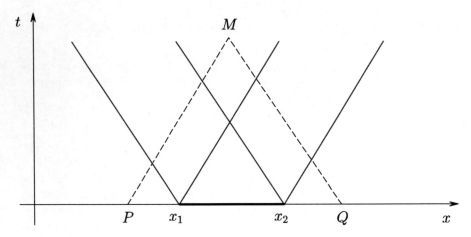

Fig. 5.6 The initial velocity is not identically zero on $[x_1, x_2]$

Hence the picture of the wave propagation looks as it is shown on Fig. 5.7. Over some time every point of the string is shifted to the same height $u = \frac{1}{2a} \int_{x_1}^{x_2} \varphi_2(z) dz$ and remains in that position all the time.

Such phenomenon is called the **residual effect**. The wave rises farther and farther.

5.3   Uniqueness Theorem for the Wave Equation

We shall prove the uniqueness theorem for the bidimensional wave equation

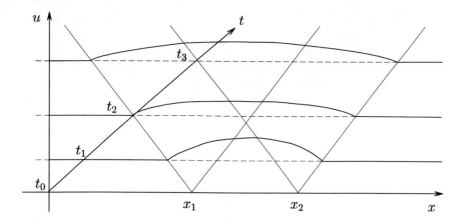

Fig. 5.7 The residual effect. The wave rises farther and farther

$$\frac{\partial^2 u}{\partial t^2} = a^2 \Delta u + f.$$

Additionally we suppose that $a = 1$. Note that if $a \neq 1$, the change of the variable $\tilde{t} = at$ leads to the equation

$$\frac{\partial^2 u}{\partial \tilde{t}^2} = \frac{\partial^2 u}{\partial x^2} + \frac{\partial^2 u}{\partial y^2} + f,$$

where the coefficient equals already one.

Thus, consider the equation

$$\frac{\partial^2 u}{\partial t^2} = \frac{\partial^2 u}{\partial x^2} + \frac{\partial^2 u}{\partial y^2} + f(x, y, t), \tag{5.4}$$

and the Cauchy problem for it

$$u(x, y, 0) = \varphi_1(x, y),$$

$$u_t(x, y, 0) = \varphi_2(x, y).$$

Let us choose an arbitrary point $M_0 = (x_0, y_0, t_0)$ in the phase space (x, y, t), $t_0 > 0$, and consider the cone $(x - x_0)^2 + (y - y_0)^2 - (t - t_0)^2 = 0$ with the apex at that point, the axis being parallel to the t-axis and the generating lines being inclined to the plane at an angle $\alpha = 45°$ (Fig. 5.8).

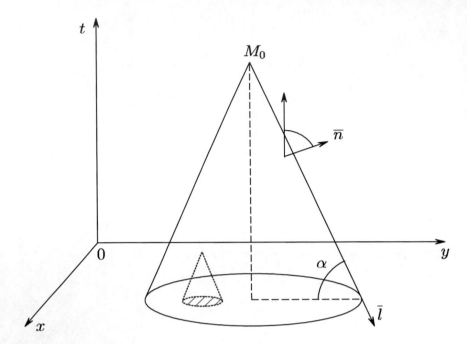

Fig. 5.8 Characteristic cone

The cone constructed in such a way is called **characteristic** (in the case $a \neq 1$ the generating lines should be drawn at an angle $\alpha = \arctan a$).

Let us prove that the solution of the Cauchy problem is unique in the class $C^2(\Omega)$, where $\Omega = \mathbb{R}^2 \times (0, T) = \{(x, y, t) \mid (x, y) \in \mathbb{R}^2, \, t \in (0, T)\}$.

Theorem 44 *The solution of the Cauchy problem for the wave equation (5.4) is uniquely determined by the values of $u(x, y, 0)$ and $u_t(x, y, 0)$ on the base of the characteristic cone.*

Proof Let us suppose that the Cauchy problem possesses two solutions $u_1(x, y, t)$ and $u_2(x, y, t)$. Then the function $u = u_1 - u_2$ satisfies the homogeneous wave equation

$$\frac{\partial^2 u}{\partial t^2} = \frac{\partial^2 u}{\partial x^2} + \frac{\partial^2 u}{\partial y^2} \tag{5.5}$$

with the homogeneous initial conditions

$$u(x, y, 0) = 0, \tag{5.6}$$

$$u_t(x, y, 0) = 0. \tag{5.7}$$

Hence for the proof of the theorem it is sufficient to show that a solution of (5.5), satisfying (5.6) and (5.7), equals zero at the apex of the characteristic cone, $u(x_0, y_0, t_0) = 0$.

We begin with an easily verifiable identity

$$\frac{\partial u}{\partial t}\left(\frac{\partial^2 u}{\partial t^2} - \frac{\partial^2 u}{\partial x^2} - \frac{\partial^2 u}{\partial y^2}\right) = \frac{1}{2}\frac{\partial}{\partial t}\left(\left(\frac{\partial u}{\partial t}\right)^2 + \left(\frac{\partial u}{\partial x}\right)^2 + \left(\frac{\partial u}{\partial y}\right)^2\right)$$

$$-\frac{\partial}{\partial x}\left(\frac{\partial u}{\partial t}\frac{\partial u}{\partial x}\right) - \frac{\partial}{\partial y}\left(\frac{\partial u}{\partial t}\frac{\partial u}{\partial y}\right).$$

In the left hand side we have an identical zero. Let us integrate this equality over the volume V enclosed by the conical surface and the plane $t = 0$. We obtain

$$\int_V \left\{\frac{\partial}{\partial t}\left(\left(\frac{\partial u}{\partial t}\right)^2 + \left(\frac{\partial u}{\partial x}\right)^2 + \left(\frac{\partial u}{\partial y}\right)^2\right) - 2\frac{\partial}{\partial x}\left(\frac{\partial u}{\partial t}\frac{\partial u}{\partial x}\right) - 2\frac{\partial}{\partial y}\left(\frac{\partial u}{\partial t}\frac{\partial u}{\partial y}\right)\right\} dV = 0.$$

Let us transform the volume integral into a surface integral with the aid of the Gauss-Ostrogradsky theorem. The boundary of the volume V consists of the lateral conical surface Γ and the base S.

$$0 = \int_S \left\{\left(\left(\frac{\partial u}{\partial t}\right)^2 + \left(\frac{\partial u}{\partial x}\right)^2 + \left(\frac{\partial u}{\partial y}\right)^2\right)\cos(n,t) - 2\frac{\partial u}{\partial t}\frac{\partial u}{\partial x}\cos(n,x) - 2\frac{\partial u}{\partial t}\frac{\partial u}{\partial y}\cos(n,y)\right\} d\sigma$$

$$+ \int_\Gamma \left\{\left(\left(\frac{\partial u}{\partial t}\right)^2 + \left(\frac{\partial u}{\partial x}\right)^2 + \left(\frac{\partial u}{\partial y}\right)^2\right)\cos(n,t) - 2\frac{\partial u}{\partial t}\frac{\partial u}{\partial x}\cos(n,x) - 2\frac{\partial u}{\partial t}\frac{\partial u}{\partial y}\cos(n,y)\right\} d\sigma.$$

At the points of the base, due to the initial conditions, we have that $u_t = u_x = u_y = 0$, and hence for the first integral above we have $\int_S = 0$. Thus, there remains the integral over Γ:

$$\int_\Gamma \left\{\left(\left(\frac{\partial u}{\partial t}\right)^2 + \left(\frac{\partial u}{\partial x}\right)^2 + \left(\frac{\partial u}{\partial y}\right)^2\right)\cos(n,t) - 2\frac{\partial u}{\partial t}\frac{\partial u}{\partial x}\cos(n,x) - 2\frac{\partial u}{\partial t}\frac{\partial u}{\partial y}\cos(n,y)\right\} d\sigma = 0,$$

$$(5.8)$$

where n stands for the normal vector of the lateral conical surface Γ.

The equation of the cone

$$F \equiv (t - t_0)^2 - (x - x_0)^2 - (y - y_0)^2 = 0$$

implies that on the lateral surface the generating lines of the cone are related by the equality

$$\cos^2(n,t) = \cos^2(n,x) + \cos^2(n,y).$$

$$(5.9)$$

Indeed, since

$$\cos{(n, x)} = \frac{F_x}{\pm\sqrt{F_x^2 + F_y^2 + F_t^2}} = \mp\frac{(x - x_0)}{\sqrt{(t - t_0)^2 + (x - x_0)^2 + (y - y_0)^2}},$$

$$\cos{(n, y)} = \frac{F_y}{\pm\sqrt{F_x^2 + F_y^2 + F_t^2}} = \mp\frac{(y - y_0)}{\sqrt{(t - t_0)^2 + (x - x_0)^2 + (y - y_0)^2}},$$

$$\cos{(n, t)} = \frac{F_t}{\pm\sqrt{F_x^2 + F_y^2 + F_t^2}} = \pm\frac{(t - t_0)}{\sqrt{(t - t_0)^2 + (x - x_0)^2 + (y - y_0)^2}},$$

then taking into account the cone equation, we obtain (5.9).

Substitution of (5.9) into (5.8) leads to the equality

$$\int_\Gamma \frac{1}{\cos{(n, t)}}(u_t^2 \cos^2{(n, x)} + u_t^2 \cos^2{(n, y)} + u_x^2 \cos^2{(n, t)} + u_y^2 \cos^2{(n, t)}$$

$$- 2u_x u_t \cos{(n, x)} \cos{(n, t)} - 2u_y u_t \cos{(n, y)} \cos{(n, t)})ds$$

$$= \int_\Gamma \frac{1}{\cos{(n, t)}} \left((u_t \cos{(n, x)} - u_x \cos{(n, t)})^2 + (u_t \cos{(n, y)} - u_y \cos{(n, t)})^2 \right) ds$$

$$= 0.$$

On the surface of the cone we have $\cos{(n, t)} = \frac{\sqrt{2}}{2}$, and in the last integral the integrand is positive. Hence it equals zero, thus,

$$u_t \cos{(n, x)} - u_x \cos{(n, t)} = 0$$

and

$$u_t \cos{(n, y)} - u_y \cos{(n, t)} = 0$$

on Γ. Hence

$$\frac{u_x}{\cos{(n, x)}} = \frac{u_t}{\cos{(n, t)}} = \frac{u_y}{\cos{(n, y)}} = \lambda,$$

where λ is a constant.

Denote by l the direction of some generating line of the cone. Then

$$\frac{\partial u}{\partial l} = \frac{\partial u}{\partial x} \cos{(x, l)} + \frac{\partial u}{\partial y} \cos{(y, l)} + \frac{\partial u}{\partial t} \cos{(t, l)}$$

$$= \lambda \left(\cos{(n, x)} \cos{(x, l)} + \cos{(n, y)} \cos{(y, l)} + \cos{(n, t)} \cos{(t, l)} \right)$$

$$= \lambda \cos{(n, l)} = 0$$

due to the orthogonality of n to l. Here we used the relation

$$\cos(n, l) = \frac{\vec{n} \cdot \vec{l}}{|\vec{n}| \cdot |\vec{l}|}$$

$$= \vec{n} \cdot \vec{l} = \cos(n, x)\cos(x, l) + \cos(n, y)\cos(y, l) + \cos(n, t)\cos(t, l).$$

Thus, $\frac{\partial u}{\partial l} = 0$ and hence $u = $ Const along every generating line. Since $u = 0$ on the base of the cone we conclude that $u = 0$ along the whole generating line, in particular, at the apex, $u(M) = u(x_0, y_0, t_0) = 0$. ∎

Corollary 45 *If the initial conditions $u(x, y, 0) = \varphi_1(x, y)$, $u_t(x, y, 0) = \varphi_2(x, y)$ are imposed on a part of the plane $t = 0$, on some domain $\mathfrak{I}$, the solution $u = u(x, y, t)$ is defined only for such points (x_0, y_0, t_0) for which the bases of the characteristic cones belong to $\mathfrak{I}$.*

For example, if the initial data are given in a disk, the solution $u(x, y, t)$ is defined only in the cone with the base on this disk and with generating lines at the angle $45°$ (or $\arctan \alpha$ when $a \neq 1$).

Conclusion 46 *When the wave equation (5.5) is considered in a bounded domain, the initial conditions are insufficient for finding the solution for any instant of time t. The solution can be found at a point (x, y) for a small interval of time $t = t(x, y)$.*

5.4 Solution of Cauchy's Problem: Kirchhoff's Formula

Here we derive the **Kirchhoff formula** (see (5.12)) giving a solution of the Cauchy problem for the wave equation in three dimensions. But before we need the following result giving such a solution in particular case.

We will deal with surface integrals over spheres, and $d\sigma_r$ denotes the surface area element for the sphere of radius r.

Lemma 47 *For any twice continuously differentiable function $\varphi(x, y, z)$ the surface integral*

$$u_\varphi(x, y, z, t) = \frac{1}{4\pi a} \int_{\Sigma_{at}} \frac{\varphi(\xi, \eta, \zeta)}{at} d\sigma_{at}$$

over the sphere of radius $r = at$ and center at the point (x, y, z) is a solution of the homogeneous wave equation

$$\frac{\partial^2 u}{\partial t^2} = a^2 \left(\frac{\partial^2 u}{\partial x^2} + \frac{\partial^2 u}{\partial y^2} + \frac{\partial^2 u}{\partial z^2} \right)$$

and satisfies the initial conditions

$$u(x, y, z, 0) = 0$$

and

$$u_t(x, y, z, 0) = \varphi(x, y, z).$$

Proof Let $M = (x, y, z)$ and $P = (\xi, \eta, \zeta)$. Consider the sphere

$$(\xi - x)^2 + (\eta - y)^2 + (\zeta - z)^2 = a^2 t^2$$

(see Fig. 5.9). Let (α, β, γ) be the directing cosines of the vector MP. Then

$$\begin{cases} \xi = x + \alpha a t, \\ \eta = y + \beta a t, \\ \zeta = z + \gamma a t, \end{cases}$$

where

$$\alpha = \frac{\partial F}{\partial x} = \frac{\xi - x}{at}, \quad \beta = \frac{\partial F}{\partial y} = \frac{\eta - y}{at}, \quad \gamma = \frac{\partial F}{\partial z} = \frac{\zeta - z}{at}, \quad \alpha^2 + \beta^2 + \gamma^2 = 1.$$

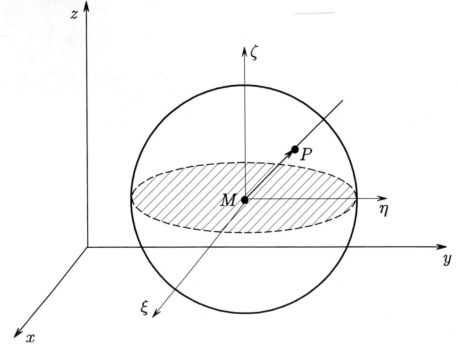

Fig. 5.9 The sphere $(\xi - x)^2 + (\eta - y)^2 + (\zeta - z)^2 = a^2 t^2$

When the point $P = (\xi, \eta, \zeta)$ runs along the sphere σ_{at}, the point (α, β, γ) runs along the sphere σ_1 of radius one and centered at the origin. Obviously,

$$d\sigma_{at} = a^2 t^2 d\sigma_1$$

and hence

$$u_\varphi(x, y, z, t) = \frac{t}{4\pi} \int_{\Sigma_1} \varphi(x + \alpha at, y + \beta at, z + \gamma at) d\sigma_1. \tag{5.10}$$

From here we obtain

$$\frac{\partial^2 u_\varphi}{\partial x^2} + \frac{\partial^2 u_\varphi}{\partial y^2} + \frac{\partial^2 u_\varphi}{\partial z^2} = \frac{t}{4\pi} \int_{\Sigma_1} \left(\frac{\partial^2 \varphi}{\partial \xi^2} + \frac{\partial^2 \varphi}{\partial \eta^2} + \frac{\partial^2 \varphi}{\partial \zeta^2} \right) d\sigma_1.$$

Returning to the sphere Σ_{at}, we have

$$\frac{\partial^2 u_\varphi}{\partial x^2} + \frac{\partial^2 u_\varphi}{\partial y^2} + \frac{\partial^2 u_\varphi}{\partial z^2} = \frac{1}{4\pi a^2 t} \int_{\Sigma_{at}} \left(\frac{\partial^2 \varphi}{\partial \xi^2} + \frac{\partial^2 \varphi}{\partial \eta^2} + \frac{\partial^2 \varphi}{\partial \zeta^2} \right) d\sigma_{at}. \tag{5.11}$$

Differentiation of (5.10) with respect to t gives us the relation

$$\frac{\partial u_\varphi}{\partial t} = \frac{1}{4\pi} \int_{\Sigma_1} \varphi(x + \alpha at, y + \beta at, z + \gamma at) d\sigma_1 + \frac{at}{4\pi} \int_{\Sigma_1} \left(\alpha \frac{\partial \varphi}{\partial \xi} + \beta \frac{\partial \varphi}{\partial \eta} + \gamma \frac{\partial \varphi}{\partial \zeta} \right) d\sigma_1$$

$$= \frac{1}{4\pi a^2 t^2} \int_{\Sigma_{at}} \varphi(\xi, \eta, \zeta) d\sigma_{at} + \frac{1}{4\pi at} \int_{\Sigma_{at}} \left(\alpha \frac{\partial \varphi}{\partial \xi} + \beta \frac{\partial \varphi}{\partial \eta} + \gamma \frac{\partial \varphi}{\partial \zeta} \right) d\sigma_{at}.$$

Application of the Gauss theorem to the last integral leads to the formula

$$\frac{\partial u_\varphi}{\partial t} = \frac{u_\varphi}{t} + \frac{1}{4\pi at} \int_{V_{at}} \left(\frac{\partial^2 \varphi}{\partial \xi^2} + \frac{\partial^2 \varphi}{\partial \eta^2} + \frac{\partial^2 \varphi}{\partial \zeta^2} \right) dV_{at},$$

where V_{at} is a ball of radius $r = at$ and centered at the same point $M = (x, y, z)$.
Denote

$$I = \int_{V_{at}} \Delta \varphi \, d\xi d\eta d\zeta,$$

and thus

$$\frac{\partial u_\varphi}{\partial t} = \frac{u_\varphi}{t} + \frac{I}{4\pi at}.$$

Fig. 5.10 Spherical
coordinates

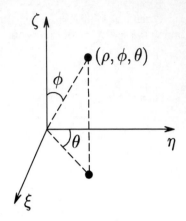

Then for the second time derivative we have

$$\frac{\partial^2 u_\varphi}{\partial t^2} = \frac{1}{t}\frac{\partial u_\varphi}{\partial t} - \frac{u_\varphi}{t^2} - \frac{I}{4\pi a t^2} + \frac{1}{4\pi a t}\frac{\partial I}{\partial t}$$

$$= \frac{1}{t}\left(\frac{u_\varphi}{\partial t} + \frac{I}{4\pi a t}\right) - \frac{1}{t}\left(\frac{u_\varphi}{t} + \frac{I}{4\pi a t}\right) + \frac{1}{4\pi a t}\frac{\partial I}{\partial t}$$

$$= \frac{1}{4\pi a t}\frac{\partial I}{\partial t}.$$

For calculating $\frac{\partial I}{\partial t}$, let us change to spherical coordinates

$$\xi = \rho \sin\phi \cos\theta, \ \eta = \rho \sin\phi \sin\theta, \ \zeta = \rho \cos\phi,$$

where $0 < \rho < \infty, 0 \le \phi < \pi, 0 \le \theta < 2\pi$ (Fig. 5.10).
Then

$$\frac{\partial I}{\partial t} = \frac{\partial}{\partial t}\left(\int_{V_{at}} \Delta\varphi \, d\xi d\eta d\zeta\right) = \frac{\partial}{\partial t}\left(\int_0^{at}\int_0^{2\pi}\int_0^\pi \Delta\varphi\rho^2 \sin\phi \, d\rho d\phi d\theta\right)$$

$$= a\int_0^{2\pi}\int_0^\pi \Delta\varphi|_{\rho=at} \, a^2t^2 \sin\phi \, d\phi d\theta = a\int_{\Sigma_{at}} \Delta\varphi \, d\sigma_{at}.$$

Hence

$$\frac{\partial^2 u_\varphi}{\partial t^2} = \frac{1}{4\pi t}\int_{\Sigma_{at}}\left(\frac{\partial^2\varphi}{\partial\xi^2} + \frac{\partial^2\varphi}{\partial\eta^2} + \frac{\partial^2\varphi}{\partial\zeta^2}\right) d\sigma_{at},$$

and using (5.11) we obtain that

$$\frac{\partial^2 u_\varphi}{\partial t^2} = a^2 \Delta u_\varphi.$$

Thus, u_φ is a solution of the wave equation. It remains to check the fulfillment of the initial conditions. We have

$$\left| u_\varphi \right| = \left| \frac{1}{4\pi a} \int_{\Sigma_{at}} \frac{\varphi(\xi, \eta, \zeta)}{at} d\sigma_{at} \right| \leq \frac{4\pi at}{4\pi a} \sup |\varphi| \to 0$$

when $t \to 0$. Thus, the first condition is fulfilled. Furthermore,

$$\frac{\partial u_\varphi}{\partial t} = \frac{1}{4\pi} \int_{\Sigma_1} \varphi(x + \alpha at, y + \beta at, z + \gamma at) d\sigma_1 + \frac{at}{4\pi} \int_{\Sigma_1} \left(\alpha \frac{\partial \varphi}{\partial \xi} + \beta \frac{\partial \varphi}{\partial \eta} + \gamma \frac{\partial \varphi}{\partial \zeta} \right) d\sigma_1.$$

Since the integral

$$\int_{\Sigma_1} \left(\alpha \frac{\partial \varphi}{\partial \xi} + \beta \frac{\partial \varphi}{\partial \eta} + \gamma \frac{\partial \varphi}{\partial \zeta} \right) d\sigma_1$$

is bounded, the second term tends to zero when $t \to 0$. For the first term we use the theorem on the mean value,

$$\lim_{t \to 0} \frac{1}{4\pi} \int_{\Sigma_1} \varphi(x + \alpha at, y + \beta at, z + \gamma at) d\sigma_1 = \frac{1}{4\pi} \varphi(x, y, z) \cdot |\Sigma_1| = \varphi(x, y, z),$$

where $|\Sigma_1|$ denotes the area of Σ_1.   ∎

With the aid of this lemma it is not difficult to construct a solution of the Cauchy problem for the wave equation

$$\begin{cases} u_{tt} = a^2 \Delta u, \\ u|_{t=0} = \varphi_1, \\ u_t|_{t=0} = \varphi_2. \end{cases}$$

The solution of this problem can be sought in the form $u = v + w$ where v and w are solutions of their respective problems

$$\begin{cases} v_{tt} = a^2 \Delta v, \\ v|_{t=0} = 0, \\ v_t|_{t=0} = \varphi_2 \end{cases} \quad \text{and} \quad \begin{cases} w_{tt} = a^2 \Delta w, \\ w|_{t=0} = \varphi_1, \\ w_t|_{t=0} = 0. \end{cases}$$

We already know the solution of the first problem $v = u_{\varphi_2}$. Let us verify that

$$w = \frac{\partial}{\partial t} \left(u_{\varphi_1} \right)$$

is the solution of the second problem. Indeed, according to Lemma 47,

$$w|_{t=0} = \varphi_1$$

and

$$\left.\frac{\partial w}{\partial t}\right|_{t=0} = \left.\frac{\partial^2 u_{\varphi_1}}{\partial t^2}\right|_{t=0}.$$

Since the function u_{φ_1} is a solution of the wave equation, we have

$$\left.\frac{\partial w}{\partial t}\right|_{t=0} = a^2 \left(\Delta u_{\varphi_1}\right)\Big|_{t=0} = a^2 \Delta \left(u_{\varphi_1}|_{t=0}\right) \equiv 0.$$

Thus,

$$v = u_{\varphi_2}, \qquad w = \frac{\partial}{\partial t}\left(u_{\varphi_1}\right)$$

and hence

$$u = u_{\varphi_2} + \frac{\partial u_{\varphi_1}}{\partial t}$$

or, more explicitly,

$$u(x, y, z, t) = \frac{1}{4\pi a^2} \int_{\Sigma_{at}} \frac{\varphi_2(\xi, \eta, \zeta)}{t}\, d\sigma_{at} + \frac{1}{4\pi a^2} \frac{\partial}{\partial t}\left(\int_{\Sigma_{at}} \frac{\varphi_1(\xi, \eta, \zeta)}{t}\, d\sigma_{at}\right). \tag{5.12}$$

This is the Kirchhoff formula giving a solution of the Cauchy problem for the wave equation in three dimensions.

5.5   Method of Descent: Poisson and d'Alembert Formulas

In the case when the initial data φ_1 and φ_2 are independent of z, it is easy to see from the Kirchhoff formula that the solution of the Cauchy problem does not change when moving the point P parallel to z-axis. Hence it is sufficient to study the oscillations of the solution occurring on the plane XoY. Thus, let us consider the equation

$$\frac{\partial^2 u}{\partial t^2} = a^2 \left(\frac{\partial^2 u}{\partial x^2} + \frac{\partial^2 u}{\partial y^2}\right)$$

with the initial conditions

$$u(x, y, 0) = \varphi_1(x, y)$$

and

$$u_t(x, y, 0) = \varphi_2(x, y).$$

The solution is given by the Kirchhoff formula. Let us transform the surface integral into an integral over its projection. Consider

$$u_\varphi = \frac{1}{4\pi a^2 t} \int_{\Sigma_{at}} \varphi(\xi, \eta, \zeta) d\sigma_{at}, \qquad \varphi(\xi, \eta, \zeta) = \varphi(\xi, \eta).$$

This integral can be represented as a sum

$$u_\varphi = \frac{1}{4\pi a^2 t} \left(\int_{\Sigma_{at}^+} \varphi(\xi, \eta, \zeta) d\sigma_{at} + \int_{\Sigma_{at}^-} \varphi(\xi, \eta, \zeta) d\sigma_{at} \right),$$

where Σ_{at}^+ and Σ_{at}^- are the upper and the lower hemispheres,

$$\Sigma_{at}^\pm: \quad \zeta - z = \pm\sqrt{(at)^2 - (\xi - x)^2 - (\eta - y)^2},$$

$$S_{at}: \quad (\xi - x)^2 + (\eta - y)^2 = (at)^2.$$

Because of the independence of the initial data of z integration over each hemisphere can be replaced by integration over the circle

$$\int_{\Sigma_{at}^\pm} \varphi(\xi, \eta, \zeta) d\sigma_{at} = \int_{S_{at}} \varphi(\xi, \eta) \frac{d\xi d\eta}{\left|\cos\left(\overrightarrow{n}, \overrightarrow{\zeta}\right)\right|}$$

and hence

$$u_\varphi = \frac{1}{2\pi a^2 t} \int_{S_{at}} \varphi(\xi, \eta) \frac{d\xi d\eta}{\left|\cos\left(\overrightarrow{n}, \overrightarrow{\zeta}\right)\right|}.$$

Let us calculate $\left|\cos\left(\overrightarrow{n}, \overrightarrow{\zeta}\right)\right|$. Denote by $Q = (\xi, \eta, 0)$ the projection of P onto the plane XoY (Fig. 5.11). Then

$$\left|\cos\left(\overrightarrow{n}, \overrightarrow{\zeta}\right)\right| = \frac{PQ}{MP} = \frac{\sqrt{(at)^2 - (\xi - x)^2 - (\eta - y)^2}}{at}$$

and

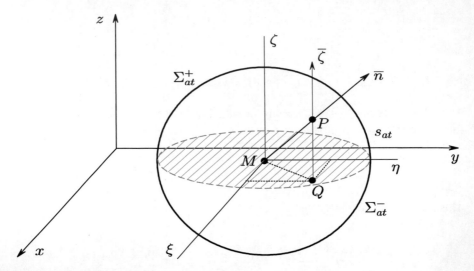

Fig. 5.11 Because of the independence of the initial data of z, integration over each hemisphere $\Sigma_{at}^{\pm}$ can be replaced by integration over the circle

$$u_\varphi = \frac{1}{2\pi a} \int_{S_{at}} \frac{\varphi(\xi, \eta) d\xi d\eta}{\sqrt{(at)^2 - (\xi - x)^2 - (\eta - y)^2}}.$$

With the aid of this formula, from the Kirchhoff formula we obtain the **Poisson formula**

$$u(x, y, t) = \frac{1}{2\pi a} \int_{S_{at}} \frac{\varphi_2(\xi, \eta) d\xi d\eta}{\sqrt{(at)^2 - (\xi - x)^2 - (\eta - y)^2}}$$

$$+ \frac{1}{2\pi a} \frac{\partial}{\partial t} \int_{S_{at}} \frac{\varphi_1(\xi, \eta) d\xi d\eta}{\sqrt{(at)^2 - (\xi - x)^2 - (\eta - y)^2}}. \qquad (5.13)$$

Finally, using the method of descent from the Poisson formula one can obtain the d'Alembert formula. If the initial data are independent also of y, then according to (5.12) (or (5.13)) the particles located in the same plane (or on a straight line parallel to the y-axis) oscillate identically. That is, the oscillations depend on the point only, and can be studied at the points of the X-axis. Consider the function

$$u_\varphi = \frac{1}{2\pi a} \int_{S_{at}} \frac{\varphi(\xi, \eta) d\xi d\eta}{\sqrt{(at)^2 - (\xi - x)^2 - (\eta - y)^2}}, \quad \varphi(\xi, \eta) = \varphi(\xi).$$

We have

$$u_\varphi = \frac{1}{2\pi a} \int\int_{S_{at}} \frac{\varphi(\xi, \eta)d\xi d\eta}{\sqrt{(at)^2 - (\xi - x)^2 - (\eta - y)^2}}$$

$$= \frac{1}{2\pi a} \int_{x-at}^{x+at} \varphi(\xi)d\xi \int_{y-\sqrt{(at)^2-(\xi-x)^2}}^{y+\sqrt{(at)^2+(\xi-x)^2}} \frac{d\eta}{\sqrt{(at)^2 - (\xi - x)^2 - (\eta - y)^2}}$$

$$= \frac{1}{2\pi a} \int_{x-at}^{x+at} \varphi(\xi)d\xi \left(\arcsin \frac{\eta - y}{\sqrt{(at)^2 - (\xi - x)^2}} \right)\Bigg|_{\eta=y-\sqrt{(at)^2-(\xi-x)^2}}^{\eta=y+\sqrt{(at)^2-(\xi-x)^2}}$$

$$= \frac{1}{2\pi a} \int_{x-at}^{x+at} \varphi(\xi)d\xi \cdot 2\arcsin 1 = \frac{1}{2a} \int_{x-at}^{x+at} \varphi(\xi)d\xi.$$

Now application of this formula to (5.13) leads to the d'Alembert formula

$$u(x, t) = \frac{1}{2a}\frac{\partial}{\partial t} \int_{x-at}^{x+at} \varphi_1(\xi)d\xi + \frac{1}{2a} \int_{x-at}^{x+at} \varphi_2(\xi)d\xi$$

$$= \frac{\varphi_1(x - at) + \varphi_1(x + at)}{2} + \frac{1}{2a} \int_{x-at}^{x+at} \varphi_2(\xi)d\xi.$$

5.6   A Closer Look at the Formulas for Solutions of Cauchy's Problems: Wave Propagation—Stability of Solution of Cauchy's Problem—Wave Diffusion

The Kirchhoff formula shows that the solution of the Cauchy problem for the wave equation is obtained by integrating the initial data, and one of the terms involves the time derivative as well. Thus, it is quite evident that the solution is stable for any finite time $t \in [0, T]$.

The wave equation does not take into account the forces of friction. As a consequence, in the ideal medium under consideration waves do not attenuate.

In the case of $n = 3$ the solution is determined by the initial data given on the surface of the sphere. When $n = 2$ the solution is determined by the values of the given functions in the interior of the whole disk. This difference in the formulas implies serious differences in the physical picture of the wave propagation.

Let $n = 3$, and suppose that in a neighborhood V_{M_0} of the point (x_0, y_0, z_0) an initial disturbance is given

$$\varphi_1(x, y, z) = \begin{cases} 0, & M = (x, y, z) \notin V_{M_0} \\ \varphi(x, y, z), & M \in V_{M_0}. \end{cases}$$

The initial velocity is supposed to equal zero.

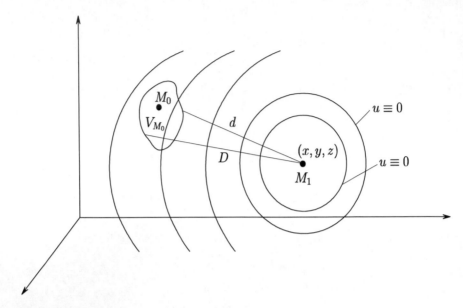

Fig. 5.12 Every point located outside the domain V_{M_0} is at rest first, then oscillates during the time $t_2 - t_1 = \frac{1}{a}(D - d)$, where D is the distance of the most remote point of V_{M_0} to M_1, and d is the distance from V_{M_0} to M_1, and eventually again goes into a state of rest

Consider an arbitrary point $M_1 = (x, y, z)$ outside the domain V_{M_0} and the spheres centered at M_1 of radii at. If the point is far enough from V_{M_0}, initially, during some time, $u = u_\varphi \equiv 0$ due to the Kirchhoff formula, since φ equals zero on near spheres, the disturbance given at M_0 is not perceived at M_1 (did not arrive there yet, Fig. 5.12). At some instant the radius of the sphere becomes sufficiently large as to intersect V_{M_0}, and at the point M_1, according to the Kirchhoff formula there will be perceived an excitation, which will last while the sphere is intersecting the domain V_{M_0}. As soon as the sphere abandons the domain V_{M_0} oscillations at the point M_1 stop. Thus, every point located outside the domain V_{M_0} is at rest first, then oscillates during the time $t_2 - t_1 = \frac{1}{a}(D - d)$, where D is the distance of the most remote point of V_{M_0} to M_1, and d is the distance from V_{M_0} to M_1, and eventually again goes into a state of rest. The wave generated in the domain V_{M_0} has two pronounced fronts, the leading and the rear. At the leading front the points from a state of rest goes into a state of oscillation, while at the rear front the opposite is observed, the oscillating points go to a state of rest. Thus, an initial disturbance, localized in space, induces at every point M_1 of space an effect, localized in time.

The considered waves are called **spherical**.

We proceed to the case $n = 2$. The integration in the corresponding Poisson formula (5.13) is performed over a disk of radius at. Similarly to the previous, a point $M_1 \notin V_{M_0}$ is first at rest. As soon as the disk "seizes" a part of the domain V_{M_0} the oscillations at the point M_1 start, but contrary to the spatial situation, these oscillations will not stop after some time, since the expanding disk will eventually

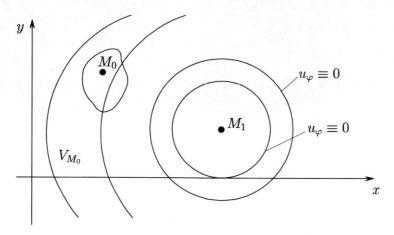

Fig. 5.13 Propagation of a cylindrical wave

seize the whole region V_{M_0}, which will always remain inside of it. In this case we are dealing with the propagation of a **cylindrical wave**, possessing a leading front only (Fig. 5.13). A pronounced rear front of such wave does not exist, although the disturbance at every point M_1 decreases to zero as $t \to \infty$. This erosion of the rear front is called **diffusion**. It has place when $n = 2$. Interestingly, it can be proved that for any odd $n > 3$ there is no wave diffusion.

Chapter 6
Fourier Method for the Wave Equation

In this chapter we study a widely used method of separation of variables for solving partial differential equations. Equations in bounded domains will be considered. First of all, we prove a theorem on uniqueness of a solution.

6.1 Uniqueness Theorem for the First Boundary Value Problem for the Wave Equation

Consider the first boundary value problem for the equation describing the vibrations of a string

$$
\begin{cases}
u_{tt} = a^2 u_{xx} + f, \\
u(x, 0) = \varphi(x), \\
u_t(x, 0) = \psi(x), \\
u(0, t) = v_1(t), \\
u(l, t) = v_2(t).
\end{cases}
\tag{6.1}
$$

Theorem 48 *The solution of the problem (6.1) is unique.*

Proof Suppose, there exist two solutions u_1 and u_2. Then, due to the linearity of the equation, the function $u = u_1 - u_2$ satisfies the homogeneous wave equation and the homogeneous initial and boundary conditions

$$
\begin{cases}
u_{tt} = a^2 u_{xx}, \\
u(x, 0) = u_t(x, 0) = 0, \\
u(0, t) = u(l, t) = 0.
\end{cases}
$$

The theorem will be proved if we show that $u \equiv 0$. The method of the proof is the method of the total energy, widely used in mechanics. Consider the expression for the total energy (see Sect. 3.2)

$$E = \frac{1}{2} \int_0^l \left(\rho u_t^2 + T_0 u_x^2 \right) dx.$$

Due to the energy conservation law, $\frac{dE}{dt} = 0$, that is, $E(t) = E(0) = \text{Const}$ for all t. Due to the initial conditions we obtain

$$E(0) = \frac{1}{2} \int_0^l \left(\rho u_t^2 + T_0 u_x^2 \right)\Big|_{t=0} dx = 0$$

and hence $E(t) \equiv 0$, that is,

$$\int_0^l \left(\rho u_t^2 + T_0 u_x^2 \right) dx = 0. \tag{6.2}$$

The integrand here is a non-negative function, $\rho u_t^2 + T_0 u_x^2 \geq 0$. Hence from (6.2) we have that in fact $\rho u_t^2 + T_0 u_x^2 \equiv 0$ and thus, $u_t \equiv 0$ and $u_x \equiv 0$. Consequently, $u(x, t) = \text{Const}$, and due to the initial conditions, $u(x, t) \equiv 0$. ∎

6.2   Fourier Method for a Fixed String

Consider the model of a string, fixed at both ends

$$\begin{cases} u_{tt} = a^2 u_{xx}, \\ u(x, 0) = \varphi_1(x) \\ u_t(x, 0) = \varphi_2(x), \\ u(0, t) = u(l, t) = 0. \end{cases} \tag{6.3}$$

We begin by solving (6.3) without taking into account the initial conditions, but the boundary conditions only. Let us look for a solution $u(x, t)$ having the form of a product of two functions depending each of one variable

$$u(x, t) = X(x)T(t).$$

Substitution of this function into the wave equation gives us the equality

$$\frac{1}{a^2} T''(t) X(x) = T(t) X''(x).$$

Dividing over $X(x)T(t)$ we obtain

$$\frac{1}{a^2}\frac{T''(t)}{T(t)} = \frac{X''(x)}{X(x)}.$$

The function on the left hand side here depends on t only, while that on the right depends only on x, and the variables t and x are independent of each other. This is possible if only both functions are equal to the same constant, which for convenience we denote by $-\lambda$,

$$\frac{1}{a^2}\frac{T''(t)}{T(t)} = \frac{X''(x)}{X(x)} = -\lambda.$$

Thus, we arrive at two equations

$$T''(t) + a^2\lambda T(t) = 0 \tag{6.4}$$

and

$$X''(x) + \lambda X(x) = 0. \tag{6.5}$$

From the boundary conditions of the problem we have

$$X(0)T(t) = 0 \quad \text{and} \quad X(l)T(t) = 0.$$

We are interested in nontrivial solutions and hence T cannot equal zero identically. Thus,

$$X(0) = 0 \quad \text{and} \quad X(l) = 0. \tag{6.6}$$

For the function $X(x)$ we obtain the problem of finding non trivial solutions of (6.5) satisfying the homogeneous boundary conditions (6.6). This is an example of a **Sturm-Liouville problem**.

More generally, a Sturm-Liouville problem is a problem of finding all values of the parameter λ, for which there exist non trivial solutions y of a Sturm-Liouville equation

$$-\left(p(x)y'\right)' + q(x)y = \lambda r(x)y, \quad x \in (a, b) \tag{6.7}$$

satisfying homogeneous boundary conditions

$$\alpha_1 y(a) + \alpha_2 y'(a) = 0 \quad \text{and} \quad \beta_1 y(b) + \beta_2 y'(b) = 0, \tag{6.8}$$

as well as the corresponding solutions. Here p, q, and r are some functions defined on $[a, b]$. Often they are required to satisfy some regularity conditions, e.g.,

$\{p, p', q, r\} \subset C[a, b]$, and $\alpha_{1,2}$, $\beta_{1,2}$ are some constants such that $|\alpha_1| + |\alpha_2| > 0$ and $|\beta_1| + |\beta_2| > 0$. The parameter λ is called the **spectral parameter**. Such values of the spectral parameter for which there exist non trivial solutions of the problem (6.7), (6.8) are called the **eigenvalues**, and the corresponding solutions are called the **eigenfunctions** of the Sturm-Liouville problem (6.7), (6.8).

Let us solve the simplest Sturm-Liouville problem (6.5), (6.6) which, as we see, arises, in particular, after applying the Fourier method of separation of variables to the first boundary value problem for the wave equation.

Equation (6.5) is a linear homogeneous equation with constant coefficients. Its characteristic equation has the form $r^2 + \lambda = 0$. From here we find a general solution of (6.5).

1. When $\lambda < 0$, a general solution of (6.5) can be written in the form

$$X(x) = c_1 e^{\sqrt{-\lambda} x} + c_2 e^{-\sqrt{-\lambda} x}.$$

Substituting this expression into the boundary conditions (6.6) we obtain the following equations for the constants c_1 and c_2,

$$\begin{cases} c_1 + c_2 = 0, \\ c_1 e^{\sqrt{-\lambda} l} + c_2 e^{-\sqrt{-\lambda} l} = 0, \end{cases}$$

and hence $c_1 = c_2 = 0$. Thus, when $\lambda < 0$, the Sturm-Liouville problem (6.5), (6.6) does not have a non-zero solution.

2. If $\lambda = 0$, then $X(x) = c_1 + c_2 x$, and due to (6.6), again $c_1 = c_2 = 0$. Hence no non zero solution of the Sturm-Liouville problem (6.5), (6.6) exists for $\lambda \leq 0$.

3. When $\lambda > 0$, a general solution of (6.5) can be written in the form

$$X(x) = c_1 \cos \sqrt{\lambda} x + c_2 \sin \sqrt{\lambda} x.$$

Substitution of this expression into (6.6) leads to the following couple of equations for the constants c_1 and c_2,

$$c_1 = 0, \quad c_2 \sin \sqrt{\lambda} l = 0.$$

If $c_2 = 0$, then $X(x) \equiv 0$ (a trivial solution). It remains to set

$$\sin \sqrt{\lambda} l = 0,$$

from where we find that $\sqrt{\lambda} l = k\pi$, $k = \pm 1, \pm 2, \dots$. Thus,

$$\lambda = \lambda_k = \left(\frac{k\pi}{l}\right)^2, \tag{6.9}$$

and hence when $\lambda > 0$, there exist eigenvalues, and they can be computed by (6.9). The corresponding eigenfunctions have the form

$$X_k(x) = \sin \frac{k\pi x}{l}.$$

They are completely determined up to a multiplicative constant. Note that since for $k < 0$ the eigenfunctions just change the sign (which is a multiplicative constant -1), it is sufficient to consider $k > 0$ ($k = 1, 2, \ldots$).

Observe that the eigenvalues of the Sturm-Liouville problem form a countable set, that is, their sequence can be associated with the sequence of natural numbers. When $\lambda = \lambda_k$ the solution of equation (6.4) has the form

$$T_k(t) = A_k \cos \frac{k\pi a}{l}t + B_k \sin \frac{k\pi a}{l}t,$$

where A_k and B_k are arbitrary constants. Thus, the functions

$$u_k(x, t) = X_k(x)T_k(t) = \left(A_k \cos \frac{k\pi a}{l}t + B_k \sin \frac{k\pi a}{l}t \right) \sin \frac{k\pi}{l}x$$

satisfy the wave equation

$$u_{tt} = a^2 u_{xx} \tag{6.10}$$

together with the boundary conditions

$$u(0, t) = u(l, t) = 0.$$

For a linear equation the **superposition principle** is valid.

The **superposition principle:** if the functions $u_1, u_2, \ldots, u_n$ are solutions of (6.10) and fulfill homogeneous boundary conditions, then any their linear combination

$$\alpha_1 u_1 + \alpha_2 u_2 + \ldots + \alpha_n u_n$$

is a solution of (6.10) as well and fulfills the homogeneous boundary conditions.

A **generalized superposition principle:** If the number of functions $u_1, u_2, \ldots$ is infinite, and all of them satisfy (6.10) with homogeneous boundary conditions, then

$$u = \sum_n \alpha_n u_n$$

with arbitrary constants α_n satisfies (6.10) and the homogeneous boundary conditions, if the series converges uniformly and can be twice differentiated termwise. We formulate and prove this principle in the following lemma.

Lemma 49 *Let* $L := \frac{\partial^2}{\partial t^2} - a^2 \frac{\partial^2}{\partial x^2}$ *be the one-dimensional wave operator. Let* $u_1, u_2, \ldots$ *be an infinite sequence of twice differentiable functions of variables* x *and* t, *satisfying*

$$Lu_k = 0, \quad k = 1, 2 \ldots,$$

and

$$u(0, t) = u(l, t) = 0.$$

Also let the series

$$u = \sum_n \alpha_n u_n$$

converge uniformly and be twice differentiable termwise. Then the sum of the series u *is a twice differentiable function, satisfying*

$$Lu = 0 \quad \text{and} \quad u(0, t) = u(l, t) = 0.$$

Proof Due to the assumptions, for the function $u = \sum_n \alpha_n u_n$ we have

$$L\left(\sum_{n=1}^{\infty} \alpha_n u_n\right) = \sum_{n=1}^{\infty} \alpha_n Lu_n = 0.$$

Since the series converges uniformly, its sum is continuous and hence its values for $x = 0$ and $x = l$ can be calculated by substituting $x = 0$ or $x = l$ termwise,

$$\sum_{n=1}^{\infty} \alpha_n \, u_n|_{x=0} = \sum_{n=1}^{\infty} \alpha_n u_n(0, t) = 0$$

and

$$\sum_{n=1}^{\infty} \alpha_n \, u_n|_{x=l} = \sum_{n=1}^{\infty} \alpha_n u_n(l, t) = 0.$$

∎

Until now the initial conditions of the problem (6.3) have not been used. Let us clarify under which conditions on the initial data the generalized superposition principle is applicable for solving the problem, or in other words, the problem admits a solution of the form

$$u(x, t) = \sum_{k=1}^{\infty} u_k(x, t) = \sum_{k=1}^{\infty} \left(A_k \cos \frac{k \pi a}{l} t + B_k \sin \frac{k \pi a}{l} t \right) \sin \frac{k \pi}{l} x.$$

Let us determine the constants A_k and B_k in order to satisfy the initial conditions

$$u(x, 0) = \varphi_1(x), \quad u_t(x, 0) = \varphi_2(x).$$

We have

$$u(x, 0) = \sum_{k=1}^{\infty} A_k \sin \frac{k \pi}{l} x = \varphi_1(x),$$

$$u_t(x, 0) = \sum_{k=1}^{\infty} \frac{k \pi a}{l} B_k \sin \frac{k \pi}{l} x = \varphi_2(x).$$

Expanding the functions $\varphi_1(x)$ and $\varphi_2(x)$ into series in terms of sines

$$\varphi_1(x) = \sum_{k=1}^{\infty} \varphi_k^{(1)} \sin \frac{k \pi}{l} x \quad \text{and} \quad \varphi_2(x) = \sum_{k=1}^{\infty} \varphi_k^{(2)} \sin \frac{k \pi}{l} x,$$

and using the uniqueness theorem for the Fourier series, we find that

$$A_k = \varphi_k^{(1)} \quad \text{and} \quad B_k = \frac{l}{k \pi a} \varphi_k^{(2)}$$

or, more explicitly,

$$A_k = \frac{2}{l} \int_0^l \varphi_1(x) \sin \frac{k \pi}{l} x \, dx \quad \text{and} \quad B_k = \frac{2}{k \pi a} \int_0^l \varphi_2(x) \sin \frac{k \pi}{l} x \, dx. \quad (6.11)$$

In order to verify the applicability of the superposition principle, let us consider the series

$$|u(x, t)| \leq \sum_{k=1}^{\infty} \left(|A_k| + |B_k| \right),$$

$$|u_t(x, t)| \leq \sum_{k=1}^{\infty} \frac{k \pi a}{l} \left(|A_k| + |B_k| \right),$$

$$|u_x(x, t)| \leq \sum_{k=1}^{\infty} \frac{k\pi}{l} \left(|A_k| + |B_k| \right),$$

$$|u_{tt}(x, t)| \leq \sum_{k=1}^{\infty} \frac{k^2 \pi^2 a^2}{l^2} \left(|A_k| + |B_k| \right),$$

$$|u_{xx}(x, t)| \leq \sum_{k=1}^{\infty} \frac{k^2 \pi^2}{l^2} \left(|A_k| + |B_k| \right).$$

If all these majorizing series converge then the corresponding functional series converge uniformly and the operations of termwise differentiation make sense. Thus, it would be sufficient to require the convergence of the series

$$\sum_{k=1}^{\infty} k^j \left(\left| \varphi_k^{(1)} \right| + \frac{2}{k\pi a} \left| \varphi_k^{(2)} \right| \right), \quad \text{for } j = 0, 1, 2.$$

Since the terms of the series are all non-negative, this is possible if only the following series converge

$$\begin{cases} \sum_{k=1}^{\infty} k^j \left| \varphi_k^{(1)} \right|, \\ \sum_{k=1}^{\infty} k^{j-1} \left| \varphi_k^{(2)} \right|, \end{cases} \quad j = 0, 1, 2.$$

The following theorem is known from the calculus.

Theorem 50 *If the function $\varphi(x)$ possesses an m-th piecewise continuous derivative, the series*

$$\sum_{k=1}^{\infty} k^{m-1} |\varphi_k|$$

converges.

Hence it is sufficient to require from $\varphi_1(x)$ the existence of piecewise continuous derivatives up to the third order, and from $\varphi_2(x)$ up to the second order. Under such conditions all the majorizing series involved converge.

Finally, let us notice that we expanded the functions $\varphi_1(x)$ and $\varphi_2(x)$ into the series in terms of the sines. In this case, as it is well known, the function has to be continued as an odd function. This continuation can lead to the loss of the regularity of the lower derivatives. Consider $\varphi_1(x)$. Its continuation as an odd function gives us a function $\widetilde{\varphi}_1(x)$ defined as follows (see Fig. 6.1)

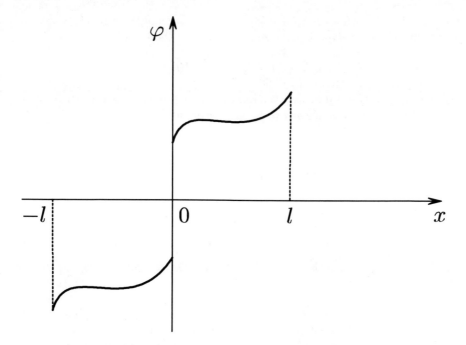

Fig. 6.1 Odd continuation of a function

$$\widetilde{\varphi}_1(x) = \begin{cases} \varphi_1(x), & 0 \le x \le l, \\ -\varphi_1(-x), & -l \le x \le 0. \end{cases}$$

It will no longer be continuous and continuously differentiable at the origin unless the following condition is fulfilled

$$\varphi_1(0) = \varphi_1(l) = 0. \tag{6.12}$$

Thus, the necessary condition for the uniform convergence of the corresponding trigonometric series $\sum_{k=1}^{\infty} \varphi_k^{(1)} \sin \frac{k\pi}{l} x$ on the whole segment is the equality to zero of the values of the function $\varphi_1(x)$ at both ends of the segment $[0, l]$. In this case, if the second derivative exists, it will be continuous as well. For the existence of the third derivative one needs to require the validity of the equalities

$$\varphi_1''(0) = \varphi_1''(l) = 0.$$

Analogously, we obtain the necessity of the condition

$$\varphi_2(0) = \varphi_2(l) = 0. \tag{6.13}$$

Finally, the following theorem is valid.

Theorem 51 *Let $\varphi_1(x)$ be twice continuously differentiable on $[0, l]$, possess a piecewise continuous third derivative and satisfy the conditions*

$$\varphi_1(0) = \varphi_1(l) = \varphi_1''(0) = \varphi_1''(l) = 0,$$

and let the function $\varphi_2(x)$ be continuously differentiable, possess a piecewise continuous second derivative and satisfy the condition

$$\varphi_2(0) = \varphi_2(l) = 0.$$

Then the function $u(x, t)$ defined by the series

$$u(x, t) = \sum_{k=1}^{\infty} \left(A_k \cos \frac{k\pi a}{l} t + B_k \sin \frac{k\pi a}{l} t \right) \sin \frac{k\pi}{l} x \qquad (6.14)$$

with the coefficients A_k and B_k defined by (6.11) is a regular solution of the problem (6.3).

Remark 52 The conditions (6.12) and (6.13) follow in fact from the compatibility of the boundary and initial conditions of the problem (6.3). Substituting $t = 0$ into the boundary conditions $u(0, t) = u(l, t) = 0$ we find that (6.12) is fulfilled. Differentiating the boundary conditions with respect to t and substituting $t = 0$ we find that (6.13) is fulfilled as well.

Remark 53 In practice it is often necessary to impose minimal possible conditions on the initial data. In this case the notion of the generalized solution is useful, and for the existence of a generalized solution it is enough to require the continuous differentiability of $\varphi_1(x)$ and the fulfillment of (6.12) and (6.13). Then the series (6.14) converges uniformly for $0 \le x \le l$ and any t fixed, and defines a continuous solution.

Let us discuss the physical meaning of the obtained formulas. The eigenfunction $u_k(x, t)$ can be written in the form

$$u_k(x, t) = \sqrt{A_k^2 + B_k^2} \, \sin \frac{k\pi}{l} x \, \sin \left(\frac{k\pi a}{l} t + \phi_k \right), \qquad (6.15)$$

where

$$\tan \phi_k = \frac{A_k}{B_k}.$$

From (6.15) it can be seen that the points of the string oscillate at the frequency $\omega_k = \frac{k\pi a}{l}$ and with the phase ϕ_k. The amplitude of the oscillations depends on the point

x and equals $F_k = \sqrt{A_k^2 + B_k^2} \sin\frac{k\pi}{l}x$. This magnitude decreases when $k \to \infty$. These are so-called **standing waves**. When $k = 1$, there are two motionless points, the ends of the string. When $k = 2$, there appears the third such point $x = l/2$. The motionless points are called the **nodes** of the standing wave. The standing wave $u_k(x, t)$ possesses $(k + 1)$ nodes $0, \frac{l}{k}, \frac{2l}{k}, \ldots, \frac{k-1}{k}l, l$. In the middle points between the nodes the amplitude of the vibrations achieves its maximum. Such points are called the **crests** (the antinodes). The frequency of the fundamental (the lowest) tone equals $\omega_1 = \frac{\pi a}{l} = \frac{\pi}{l}\sqrt{\frac{T_0}{\rho}}$. In general, the frequencies ω_k are called the harmonics. The higher tones, corresponding to ω_k, $k = 2, 3, \ldots$ are called **overtones**. The amplitude of $u_k(x, t)$ decreases quite rapidly for large k, and hence the combined effect of the higher harmonics influences the quality (the timbre) of the sound.

6.3 General Scheme of the Fourier Method

Let us consider a hyperbolic equation

$$A(t)\frac{\partial^2 u}{\partial t^2} + C(x)\frac{\partial^2 u}{\partial x^2} + D(t)\frac{\partial u}{\partial t} + E(x)\frac{\partial u}{\partial x} + (F(t) + H(x))u = 0, \qquad (6.16)$$

where we assume that $A(t) > 0$ and $C(x) < 0$. The coefficients in the equation admit the separation of variables. We will look for the solutions of (6.16) satisfying additionally the initial conditions

$$\begin{cases} u(x, 0) = \varphi(x), \\ u_t(x, 0) = \psi(x), \end{cases}$$

and the boundary conditions

$$\begin{cases} (\alpha u + \beta u_x)|_{x=0} = 0, & |\alpha| + |\beta| > 0, \\ (au + bu_x)|_{x=l} = 0, & |a| + |b| > 0. \end{cases} \qquad (6.17)$$

Here α, β, a, and b are constants, and the inequalities just tell us that at least one of the constants from each pair is different from zero.

Substituting $u(x, t) = X(x)T(t)$ into (6.16) we obtain

$$A(t)T''X + C(x)X''T + D(t)T'X + E(x)TX' + F(t)TX + H(x)TX = 0.$$

Dividing over $X(x)T(t)$ leads to the equality

$$A(t)\frac{T''}{T} + D(t)\frac{T'}{T} + F(t) = -C(x)\frac{X''}{X} - E(x)\frac{X'}{X} - H(x).$$

Since both sides of the equality are functions of different independent variables, we conclude (e.g., by differentiating the equality first with respect to t and then with respect to x) that both functions must be equal to the same constant which we denote by $-\lambda$. Thus,

$$A(t)\frac{T''}{T} + D(t)\frac{T'}{T} + F(t) = -C(x)\frac{X''}{X} - E(x)\frac{X'}{X} - H(x) = -\lambda,$$

which gives us two linear ordinary differential equations

$$C(x)X'' + E(x)X' + H(x)X - \lambda X = 0 \tag{6.18}$$

and

$$A(t)T'' + D(t)T' + F(t)T + \lambda T = 0. \tag{6.19}$$

From the boundary conditions (6.17) we obtain the boundary conditions for the Eq. (6.18)

$$\begin{cases} \alpha X(0) + \beta X'(0) = 0, \\ a X(l) + b X'(l) = 0. \end{cases} \tag{6.20}$$

Definition 54 A linear ordinary differential equation of second order

$$\frac{d}{dx}\left(p_1(x)\frac{dy}{dx}\right) + p_2(x)y = 0$$

is said to be of **self-adjoint** form.

Let us reduce Eq. (6.18) to a self-adjoint form. To this purpose we multiply it by a function $(-r(x))$ chosen in such a way that

$$(Er) = (Cr)' = C'r + Cr', \quad \frac{dr}{r} = \frac{(E - C')}{C}dx$$

and hence

$$r = \exp\left(\int_{x_0}^x \frac{(E - C')}{C}dx\right).$$

Thus, $p_1 = -Cr$, $p_2 = -Hr + \lambda r$. Denote $p := p_1$ and $q := Hr$. Then (6.18) is written in the form

$$\frac{d}{dx}\left(p(x)\frac{dX}{dx}\right) - q(x)X(x) + \lambda r(x)X(x) = 0$$

or

$$L(X) + \lambda r X = 0, \quad \text{where } L(X) := \frac{d}{dx}\left(p\frac{dX}{dx}\right) - qX.$$

6.4 Properties of Eigenfunctions and Eigenvalues

In this section we consider the Sturm-Liouville equation

$$\frac{d}{dx}\left(p(x)\frac{dX}{dx}\right) - q(x)X(x) + \lambda r(x)X(x) = 0, \quad 0 < x < l \tag{6.21}$$

with the boundary conditions of the form

$$\begin{cases} \alpha X(0) + \beta X'(0) = 0, \\ a X(l) + b X'(l) = 0 \end{cases} \tag{6.22}$$

and assume that the following conditions are fulfilled. The coefficients in Eq. (6.21) are real valued continuous functions together with $p'(x)$. The functions p and r are strictly positive on $[0, l]$. The constants in the boundary conditions are real and satisfy the conditions $|\alpha| + |\beta| > 0$, $|a| + |b| > 0$. Such Sturm-Liouville problem is called **regular**.

Below we discuss the main properties of regular Sturm-Liouville problems.

Theorem 55 *The eigenfunctions belonging to the same eigenvalue are linearly dependent, that is, the multiplicity of any eigenvalue of a regular Sturm-Liouville problem equals one.*

Proof Suppose that two functions X_1 and X_2 satisfy (6.21) and (6.22) with the same value of λ. In particular, this means that the following equalities coming from (6.22) are valid

$$\begin{cases} \alpha X_1(0) + \beta X_1'(0) = 0, \\ \alpha X_2(0) + \beta X_2'(0) = 0. \end{cases} \tag{6.23}$$

Since, due to the condition $|\alpha| + |\beta| > 0$, α and β cannot both equal zero, the necessary condition for (6.23) is that the determinant of the system be zero,

$$\begin{vmatrix} X_1(0) & X_1'(0) \\ X_2(0) & X_2'(0) \end{vmatrix} = W[X_1, X_2]|_{x=0} = 0.$$

That is, the Wronskian of the solutions X_1 and X_2 vanishes at $x = 0$. Hence, as it is known from the course of differential equations (see, e.g., [34, Chapter 4, Sect. 8]),

$W[X_1, X_2] = 0$ on the whole interval, which means that X_1 and X_2 are linearly dependent.   ∎

Theorem 56 *The eigenfunctions belonging to different eigenvalues are orthogonal on $[0, l]$ with the weight $r(x)$. That is,*

$$\int_0^l r(x)X_1(x)X_2(x)\,dx = 0. \qquad (6.24)$$

Proof Let X_1 and X_2 be eigenfunctions belonging to the eigenvalues λ_1 and λ_2, respectively, and $\lambda_1 \neq \lambda_2$. We have then

$$L(X_1) + \lambda_1 r X_1 = 0 \qquad (6.25)$$

and

$$L(X_2) + \lambda_2 r X_2 = 0. \qquad (6.26)$$

Let us show that these equalities together with (6.22) satisfied by both eigenfunctions imply (6.24). Multiplying (6.25) by X_2 and (6.26) by X_1, and resting the second equality from the first one, we obtain

$$X_2 L(X_1) - X_1 L(X_2) + (\lambda_1 - \lambda_2) r X_1 X_2 = 0.$$

Integration of this equality from 0 to l gives

$$\int_0^l X_2(pX_1')'\,dx - \int_0^l X_1(pX_2')'\,dx + \int_0^l q\,(X_1 X_2 - X_2 X_1)\,dx$$

$$+ (\lambda_1 - \lambda_2) \int_0^l r X_1 X_2\,dx = 0.$$

Hence

$$(\lambda_1 - \lambda_2) \int_0^l r X_1 X_2\,dx = \int_0^l \left(X_1(pX_2')' - X_2(pX_1')' \right)\,dx.$$

Integration by parts on the right hand side leads to the equalities

$$(\lambda_1 - \lambda_2) \int_0^l r X_1 X_2\,dx = \left(X_1 X_2' - X_2 X_1' \right) p\Big|_0^l - \int_0^l \left(pX_1' X_2' - pX_2' X_1' \right)\,dx$$

$$= p(l) \begin{vmatrix} X_1(l) & X_2(l) \\ X_1'(l) & X_2'(l) \end{vmatrix} - p(0) \begin{vmatrix} X_1(0) & X_2(0) \\ X_1'(0) & X_2'(0) \end{vmatrix}.$$

Due to the boundary conditions, the columns in the determinants are proportional, and hence the determinants equal zero. Thus,

$$(\lambda_1 - \lambda_2) \int_0^l r X_1 X_2 \, dx = 0.$$

Since $\lambda_1 \neq \lambda_2$, this finishes the proof.   ∎

Thus, to every eigenvalue λ_0 there corresponds exactly one eigenfunction X_0 determined up to an arbitrary multiplicative constant. The eigenfunction can be normalized by the condition

$$\|X_0\| = \sqrt{\int_0^l r X_0^2 dx} = 1.$$

If an eigenfunction X_0 does not satisfy this equality, it can be easily normalized. Indeed, denote

$$\|X_0\| := \sqrt{\int_0^l r X_0^2 dx}. \tag{6.27}$$

Consider the eigenfunction differing from X_0 by a constant factor: $\widetilde{X}_0 = \frac{X_0}{\|X_0\|}$. Obviously,

$$\|\widetilde{X}_0\| = \sqrt{\int_0^l \frac{r X_0^2}{\|X_0\|^2} dx} = \frac{1}{\|X_0\|} \sqrt{\int_0^l r X_0^2 dx} = 1,$$

and $\widetilde{X}_0$ is the normalized eigenfunction.

Let X_i be the normalized eigenfunctions corresponding to the eigenvalues λ_i. Their orthogonality means that

$$\int_0^l r X_i X_j \, dx = \delta_{ij} = \begin{cases} 0, & i \neq j, \\ 1, & i = j. \end{cases}$$

Theorem 57 *If $p(x)$, $p'(x)$, and $q(x)$ are continuous, and $p(x) > 0$, $q(x) \geq 0$, all eigenvalues of the Sturm-Liouville problem are positive under the condition*

$$\left(p(x) X(x) X'(x) \right) |_{x=l} - \left(p(x) X(x) X'(x) \right) |_{x=0} \leq 0.$$

Remark 58 The last condition is fulfilled in the most frequently encountered Sturm-Liouville problems. For example, it is valid if

1. $\frac{\alpha}{\beta} \leq 0$, $\frac{a}{b} \geq 0$;

2. $\frac{\alpha}{\beta} \leq 0, b = 0$;
3. $\beta = 0, \frac{a}{b} \geq 0$;
4. $\beta = 0, b = 0$.

Proof Let λ_0 be an eigenvalue and X_0 the corresponding normalized eigenfunction. Thus,

$$L(X_0) = -\lambda_0 r X_0.$$

Multiplying this equation by X_0 and integrating from 0 to l gives

$$-\int_0^l \lambda_0 r X_0^2 dx = -\lambda_0 = \int_0^l L(X_0) X_0 dx$$

that leads to the chain of relations

$$\lambda_0 = -\int_0^l L(X_0) X_0 dx = -\int_0^l X_0 \left(pX_0'\right)' dx + \int_0^l q X_0^2 dx$$

$$= -pX_0 X_0'\big|_0^l + \int_0^l p\left(X_0'\right)^2 dx + \int_0^l q X_0^2 dx > 0.$$

∎

Remark 59 Often the case 1 arises in real world problems because the forces acting at the ends of the interval are directed in opposite directions.

Let us discuss the question: how many eigenvalues a Sturm-Liouville problem can have. In general, it is far from being obvious whether a given Sturm-Liouville problem possesses at least one eigenvalue. In 1836 J. Sturm and J. Liouville independently from each other published papers in a same journal in which precisely that question was asked. The following example shows that a Sturm-Liouville problem can have no eigenvalue.

Example 60 Consider the Sturm-Liouville problem

$$y'' - \lambda y = 0, \quad x \in \mathbb{R},$$

$$y(-\infty) = y(\infty) = 0.$$

Thus, instead of a finite interval, we consider a problem on the whole line. This problem turns out to have no eigenvalue. Indeed, a general solution of the equation has the form $y(x) = c_1 e^{\sqrt{\lambda}x} + c_2 e^{-\sqrt{\lambda}x}$ where both linearly independent (when $\lambda \neq 0$) solutions decay at one infinity but grow at the other. The value $\lambda = 0$ is not an eigenvalue either, because no linear combination of the linearly independent solutions $y_1(x) = 1$ and $y_2(x) = x$ in this case satisfies any of the boundary conditions at infinity.

It is all the more interesting that in the case of regular Sturm-Liouville problems there always exists not just one but an infinite sequence of eigenvalues, see the following theorem which we give here without proof (that can be found, e.g., in [22, Chapter VII, Sect. 2]).

Theorem 61 *The set of all eigenvalues of a regular Sturm-Liouville problem represents an infinite, unbounded, strictly monotonous sequence of numbers*

$$\lambda_1 < \lambda_2 < \ldots < \lambda_n < \ldots.$$

Remark 62 Quite often in applications so-called **periodic Sturm-Liouville problems** arise

$$\frac{d}{dx}\left(p(x)\frac{dX}{dx}\right) - q(x)X(x) + \lambda r(x)X(x) = 0, \quad 0 < x < l, \tag{6.28}$$

$$X(0) = X(l), \qquad X'(0) = X'(l), \tag{6.29}$$

where the coefficients p, q, and r are periodic functions of a period l. The periodic boundary conditions (6.29) and the Sturm-Liouville equation (6.28) imply that an eigenfunction can be extended to a periodic function on the real line. Theory of the periodic Sturm-Liouville problems with the coefficients p, q, and r satisfying the conditions of a regular Sturm-Liouville problem is analogous to the presented theory of regular Sturm-Liouville problems with the only substantial difference that the multiplicity of the eigenvalues can be greater than one. We refer the interested reader, e.g., to [72] for more information on this subject.

Let us return to Eq. (6.16). Application of the Fourier method to it led to a couple of Eqs. (6.18) and (6.19). Let $\lambda_1, \lambda_2,\ldots,\lambda_n,\ldots$ be the eigenvalues of the Sturm-Liouville problem (6.18), (6.20). Every λ_n defines a pair of linearly independent solutions $T_n^1(t)$ and $T_n^2(t)$ of Eq. (6.19) which can be normalized as follows

$$T_n^1(0) = 1, \quad \left[T_n^1(t)\right]'\bigg|_{t=0} = 0,$$

$$T_n^2(0) = 0, \quad \left[T_n^2(t)\right]'\bigg|_{t=0} = 1.$$

Consider a series

$$u(x,t) = \sum_{n=1}^{\infty}\left(A_n T_n^1(t) + B_n T_n^2(t)\right) X_n(x),$$

where the coefficients A_n and B_n are to be chosen in such a way that the initial conditions be fulfilled

$$u(x, 0) = \sum_{n=1}^{\infty} A_n X_n(x) = \varphi(x),$$

$$u_t(x, 0) = \sum_{n=1}^{\infty} B_n X_n(x) = \psi(x).$$

The coefficients A_n and B_n are then well defined if the functions $\varphi(x)$ and $\psi(x)$ admit generalized Fourier series expansions in terms of the eigenfunctions of the Sturm-Liouville problem (6.18), (6.20),

$$\varphi(x) = \sum_{n=1}^{\infty} \varphi_n X_n(x), \quad \psi(x) = \sum_{n=1}^{\infty} \psi_n X_n(x).$$

Due to the orthonormality of the functions X_n we have that

$$\varphi_n = \int_0^l r(x)\varphi(x)X_n(x)\,dx \quad \text{and} \quad \psi_n = \int_0^l r(x)\psi(x)X_n(x)\,dx.$$

The following theorem is valid which we give here without proof (see, e.g., [83, p. 277]).

Theorem 63 (Steklov's Theorem) *If the function $\varphi(x)$ is three times continuously differentiable, satisfies the boundary conditions (6.20), and the coefficients p and q are twice continuously differentiable, then $\varphi(x)$ admits a generalized Fourier series expansion in terms of eigenfunctions of the Sturm-Liouville problem, and the series can be twice differentiated termwise.*

Often in practice less smooth initial data than those mentioned in Steklov's theorem should be considered, as well as generalized solutions of the initial boundary value problem. Then the following theorem becomes crucial.

Theorem 64 (The Completeness of the System of Eigenfunctions) *The orthonormal system $\{X_n\}$ of all eigenfunctions of a regular Sturm-Liouville problem is complete in the space $L_2(0, l; r)$ (the space of all functions square integrable with the weight $r(x)$).*

We recall that the norm of a function φ in $L_2(0, l; r)$ is defined by the equality

$$\|\varphi\|_{L_2(0,l;r)} := \left(\int_0^l r(x)\varphi^2(x)dx \right)^{\frac{1}{2}}. \tag{6.30}$$

We encountered this definition above, see (6.27).

The completeness of the system $\{X_n\}$ in $L_2(0, l; r)$ implies that any function $\varphi \in L_2(0, l; r)$ admits a series expansion in terms of the eigenfunctions:

$$\varphi(x) = \sum_{n=1}^{\infty} \varphi_n X_n(x), \tag{6.31}$$

which converges in the sense of the norm (6.30). It is worth mentioning that the space $L_2(0, l; r)$ ($r \in C[0, l]$, $r(x) > 0$, $\forall x \in [0, l]$) is much wider than the set of smooth functions. It contains all continuous functions, as well as all piecewise continuous functions, admitting a finite number of jump discontinuities, besides some other subclasses of functions. The convergence with respect to the norm (6.30) does not imply even a pointwise convergence; however, if the function satisfies additional smoothness conditions as well as the boundary conditions (6.20), as we see from Steklov's theorem the convergence in a much stronger sense takes place.

Remark 65 Let φ be a piecewise continuously differentiable function, defined on $[0, l]$. Then for any $x \in (0, l)$ its generalized Fourier series (6.31) converges to $(\varphi(x_+) + (\varphi(x_-))/2$ (the mean value), which, of course, coincides with $\varphi(x)$ at the points of continuity.

If $\varphi \in C^1[0, l]$ and satisfies the boundary conditions (6.20), then its generalized Fourier series (6.31) converges uniformly to φ on $[0, l]$.

Example 66 Consider $\varphi \equiv 1$ on $[0, \pi]$ and the orthonormal system

$$\left\{ \sqrt{\frac{2}{\pi}} \sin nx \right\}_{n=1}^{\infty}$$

of eigenfunctions of the Sturm-Liouville problem

$$y'' + \lambda y = 0, \quad 0 < x < \pi,$$

$$y(0) = y(\pi) = 0. \tag{6.32}$$

The corresponding Fourier coefficients of the function φ have the form

$$\left\langle \varphi, \sqrt{\frac{2}{\pi}} \sin nx \right\rangle = \sqrt{\frac{2}{\pi}} \int_0^{\pi} \sin nx \, dx = -\sqrt{\frac{2}{\pi}} \left. \frac{\cos nx}{n} \right|_0^{\pi} = \sqrt{\frac{2}{\pi}} \frac{(1 - (-1)^n)}{n}.$$

Thus,

$$1 = \frac{2}{\pi} \sum_{n=1}^{\infty} \frac{(1 - (-1)^n)}{n} \sin nx = \frac{4}{\pi} \sum_{k=0}^{\infty} \frac{1}{2k + 1} \sin(2k + 1)x.$$

The series converges to the unity for all $x \in (0, \pi)$ but does not converge uniformly on $[0, \pi]$ and even pointwise (in the origin) to 1 because the function φ does not satisfy the boundary conditions of the problem (6.32).

6.5 Free Vibrations of a Rectangular Membrane

We consider a homogeneous rectangular membrane, fastened along the edges. The sides of the membrane are of length l_1 and l_2 (Fig. 6.2). Thus, according to Sect. 3.4, the corresponding mathematical model can be formulated as the following boundary value problem

$$\frac{\partial^2 u}{\partial t^2} = a^2 \Delta u = a^2 \left(\frac{\partial^2 u}{\partial x^2} + \frac{\partial^2 u}{\partial y^2} \right), \quad 0 < x < l_1, \quad 0 < y < l_2, \tag{6.33}$$

$$u|_\Gamma = 0, \tag{6.34}$$

$$u|_{t=0} = \varphi(x, y), \quad u_t|_{t=0} = \psi(x, y). \tag{6.35}$$

Since the boundary Γ is a rectangle, the boundary conditions can be written in the form

$$u(0, y, t) = 0, \quad u(x, 0, t) = 0,$$
$$u(l_1, y, t) = 0, \quad u(x, l_2, t) = 0.$$

Let us apply the Fourier method. We look for a solution of the form

$$U(x, y, t) = T(t)W(x, y).$$

Substitution to the wave equation gives

$$T'' W = a^2 T \, \Delta W.$$

Dividing over $a^2 T W$ we obtain

$$\frac{T''}{a^2 T} = \frac{\Delta W}{W} = -\lambda^2.$$

We have chosen $-\lambda^2$ in advance because, according to the physical meaning of the problem, the oscillations cannot be unbounded.

Thus, two separate equations are obtained

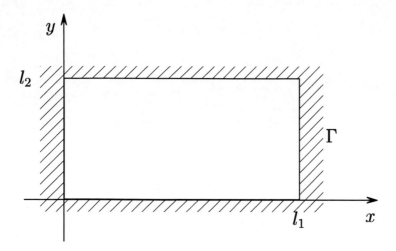

Fig. 6.2 Rectangular membrane

$$T'' + \lambda^2 a^2 T = 0$$

and

$$\Delta W + \lambda^2 W = 0.$$

The last equation is called the Helmholtz equation.

Let us apply the Fourier method to the Helmholtz equation as well. Consider $W(x, y) = X(x)Y(y)$. Then

$$X''Y + XY'' + \lambda^2 XY = 0.$$

Dividing over XY leads to the equation

$$-\frac{X''}{X} = \frac{Y''}{Y} + \lambda^2.$$

Again, this equality is possible if only both sides equal the same constant, which we denote as α^2,

$$-\frac{X''}{X} = \frac{Y''}{Y} + \lambda^2 = \alpha^2.$$

Thus,

$$X'' + \alpha^2 X = 0, \tag{6.36}$$

$$Y'' + \beta^2 Y = 0, \tag{6.37}$$

where $\lambda^2 = \alpha^2 + \beta^2$.

Let us write down the boundary conditions for the obtained equations. From the conditions

$$X(0)Y(y)T(t) = 0, \quad \text{and} \quad X(l_1)Y(y)T(t) = 0$$

it follows that

$$X(0) = X(l_1) = 0, \tag{6.38}$$

since $T(t) \neq 0$ and $Y(y) \neq 0$ identically. Similarly,

$$Y(0) = Y(l_2) = 0. \tag{6.39}$$

The eigenvalues and the corresponding eigenfunctions of the Sturm-Liouville problems (6.36), (6.38) and (6.37), (6.39), respectively, have the form

$$\alpha_k = \frac{k\pi}{l_1}, \quad k = 1, 2, \ldots, \quad X_k(x) = \sin\left(\frac{k\pi}{l_1}x\right)$$

and

$$\beta_n = \frac{n\pi}{l_2}, \quad n = 1, 2, \ldots, \quad Y_n(y) = \sin\left(\frac{n\pi}{l_2}y\right).$$

To every pair α_k, β_n there is an associated $\lambda_{kn}^2 = \alpha_k^2 + \beta_n^2$, which is a Dirichlet eigenvalue of the Laplacian in the rectangle. To every such value

$$\lambda_{kn} = \sqrt{\alpha_k^2 + \beta_n^2} = \pi\sqrt{\frac{k^2}{l_1^2} + \frac{n^2}{l_2^2}}$$

there corresponds a general solution of the equation $T'' + \lambda_{kn}^2 a^2 T = 0$:

$$T_{kn}(t) = A_{kn}\cos(a\lambda_{kn}t) + B_{kn}\sin(a\lambda_{kn}t).$$

Making use of the general superposition principle, the solution $u(x, y, t)$ of the problem (6.33)–(6.35) is sought in the form of the double series

$$u(x, y, t) = \sum_{k,n=1}^{\infty} (A_{kn}\cos(a\lambda_{kn}t) + B_{kn}\sin(a\lambda_{kn}t))\sin\left(\frac{k\pi}{l_1}x\right)\sin\left(\frac{n\pi}{l_2}y\right).$$

Whenever this series converges, and its substitution into (6.33) makes sense, it is a solution of (6.33) satisfying the boundary conditions (6.34) on the boundary of the rectangle. Thus, the coefficients A_{kn}, B_{kn} should be chosen in such a way that the initial conditions be fulfilled,

$$u(x, y, 0) = \sum_{k,n=1}^{\infty} A_{kn} \sin\left(\frac{k\pi}{l_1}x\right) \sin\left(\frac{n\pi}{l_2}y\right) = \varphi(x, y),$$

$$u_t(x, y, 0) = \sum_{k,n=1}^{\infty} B_{kn}a\lambda_{kn} \sin\left(\frac{k\pi}{l_1}x\right) \sin\left(\frac{n\pi}{l_2}y\right) = \psi(x, y).$$

Obviously, the coefficients A_{kn}, B_{kn} will be defined if the functions $\varphi(x, y)$ and $\psi(x, y)$ admit such double Fourier series expansions. Supposing that this is indeed the case, in order to calculate the coefficients A_{kn}, B_{kn} it is useful to note that the functions

$$\omega_{kn}(x, y) := \sin\left(\frac{k\pi}{l_1}x\right) \sin\left(\frac{n\pi}{l_2}y\right)$$

represent an orthogonal system of functions in the rectangle $0 \le x \le l_1, 0 \le y \le l_2$, that is,

$$\int_0^{l_1} \int_0^{l_2} \omega_{kn}(x, y)\omega_{rs}(x, y)dxdy = \begin{cases} 0, & k \ne r \quad \text{or} \quad n \ne s, \\ \frac{l_1 l_2}{4}, & k = r \quad \text{and} \quad n = s. \end{cases}$$

With the aid of this property, the coefficients A_{kn}, B_{kn} can be calculated by the formulas

$$A_{kn} = \frac{4}{l_1 l_2} \int_0^{l_1} \int_0^{l_2} \varphi(x, y) \sin\left(\frac{k\pi}{l_1}x\right) \sin\left(\frac{n\pi}{l_2}y\right) dxdy,$$

$$B_{kn} = \frac{4}{l_1 l_2 a\lambda_{kn}} \int_0^{l_1} \int_0^{l_2} \psi(x, y) \sin\left(\frac{k\pi}{l_1}x\right) \sin\left(\frac{n\pi}{l_2}y\right) dxdy.$$

6.6 Solution of a Non-homogeneous Problem

Let us study the procedure of solution of non-homogeneous problems by the Fourier method considering the model of forced vibrations of a homogeneous string with a given solution at the ends,

$$\frac{\partial^2 u}{\partial t^2} = a^2 \frac{\partial^2 u}{\partial x^2} + f(x, t),$$

$$\begin{cases} u(0, t) = \gamma_1(t), \\ u(l, t) = \gamma_2(t), \end{cases}$$

$$\begin{cases} u|_{t=0} = \varphi(x), \\ u_t|_{t=0} = \psi(x). \end{cases}$$

It is useful to highlight two auxiliary steps simplifying the task.

Step 1 First of all, we get rid of the non-homogeneity in the boundary conditions. For this purpose, let us choose a simple function $m(x, t)$ satisfying the boundary conditions. For example,

$$m(x, t) = \gamma_1(t) + (\gamma_2(t) - \gamma_1(t)) \frac{x}{l}.$$

Thus, $m(0, t) = \gamma_1(t), m(l, t) = \gamma_2(t)$.

Remark 67 In a general situation, when the boundary conditions are of the form

$$\alpha u(0, t) + \beta u_x(0, t) = \gamma_1(t), \quad |\alpha| + |\beta| > 0,$$

$$\mu u(l, t) + \nu u_x(l, t) = \gamma_2(t), \quad |\mu| + |\nu| > 0,$$

the function $m(x, t)$ can be chosen in the form

$$m(x, t) = (ax + b)\, \gamma_1(t) + (cx + d)\, \gamma_2(t)$$

if $|\alpha| + |\mu| > 0$ and in the form

$$m(x, t) = \left(ax^2 + bx\right) \gamma_1(t) + \left(cx^2 + dx\right) \gamma_2(t),$$

otherwise, i.e., when $|\alpha| + |\mu| = 0$. Here a, b, c, and d are constants guaranteeing the fulfillment of the boundary conditions by the function $m(x, t)$.

The solution u is sought in the form $u = v + m$. Then

$$\frac{\partial^2 v}{\partial t^2} = a^2 \frac{\partial^2 v}{\partial x^2} + f(x, t) - \left(\frac{\partial^2 m}{\partial t^2} - a^2 \frac{\partial^2 m}{\partial x^2}\right),$$

and hence

$$\frac{\partial^2 v}{\partial t^2} = a^2 \frac{\partial^2 v}{\partial x^2} + g(x, t),$$

where $g(x, t) := f(x, t) - \left(\frac{\partial^2 m}{\partial t^2} - a^2 \frac{\partial^2 m}{\partial x^2} \right)$. The function v satisfies the homogeneous boundary conditions,

$$\begin{cases} v(0, t) = 0, \\ v(l, t) = 0, \end{cases}$$

and the following initial conditions

$$\begin{cases} v|_{t=0} = u|_{t=0} - m|_{t=0} = \varphi(x) - m(x, 0) =: \varphi_1(x), \\ v_t|_{t=0} = u_t|_{t=0} - m_t|_{t=0} =: \psi_1(x). \end{cases}$$

Thus, for the unknown function v we obtain a problem with homogeneous boundary conditions.

Step 2 The obtained problem for v we divide into the following two separate problems, setting $v = v_1 + v_2$, where

$$\begin{cases} \frac{\partial^2 v_1}{\partial t^2} = a^2 \frac{\partial^2 v_1}{\partial x^2} \\ v_1(0, t) = v_1(l, t) = 0, \\ v_1|_{t=0} = \varphi_1(x), \\ \frac{\partial v_1}{\partial t} \Big|_{t=0} = \psi_1(x), \end{cases} \qquad \begin{cases} \frac{\partial^2 v_2}{\partial t^2} = a^2 \frac{\partial^2 v_2}{\partial x^2} + g(x, t) \\ v_2(0, t) = v_2(l, t) = 0, \\ v_2|_{t=0} = 0, \\ \frac{\partial v_2}{\partial t} \Big|_{t=0} = 0. \end{cases}$$

The wave equation in the first problem is homogeneous, but the initial conditions are non-homogeneous. Such problem was considered in Sect. 6.2.

The solution v_2 of the second problem, with the homogeneous initial conditions and a non-homogeneous wave equation can be sought in the form of a series in terms of eigenfunctions of the first problem (for v_1). We have

$$v_2(x, t) = \sum_{k=1}^{\infty} T_k(t) X_k(x).$$

In the case under consideration $X_k(x) = \sin\left(\frac{k\pi}{l} x\right)$, and hence

$$v_2(x, t) = \sum_{k=1}^{\infty} T_k(t) \sin\left(\frac{k\pi}{l} x\right).$$

Substitution of $v_2(x, t)$ into the equation gives the equality

$$\sum_{k=1}^{\infty} \left(T_k''(t) + \left(\frac{k\pi a}{l}\right)^2 T_k(t) \right) \sin\left(\frac{k\pi}{l}x\right) = g(x, t),$$

or

$$\left(T_k''(t) + \left(\frac{k\pi a}{l}\right)^2 T_k(t) \right) = g_k(t), \tag{6.40}$$

where $g_k(t)$ are the Fourier coefficients from the series expansion of the function $g(x, t)$ in terms of the eigenfunctions $X_k(x)$,

$$g_k(t) = \frac{2}{l} \int_0^l g(x, t) \sin\left(\frac{k\pi}{l}x\right) dx.$$

Due to the homogeneous initial conditions, the obtained problem for T_k should be solved under the conditions

$$T_k(0) = T_k'(0) = 0. \tag{6.41}$$

The functions $T_k(t)$ are solutions of corresponding non homogeneous differential equations with constant coefficients (6.40), and can be found by the method of variation of constants. A general solution of the homogeneous equation has the form

$$T_k^{\text{hom}}(t) = A_k \cos\left(\frac{k\pi a}{l}t\right) + B_k \sin\left(\frac{k\pi a}{l}t\right).$$

Hence, according to the method of variation of constants (also known as variation of parameters, see, e.g., [13, Sect. 3.6]) the solution of the non-homogeneous equation (6.40) is sought in the form

$$T_k(t) = A_k(t) \cos\left(\frac{k\pi a}{l}t\right) + B_k(t) \sin\left(\frac{k\pi a}{l}t\right).$$

Imposing the following condition on the functions $A_k(t)$ and $B_k(t)$:

$$A_k'(t) \cos\left(\frac{k\pi a}{l}t\right) + B_k'(t) \sin\left(\frac{k\pi a}{l}t\right) = 0, \tag{6.42}$$

we find that $T_k(t)$ is a solution of (6.40) if and only if

$$- A_k'(t) \sin\left(\frac{k\pi a}{l}t\right) + B_k'(t) \cos\left(\frac{k\pi a}{l}t\right) = \frac{l}{k\pi a} g_k(t). \tag{6.43}$$

From (6.42) and (6.43) we obtain

$$A_k'(t) = -\frac{l}{k\pi a} g_k(t) \sin\left(\frac{k\pi a}{l}t\right), \quad B_k'(t) = \frac{l}{k\pi a} g_k(t) \cos\left(\frac{k\pi a}{l}t\right).$$

Thus,

$$A_k(t) = -\frac{l}{k\pi a} \int_0^t g_k(\tau) \sin\left(\frac{k\pi a}{l}\tau\right) d\tau + \widetilde{A}_k$$

and

$$B_k(t) = \frac{l}{k\pi a} \int_0^t g_k(\tau) \cos\left(\frac{k\pi a}{l}\tau\right) d\tau + \widetilde{B}_k,$$

where $\widetilde{A}_k$ and $\widetilde{B}_k$ are arbitrary constants.
    Hence

$$T_k(t) = \widetilde{A}_k \cos\left(\frac{k\pi a}{l}t\right) + \widetilde{B}_k \sin\left(\frac{k\pi a}{l}t\right)$$

$$- \frac{l}{k\pi a} \cos\left(\frac{k\pi a}{l}t\right) \int_0^t g_k(\tau) \sin\left(\frac{k\pi a}{l}\tau\right) d\tau$$

$$+ \frac{l}{k\pi a} \sin\left(\frac{k\pi a}{l}t\right) \int_0^t g_k(\tau) \cos\left(\frac{k\pi a}{l}\tau\right) d\tau.$$

Substituting this expression into (6.41) we obtain that $\widetilde{A}_k = \widetilde{B}_k = 0$, and finally,

$$T_k(t) = \frac{l}{k\pi a} \int_0^t g_k(\tau) \sin\left(\frac{k\pi a}{l}(t - \tau)\right) d\tau.$$

6.7   Free Vibrations of a Circular Membrane

Free vibrations of a membrane are described by the wave equation $\frac{\partial^2 u}{\partial t^2} = a^2 \Delta u$ which can be written in polar coordinates as follows

$$\frac{\partial^2 u}{\partial t^2} = a^2 \left(\frac{\partial^2 u}{\partial r^2} + \frac{1}{r}\frac{\partial u}{\partial r} + \frac{1}{r^2}\frac{\partial^2 u}{\partial \varphi^2} \right). \tag{6.44}$$

Similarly to the case of a rectangular membrane, we apply the Fourier method for solving this equation. The solution is sought in the form

$$u(r, \varphi, t) = T(t)W(r, \varphi).$$

Substitution into Eq. (6.44) leads to the equation

$$\frac{T''}{a^2 T} = \frac{1}{W}\left(\frac{\partial^2 W}{\partial r^2} + \frac{1}{r}\frac{\partial W}{\partial r} + \frac{1}{r^2}\frac{\partial^2 W}{\partial \varphi^2}\right) = -\lambda^2.$$

Here the constant λ^2 is a separation constant. We have chosen $-\lambda^2$ for the same reason as before for rectangular membrane and in other similar cases. Namely, the oscillation cannot grow indefinitely over time.

Thus, we obtain two equations

$$T'' + \lambda^2 a^2 T = 0$$

and

$$\frac{\partial^2 W}{\partial r^2} + \frac{1}{r}\frac{\partial W}{\partial r} + \frac{1}{r^2}\frac{\partial^2 W}{\partial \varphi^2} + \lambda^2 W = 0.$$

To the last equation we again apply the Fourier method, setting

$$W(r, \varphi) = R(r)\Phi(\varphi).$$

Then

$$R''\Phi + \frac{1}{r}R'\Phi + \frac{1}{r^2}R\Phi'' + \lambda^2 R\Phi = 0.$$

Separation of variables in this equation leads to the equalities

$$r^2\frac{R''}{R} + r\frac{R'}{R} + \lambda^2 r^2 = -\frac{\Phi''}{\Phi} = \nu^2,$$

where ν^2 is a separation constant. From here two equations follow

$$\Phi'' + \nu^2 \Phi = 0$$

and

$$r^2 R'' + rR' + \left(\lambda^2 r^2 - \nu^2\right)R = 0. \tag{6.45}$$

In order to obtain a solution being a single-valued function we need to require $\Phi(\varphi)$ to be a periodic function with a period 2π: $\Phi(\varphi) = \Phi(\varphi + 2\pi)$ for all φ. A periodic solution Φ of the first differential equation is possible if only $\nu = n$, $n = 0, \pm 1, \pm 2, \ldots$, and the corresponding solution has the form

$$\Phi(\varphi) = A_n \cos n\varphi + B_n \sin n\varphi.$$

Obviously, it is sufficient to consider $n = 0, 1, 2, \ldots$.

In Eq. (6.45), let us change the variable $\lambda r = \rho$. Then

$$R'_r = R'_\rho \cdot \rho'_r = \lambda R'_\rho, \quad R''_{rr} = \lambda^2 R''_{\rho\rho},$$

and the equation takes the form

$$\rho^2 R''_{\rho\rho} + \rho R'_\rho + \left(\rho^2 - \nu^2\right) R = 0. \tag{6.46}$$

Thus, for solving the problem on vibrations of a circular membrane it is necessary to solve Eq. (6.46), called the **Bessel equation**. Additionally, the following natural conditions are being imposed on the solution $R(r)$,

$$R(l) = 0 \quad \text{(the membrane is fixed)}$$

and

$$R(0) < \infty \quad \text{(the vibrations in the origin are bounded)}.$$

Thus, we need to interrupt the presentation in order to provide some additional information on Bessel functions. We prefer to present it in a fairly complete context, and go slightly beyond the minimum required for this task. Thus, the subsequent sections will be devoted to the detailed study of Bessel functions and their properties, and we will return to solving the problem of oscillations of a circular membrane in Sect. 6.11. The reader who is interested in solving the problem and does not need this additional information on Bessel functions can proceed directly to Sect. 6.11.

6.8 Bessel Equation and Bessel Functions

The solution of the Bessel equation

$$x^2 y'' + xy' + (x^2 - \nu^2)y = 0 \tag{6.47}$$

can be sought in the form of a power series. Since the coefficient of y'' turns into zero in the origin, a generalized power series should be considered. Namely, we are looking for a solution in the form

$$y(x) = x^\mu \sum_{k=0}^{\infty} C_k x^k,$$

where the coefficient C_0 can be supposed to be different from zero (otherwise, factoring the corresponding power out of the sum and renumbering the series can be written in this form). Thus, $C_0 \neq 0$.

Substituting $y(x)$ in the form of the series into the equation we find two possible values of μ: $\mu = \pm\nu$. Let us find the coefficients of the series for $\mu = \nu$. We have

$$y(x) = C_0 \sum_{k=0}^{\infty} (-1)^k \frac{x^{\nu+2k}}{2^{2k} k! (\nu+k)(\nu+k-1)\ldots(\nu+1)}.$$

Usually C_0 is chosen as

$$C_0 = \frac{1}{2^\nu \Gamma(\nu+1)},$$

where Γ stands for the gamma function (see, e.g., [2, Chapter 6]). Taking into account the formula $\Gamma(s+1) = s\Gamma(s)$, the solution is written in the form

$$y(x) = \sum_{k=0}^{\infty} (-1)^k \frac{\left(\frac{x}{2}\right)^{\nu+2k}}{k! \Gamma(\nu+k+1)}. \qquad (6.48)$$

Definition 68 The solution (6.48) of Eq. (6.47) is called the **Bessel function of the first kind of order** ν and is denoted by $J_\nu(x)$,

$$J_\nu(x) = \sum_{k=0}^{\infty} (-1)^k \frac{\left(\frac{x}{2}\right)^{\nu+2k}}{k! \Gamma(\nu+k+1)}. \qquad (6.49)$$

For $\mu = -\nu$, from (6.49) we obtain another solution of (6.47)

$$J_{-\nu}(x) = \sum_{k=0}^{\infty} (-1)^k \frac{\left(\frac{x}{2}\right)^{-\nu+2k}}{k! \Gamma(-\nu+k+1)}.$$

In order to obtain a general solution we need to check that $J_\nu(x)$ and $J_{-\nu}(x)$ are linearly independent. The linear independence is valid indeed for certain values of ν, since the series expansions of $J_\nu(x)$ and $J_{-\nu}(x)$ begin with different powers of x. Hence for all noninteger ν the general solution of (6.47) has the form

$$y(x) = C_1 J_\nu(x) + C_2 J_{-\nu}(x).$$

If $\nu = n$ is an integer number, then taking into account that $\Gamma(-m) = \infty$ for $m = 0, 1, 2, \ldots$ and $\Gamma(m) = \frac{\Gamma(m+1)}{m}$, we have

$$J_{-n}(x) = \sum_{k=n}^{\infty}(-1)^k \frac{\left(\frac{x}{2}\right)^{-n+2k}}{k!\,\Gamma(-n+k+1)}.$$

Making the change in this sum $k - n \to k$ we obtain

$$J_{-n}(x) = \sum_{k=0}^{\infty}(-1)^{k+n}\frac{\left(\frac{x}{2}\right)^{n+2k}}{(k+n)!\,\Gamma(k+1)} = (-1)^n\sum_{k=0}^{\infty}(-1)^k\frac{\left(\frac{x}{2}\right)^{n+2k}}{(k+n)!\,k!}.$$

Thus,

$$J_{-n}(x) = (-1)^n J_n(x),$$

and hence for integer n the functions J_n and J_{-n} are linearly dependent. In order to obtain a second solution in this case the following function is composed

$$Y_\nu(x) = \frac{J_\nu(x)\cos\nu\pi - J_{-\nu}(x)}{\sin\nu\pi}. \tag{6.50}$$

Obviously, for a noninteger ν the function $Y_\nu(x)$ is a solution of (6.47). Expanding the indeterminacy when $\nu \to n$ (which is quite complicated in this case, see, e.g., [70]) we obtain the second solution of (6.47) in the form

$$Y_n(x) = \frac{2}{\pi}J_n(x)\ln\frac{x}{2} - \frac{1}{\pi}\sum_{k=0}^{n-1}\frac{\left(\frac{x}{2}\right)^{-n+2k}(n-k-1)!}{k!} \tag{6.51}$$

$$-\frac{1}{\pi}\sum_{k=0}^{\infty}\frac{(-1)^k\left(\frac{x}{2}\right)^{n+2k}}{k!(k+n)!}\left(\frac{\Gamma'(k+1)}{\Gamma(k+1)} + \frac{\Gamma'(k+n+1)}{\Gamma(k+n+1)}\right).$$

Definition 69 A function $Y_\nu(x)$ as defined above is called the **Bessel function of the second kind of order ν or the Neumann function**

The functions $J_\nu(x)$ and $Y_\nu(x)$ are linearly independent. Thus, the general solution of (6.47) can be written in the form

$$y(x) = C_1 J_\nu(x) + C_2 Y_\nu(x).$$

6.9 Properties of Bessel Functions

Property 1: Behavior at the Origin From (6.49) it is obvious that $J_\nu(0) = 0$ for $\nu > 0$, $J_\nu(0) = 1$ when $\nu = 0$ and $J_\nu(0) = \infty$ for $\nu < 0$.

Property 2 Derivatives of Bessel Functions First, let us notice that the Bessel function J_ν admits a power series expansion which is absolutely convergent for all x, according to the d'Alembert criterion. The interval of its uniform convergence is the whole axis, and thus for all x the series can be differentiated termwise any number of times. Let us calculate the derivative

$$\frac{d}{dx}\left(\frac{J_\nu(x)}{x^\nu}\right) = \frac{d}{dx}\sum_{k=0}^{\infty}(-1)^k\frac{x^{2k}}{2^{\nu+2k}k!\Gamma(\nu+k+1)}$$

$$= \sum_{k=1}^{\infty}(-1)^k\frac{2kx^{2k-1}}{2^{\nu+2k}k!\Gamma(\nu+k+1)}$$

$$= \sum_{k=1}^{\infty}(-1)^k\frac{x^{2k+\nu-1}}{(k-1)!\Gamma(\nu+k+1)}\frac{1}{x^\nu 2^{\nu+2k-1}}$$

$$= \frac{1}{x^\nu}\sum_{k=0}^{\infty}\frac{(-1)^{k+1}}{k!\Gamma(k+\nu+2)}\left(\frac{x}{2}\right)^{\nu+2k+1}$$

$$= -\frac{1}{x^\nu}J_{\nu+1}(x).$$

Thus,

$$\frac{d}{dx}\left(\frac{J_\nu(x)}{x^\nu}\right) = -\frac{1}{x^\nu}J_{\nu+1}(x), \tag{6.52}$$

or in a more symmetric form,

$$\left(\frac{1}{x}\frac{d}{dx}\right)\left(\frac{J_\nu(x)}{x^\nu}\right) = -\frac{1}{x^{\nu+1}}J_{\nu+1}(x).$$

Moreover, the following formula is valid

$$\left(\frac{1}{x}\frac{d}{dx}\right)^l\left(\frac{J_\nu(x)}{x^\nu}\right) = (-1)^l\frac{1}{x^{\nu+l}}J_{\nu+l}(x).$$

Analogously,

$$\frac{d}{dx}\left(x^\nu J_\nu(x)\right) = x^\nu J_{\nu-1}(x). \tag{6.53}$$

Indeed,

$$\frac{d}{dx}\left(x^{\nu}J_{\nu}(x)\right) = \frac{d}{dx}\sum_{k=0}^{\infty}(-1)^{k}\frac{x^{2\nu+2k}}{2^{\nu+2k}k!\Gamma(\nu+k+1)}$$

$$= \sum_{k=0}^{\infty}(-1)^{k}\frac{2(\nu+k)x^{2\nu+2k-1}}{2^{\nu+2k}k!\Gamma(\nu+k+1)}$$

$$= x^{\nu}\sum_{k=0}^{\infty}(-1)^{k}\frac{x^{\nu+2k-1}}{2^{\nu+2k-1}k!\Gamma(\nu+k)}$$

$$= x^{\nu}J_{\nu-1}(x),$$

where we used $\Gamma(\nu+k+1) = (\nu+k)\Gamma(\nu+k)$. More generally,

$$\left(\frac{1}{x}\frac{d}{dx}\right)^{m}\left(x^{\nu}J_{\nu}(x)\right) = x^{\nu-m}J_{\nu-m}(x).$$

Let us differentiate (6.52) and multiply it by $x^{2\nu}$. We have

$$x^{\nu}J_{\nu}'(x) - \nu x^{\nu-1}J_{\nu}(x) = -x^{\nu}J_{\nu+1}(x),$$

or

$$J_{\nu}'(x) = \frac{\nu}{x}J_{\nu}(x) - J_{\nu+1}(x). \tag{6.54}$$

Differentiating (6.53), we obtain

$$x^{\nu}J_{\nu}'(x) + \nu x^{\nu-1}J_{\nu}(x) = x^{\nu}J_{\nu-1}(x),$$

or

$$J_{\nu}'(x) = -\frac{\nu}{x}J_{\nu}(x) + J_{\nu-1}(x). \tag{6.55}$$

The recurrent formulas (6.54) and (6.55) allow us to calculate the derivative of the Bessel function in terms of the same Bessel function and a Bessel function of a neighboring index. Besides, from these formulas an important recurrent relation follows

$$J_{\nu+1}(x) = \frac{2\nu}{x}J_{\nu}(x) - J_{\nu-1}(x). \tag{6.56}$$

Property 3: The Wronskian of Bessel Functions $W[J_{\nu}(x), J_{-\nu}(x)]$ According to the Liouville-Ostrogradsky formula (or Abel's identity),

$$W(x) = Ce^{-\int_{x_0}^{x} p(t)dt},$$

where $p(t)$ is the coefficient of the first derivative in the second order differential equation written in its reduced form. In our case $p(t) = 1/t$ and hence

$$W(x) = \frac{C}{x}.$$

Let us compute C,

$$C = \lim_{x \to 0} (x \cdot W(x)) = \lim_{x \to 0} \left(x \cdot \begin{vmatrix} J_\nu(x) & J_{-\nu}(x) \\ J_\nu'(x) & J_{-\nu}'(x) \end{vmatrix}\right)$$

$$= \lim_{x \to 0} (x \cdot (J_\nu(x) J_{-\nu}'(x) - J_{-\nu}(x) J_\nu'(x))). \qquad (6.57)$$

Again, it is convenient to use the series expansions of the Bessel functions

$$J_\nu(x) = \sum_{k=0}^{\infty} (-1)^k \frac{\left(\frac{x}{2}\right)^{\nu+2k}}{k! \Gamma(\nu+k+1)} \quad \text{and} \quad J_{-\nu}(x) = \sum_{k=0}^{\infty} (-1)^k \frac{\left(\frac{x}{2}\right)^{-\nu+2k}}{k! \Gamma(-\nu+k+1)}.$$

In the products under the sign of a limit in (6.57) the only summand different from zero corresponds to $k = 0$, and thus,

$$C = \lim_{x \to 0} x \cdot \left(\frac{\left(\frac{x}{2}\right)^\nu}{\Gamma(\nu+1)} \cdot \frac{-\nu x^{-\nu-1}}{2^{-\nu}\Gamma(-\nu+1)} - \frac{\nu x^{\nu-1}}{2^\nu \Gamma(\nu+1)} \cdot \frac{\left(\frac{x}{2}\right)^{-\nu}}{\Gamma(-\nu+1)} \right)$$

$$= -\frac{2\nu}{\Gamma(\nu+1)\Gamma(-\nu+1)} = -\frac{2 \sin \nu\pi}{\pi}.$$

Here we used the identities

$$\Gamma(\nu+1) = \nu\Gamma(\nu) \quad \text{and} \quad \Gamma(\nu)\Gamma(1-\nu) = \frac{\pi}{\sin \nu\pi}.$$

Thus,

$$W[J_\nu(x), J_{-\nu}(x)] = -\frac{2 \sin \nu\pi}{\pi} \cdot \frac{1}{x}.$$

Corollary 70 *For integer ν the Wronskian equals zero, and the functions J_ν, $J_{-\nu}$ are linearly dependent. For noninteger ν the functions J_ν, $J_{-\nu}$ are linearly independent, since $W[J_\nu(x), J_{-\nu}(x)] \neq 0$.*

Property 4: The Asymptotics of Bessel Functions at Infinity Let us obtain the asymptotic formula for half integer orders $\nu = n + 1/2$. The obtained formula remains valid for all ν.

Consider the Bessel function

$$J_{1/2}(x) = \sum_{k=0}^{\infty} (-1)^k \frac{\left(\frac{x}{2}\right)^{1/2+2k}}{k!\Gamma(k+1+1/2)}.$$

Taking into account the property of the gamma function

$$\Gamma\left(k+1+\frac{1}{2}\right) = \left(k+\frac{1}{2}\right)\Gamma\left(k+\frac{1}{2}\right) = \ldots = \left(k+\frac{1}{2}\right)\left(k+\frac{1}{2}-1\right)\ldots\frac{1}{2}\Gamma\left(\frac{1}{2}\right)$$

and the identity $\Gamma\left(\frac{1}{2}\right) = \sqrt{\pi}$, that is,

$$\Gamma\left(k+1+\frac{1}{2}\right) = \frac{(2k+1)!!}{2^{k+1}}\sqrt{\pi},$$

we have

$$J_{1/2}(x) = \sum_{k=0}^{\infty} \sqrt{\frac{2}{x}} \frac{(-1)^k x^{2k+1} 2^{k+1}}{k!2^{2k+1}(2k+1)!!\sqrt{\pi}} = \sqrt{\frac{2}{\pi x}} \sum_{k=0}^{\infty} \frac{(-1)^k x^{2k+1}}{(2k+1)!} = \sqrt{\frac{2}{\pi x}}\sin x,$$

where

$$n!! = \begin{cases} 2 \cdot 4 \cdot 6 \cdot \ldots \cdot n, & n\text{ - even}; \\ 1 \cdot 3 \cdot 5 \cdot \ldots \cdot n, & n\text{ - odd}. \end{cases}$$

Here we took into account that

$$k!\,(2k+1)!!\,2^k = (2k+1)!!\,(2k)!! = (2k+1)!.$$

In a similar way one can show that

$$J_{-1/2}(x) = \sqrt{\frac{2}{\pi x}}\cos x.$$

Thus,

$$J_{1/2}(x) = \sqrt{\frac{2}{\pi x}}\sin x, \quad J_{-1/2}(x) = \sqrt{\frac{2}{\pi x}}\cos x. \tag{6.58}$$

Using the recurrent formula (6.56) we obtain

$$J_{3/2}(x) = \frac{1}{x}J_{1/2}(x) - J_{-1/2}(x) = \sqrt{\frac{2}{\pi x}}\left(\frac{\sin x}{x} - \cos x\right)$$

$$= \sqrt{\frac{2}{\pi x}}\left(\sin\left(x - \frac{\pi}{2}\right) + \frac{1}{x}\cos\left(x - \frac{\pi}{2}\right)\right),$$

$$J_{5/2}(x) = \frac{3}{x}J_{3/2}(x) - J_{1/2}(x)$$

$$= \sqrt{\frac{2}{\pi x}}\left(\frac{3}{x}\sin\left(x - \frac{\pi}{2}\right) + \frac{3}{x^2}\cos\left(x - \frac{\pi}{2}\right) - \sin x\right)$$

$$= \sqrt{\frac{2}{\pi x}}\left(\sin(x - \pi) - \frac{3}{x}\cos(x - \pi) - \frac{3}{x^2}\sin(x - \pi)\right)$$

$$= \sqrt{\frac{2}{\pi x}}\left(\left(1 - \frac{3}{x^2}\right)\sin(x - \pi) - \frac{3}{x}\cos(x - \pi)\right).$$

One can prove by induction that in general

$$J_{n+1/2}(x) = \sqrt{\frac{2}{\pi x}}\left(P_n\left(\frac{1}{x}\right)\sin\left(x - \frac{n\pi}{2}\right) + Q_n\left(\frac{1}{x}\right)\cos\left(x - \frac{n\pi}{2}\right)\right),$$
(6.59)

where

$$P_n(0) = 1 \quad \text{and} \quad Q_n(0) = 0.$$

From (6.59) we obtain

$$J_{n+1/2}(x) \sim \sqrt{\frac{2}{\pi x}}\sin\left(x - \frac{n\pi}{2}\right), \quad x \to \infty,$$

or, more generally, replacing $n + 1/2$ by ν,

$$J_\nu(x) \sim \sqrt{\frac{2}{\pi x}}\sin\left(x - \frac{\nu\pi}{2} + \frac{\pi}{4}\right) \quad x \to \infty.$$

Let us emphasize that this formula is valid for all ν, and more precisely,

$$J_\nu(x) = \sqrt{\frac{2}{\pi x}}\left(\sin\left(x - \frac{\nu\pi}{2} + \frac{\pi}{4}\right) + O\left(x^{-1}\right)\right).$$
(6.60)

Property 5: Zeros of Bessel Functions

Theorem 71 *All zeros of the Bessel function J_v for $v \geq 0$ are real.*

Proof The function J_v is a solution of the Bessel equation

$$x^2 J_v''(x) + x J_v'(x) + (x^2 - v^2) J_v(x) = 0.$$

Setting $x = \lambda t$ we find that $J_v(\lambda t)$ satisfies the equation

$$t^2 \frac{d^2 J_v}{dt^2} + t \frac{d J_v}{dt} + (\lambda^2 t^2 - v^2) J_v = 0.$$

Let us write this equation in a self-adjoint form (preliminarily dividing it by t)

$$\frac{d}{dt}\left(t \frac{d J_v}{dt}\right) - \frac{v^2}{t} J_v + \lambda^2 t J_v = 0, \quad p = t, \quad q = -\frac{v^2}{t}. \tag{6.61}$$

Take two different values of λ: λ_1 and λ_2, and write down the corresponding equations

$$\frac{d}{dt}\left(t \frac{d J_v(\lambda_1 t)}{dt}\right) - \frac{v^2}{t} J_v(\lambda_1 t) + \lambda_1^2 t J_v(\lambda_1 t) = 0,$$

$$\frac{d}{dt}\left(t \frac{d J_v(\lambda_2 t)}{dt}\right) - \frac{v^2}{t} J_v(\lambda_2 t) + \lambda_2^2 t J_v(\lambda_2 t) = 0.$$

Multiplying the first of them by $J_v(\lambda_2 t)$, the second one by $J_v(\lambda_1 t)$ and subtracting the first from the second, we obtain

$$J_v(\lambda_1 t) \frac{d}{dt}\left(t \frac{d J_v(\lambda_2 t)}{dt}\right) - J_v(\lambda_2 t) \frac{d}{dt}\left(t \frac{d J_v(\lambda_1 t)}{dt}\right) + \left(\lambda_2^2 - \lambda_1^2\right) t J_v(\lambda_1 t) J_v(\lambda_2 t) = 0. \tag{6.62}$$

Notice that

$$J_v(\lambda_1 t) \frac{d}{dt}\left(t \frac{d J_v(\lambda_2 t)}{dt}\right) - J_v(\lambda_2 t) \frac{d}{dt}\left(t \frac{d J_v(\lambda_1 t)}{dt}\right)$$

$$= \frac{d}{dt}\left(t\left(J_v(\lambda_1 t) \frac{d J_v(\lambda_2 t)}{dt} - J_v(\lambda_2 t) \frac{d J_v(\lambda_1 t)}{dt}\right)\right).$$

Integration of (6.62) from 0 to 1 gives then

$$\int_0^1 \frac{d}{dt}\left(t\left(J_v(\lambda_1 t) \frac{d J_v(\lambda_2 t)}{dt} - J_v(\lambda_2 t) \frac{d J_v(\lambda_1 t)}{dt}\right)\right) dt = -\left(\lambda_2^2 - \lambda_1^2\right) \int_0^1 t J_v(\lambda_1 t) J_v(\lambda_2 t) dt$$

or

$$t \left(J_\nu (\lambda_1 t) \frac{d J_\nu (\lambda_2 t)}{dt} - J_\nu (\lambda_2 t) \frac{d J_\nu (\lambda_1 t)}{dt} \right) \Big|_0^1 = - \left(\lambda_2^2 - \lambda_1^2 \right) \int_0^1 t J_\nu (\lambda_1 t) J_\nu (\lambda_2 t) \, dt.$$

From the power series representation of the Bessel function J_ν we deduce that the expression on the left hand side tends to zero when $t \to 0$, and hence

$$\left(\lambda_2^2 - \lambda_1^2 \right) \int_0^1 t J_\nu (\lambda_1 t) J_\nu (\lambda_2 t) \, dt = \left(\lambda_1 J_\nu (\lambda_2) J_\nu' (\lambda_1) - \lambda_2 J_\nu' (\lambda_2) J_\nu (\lambda_1) \right),$$

$$(6.63)$$

where we took into account that $\frac{d J_\nu (\lambda t)}{dt} = \lambda J_\nu' (\lambda t)$.

Let us notice that integrating (6.62) from 0 to some $l > 0$ and repeating the above procedure leads to a generalization of (6.63) which will be useful below:

$$\left(\lambda_2^2 - \lambda_1^2 \right) \int_0^l t J_\nu (\lambda_1 t) J_\nu (\lambda_2 t) \, dt = l \left(\lambda_1 J_\nu (\lambda_2 l) J_\nu' (\lambda_1 l) - \lambda_2 J_\nu' (\lambda_2 l) J_\nu (\lambda_1 l) \right).$$

$$(6.64)$$

Since the coefficients in the power series representation of the Bessel function are all real, if $J_\nu (\lambda) = 0$ then $J_\nu (\overline{\lambda}) = 0$ as well. This is because $J_\nu (\overline{\lambda}) = \overline{J_\nu (\lambda)}$. Suppose now that λ_1 is a complex zero of J_ν. Take $\lambda_2 := \overline{\lambda_1}$. Then from (6.63) we obtain

$$\left(\left(\overline{\lambda_1} \right)^2 - \lambda_1^2 \right) \int_0^1 t J_\nu (\lambda_1 t) J_\nu (\overline{\lambda_1} t) \, dt = 0$$

or

$$\int_0^1 t J_\nu (\lambda_1 t) \, \overline{J_\nu (\lambda_1 t)} dt = 0$$

which is impossible because the product $J_\nu (\lambda_1 t) \overline{J_\nu (\lambda_1 t)} = |J_\nu (\lambda_1 t)|^2$ is positive. Thus, J_ν does not possess zeros with a nontrivial imaginary part. ∎

From the asymptotics of the Bessel function at infinity (formula (6.60)) we see that J_ν changes its sign an infinite number of times. Since $J_\nu(x)$ is a continuous function, this means that it has an infinite number of real zeros (Fig. 6.3), and the absolute values of negative zeros are equal to their positive counterparts (this is due to the presence of even powers only in the power series representation of the Bessel function).

Property 6: Orthogonality of Bessel Functions Let $\lambda_1 = \frac{\mu_1}{l}$ and $\lambda_2 = \frac{\mu_2}{l}$, where μ_1 and μ_2 are zeros of the Bessel function, $J_\nu(\mu_1) = J_\nu(\mu_2) = 0$. Then from (6.64) we obtain the equality

$$\int_0^l t J_\nu \left(\frac{\mu_1}{l} t \right) J_\nu \left(\frac{\mu_2}{l} t \right) dt = 0 \qquad (6.65)$$

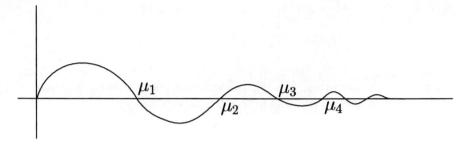

Fig. 6.3 Schematic representation of the graph of a Bessel function $J_\nu(x)$, $\nu > 0$. μ_1, μ_2, indicate its positive zeros

which means that the Bessel functions $J_\nu\left(\frac{\mu_1}{l}t\right)$ and $J_\nu\left(\frac{\mu_2}{l}t\right)$ are orthogonal on the interval $(0, l)$ with the weight t for any pair of zeros of the Bessel function $\mu_1 \neq \mu_2$.

Now suppose that μ_1 is a zero of the Bessel function J_ν, and μ_2 is not, but a sufficiently close number to μ_1. From (6.64) we have then

$$\int_0^l t J_\nu\left(\frac{\mu_1}{l}t\right) J_\nu\left(\frac{\mu_2}{l}t\right) dt = \frac{l^2}{\mu_2^2 - \mu_1^2}\mu_1 J_\nu'(\mu_1) J_\nu(\mu_2).$$

Let us pass to the limit when $\mu_2 \to \mu_1$, applying L'Hôpital's rule on the right hand side. We obtain

$$\int_0^l t J_\nu^2\left(\frac{\mu_1}{l}t\right) dt = \lim_{\mu_2 \to \mu_1} \frac{l^2\mu_1 J_\nu'(\mu_1) J_\nu(\mu_2)}{\mu_2^2 - \mu_1^2}$$

$$= \lim_{\mu_2 \to \mu_1} \frac{l^2 J_\nu'(\mu_1) J_\nu'(\mu_2) \mu_1}{2\mu_2} = \frac{l^2\left(J_\nu'(\mu_1)\right)^2}{2}.$$

Substituting μ_1 into the formula $J_\nu'(x) = \frac{\nu}{x}J_\nu(x) - J_{\nu+1}(x)$ and recalling that $J_\nu(\mu_1) = 0$, we obtain $J_\nu'(\mu_1) = -J_{\nu+1}(\mu_1)$ and thus,

$$\left\| J_\nu\left(\frac{\mu_1}{l}t\right) \right\|^2 := \int_0^l t J_\nu^2\left(\frac{\mu_1}{l}t\right) dt = \frac{l^2 J_{\nu+1}^2(\mu_1)}{2}.$$

Hence if μ_i and μ_j are zeros of the Bessel function J_ν, then

$$\int_0^l t J_\nu\left(\frac{\mu_i}{l}t\right) J_\nu\left(\frac{\mu_j}{l}t\right) dt = \begin{cases} 0, & i \neq j, \\ \frac{l^2 J_{\nu+1}^2(\mu_j)}{2}, & i = j. \end{cases}$$

The fact that the Bessel functions $J_\nu\left(\frac{\mu_i}{l}t\right)$ and $J_\nu\left(\frac{\mu_j}{l}t\right)$ are orthogonal on $(0, l)$ with the weight t is a consequence of the self-adjointness of the Bessel equation (6.61), and of the fact that we found its solutions satisfying the boundary condition

$J_\nu(\lambda t) = 0$. The usual boundary conditions for (6.61) have the form

$$J_\nu(0) = 0, \quad J_\nu(\lambda l) = J_\nu(\mu) = 0,$$

the first of which is fulfilled by the function J_ν for $\nu > 0$. The fulfillment of the second condition leads to the eigenfunctions considered above. Notice that instead of a Dirichlet type second condition, a more general condition

$$\alpha J_\nu(\lambda l) + \beta J'_\nu(\lambda l) = 0, \quad \lambda l = \mu \tag{6.66}$$

can be considered with $\alpha, \beta \in \mathbb{R}$, $|\alpha| + |\beta| > 0$. In this case the result on the existence of a countable set of eigenvalues and the orthogonality of the corresponding eigenfunctions remains valid, though with a slightly more complicated expression for their norm.

The use of the above mentioned properties allows one to expand arbitrary functions into series in terms of Bessel functions of certain order. We consider this topic in the next section.

6.10   Expansion in Series of Bessel Functions

Let us suppose that a function $\varphi(x)$ admits the following series representation

$$\varphi(x) = \sum_{j=1}^{\infty} J_\nu\left(\frac{\mu_j}{l}x\right)\varphi_j \tag{6.67}$$

where μ_j are zeros of the Bessel function. Let us obtain a formula for the coefficients φ_j. Multiplication of the series by $x J_\nu\left(\frac{\mu_k}{l}x\right)$ and subsequent integration from 0 to l, due to the orthogonality of the Bessel functions (6.65) gives

$$\varphi_k = \frac{1}{\left\|J_\nu\left(\frac{\mu_k}{l}x\right)\right\|^2} \int_0^l x\varphi(x) J_\nu\left(\frac{\mu_k}{l}x\right) dx, \tag{6.68}$$

where

$$\left\|J_\nu\left(\frac{\mu_k}{l}x\right)\right\|^2 = \frac{l^2 J_{\nu+1}^2(\mu_k)}{2}. \tag{6.69}$$

If μ_k are zeros of a more complicated expression (6.66), the series expansion (6.67) and the formula for the coefficients (6.68) remain valid with the only difference that for $\left\|J_\nu\left(\frac{\mu_k}{l}x\right)\right\|^2$ the formula (6.69) needs to be replaced with a corresponding, more complicated expression.

Thus, if $\varphi(x)$ admits a series expansion of the form (6.67), then the coefficients φ_k are determined uniquely. The series (6.67) is a series in terms of eigenfunctions of a Sturm-Liouville problem. Referring the reader to [83] for corresponding details, we only mention here that if $\varphi(x) \in C^2[0, l]$ satisfies the boundary conditions $\varphi(x) = O(x^\nu)$, $x \to 0$ and $\alpha\varphi(l) + \beta\varphi'(l) = 0$, $|\alpha| + |\beta| > 0$, and $x^{-1/2}\left(\frac{d}{dx}\left(x\frac{d\varphi}{dx}\right) - \frac{\nu^2}{x}\varphi\right) \in L_2(0, l)$, then it admits a regularly convergent series expansion of the form (6.67), which means that the series is uniformly convergent on any segment which is strictly inside of $(0, l)$. More on the convergence of such series can be found in [84].

6.11 Free Vibrations of a Circular Membrane (Continuation)

We now return to solving the problem of free vibrations a circular membrane, started in Sect. 6.7. We have the following equations

$$\begin{cases} r^2 R'' + r R' + \left(\lambda^2 r^2 - n^2\right) R = 0, & 0 < r < l, \\ \Phi'' + n^2\Phi = 0, \\ T'' + a^2\lambda^2 T = 0. \end{cases}$$

As we have seen, the equation for $\Phi(\varphi)$ admits a general solution of the form

$$\Phi(\varphi) = C_n^1 \cos n\varphi + C_n^2 \sin n\varphi.$$

A general solution of the first equation has the form

$$R_n(r) = \widetilde{A}_n J_n(\lambda r) + \widetilde{B}_n Y_n(\lambda r).$$

Since the boundary of the membrane is fixed, the condition

$$R(l) = 0 \tag{6.70}$$

must be fulfilled. Moreover, obviously the solution must be finite, in particular, the inequality $|R(0)| < \infty$ must be satisfied. Hence $\widetilde{B}_n = 0$, since $Y_n(0) = \infty$. The condition (6.70) leads to the equality

$$J_n(\lambda l) = 0.$$

Let μ_{nk}, $k = 1, 2, \ldots$ be all positive roots of the equation $J_n(\mu) = 0$, and thus, for every n we have a complete set of the eigenvalues $\lambda_{nk} = \frac{\mu_{nk}}{l}$ and eigenfunctions $J_n(\lambda_{nk}r)$. Finally, from the third equation we obtain

$$T_{nk}(t) = M_{nk} \cos a\lambda_{nk}t + L_{nk} \sin a\lambda_{nk}t.$$

Applying the superposition principle, we find

$$u(r, \varphi, t) = \sum_{n=0}^{\infty} \sum_{k=1}^{\infty} \left(C_n^1 \cos n\varphi + C_n^2 \sin n\varphi \right) (M_{nk} \cos a\lambda_{nk}t + L_{nk} \sin a\lambda_{nk}t) J_n(\lambda_{nk}r),$$

or

$$u(r, \varphi, t) = \sum_{n=0}^{\infty} \sum_{k=1}^{\infty} \Bigg((A_{nk} \cos a\lambda_{nk}t + B_{nk} \sin a\lambda_{nk}t) \cos n\varphi$$
$$+ (C_{nk} \cos a\lambda_{nk}t + D_{nk} \sin a\lambda_{nk}t) \sin n\varphi \Bigg) J_n(\lambda_{nk}r),$$

where A_{nk}, B_{nk}, C_{nk}, and D_{nk} are new constants.
Let us use the initial conditions

$$u|_{t=0} = f(r, \varphi),$$

$$\frac{\partial u}{\partial t}\bigg|_{t=0} = g(r, \varphi).$$

We have

$$u(r, \varphi, 0) = \sum_{n=0}^{\infty} \sum_{k=1}^{\infty} (A_{nk} \cos n\varphi + C_{nk} \sin n\varphi) J_n(\lambda_{nk}r) = f(r, \varphi).$$

For every r fixed this is a Fourier series expansion of the function $f(r, \varphi)$ on $[0, 2\pi]$:

$$f(r, \varphi) = \frac{a_0(r)}{2} + \sum_{n=1}^{\infty} (a_n(r) \cos n\varphi + b_n(r) \sin n\varphi).$$

Hence the coefficients are calculated according to the formulas

$$\frac{a_0(r)}{2} = \sum_{k=1}^{\infty} A_{0k} J_0(\frac{\mu_{0k}}{l}r) = \frac{1}{2\pi} \int_0^{2\pi} f(r, \varphi)d\varphi, \quad \text{for } n = 0$$

and

$$a_n(r) = \sum_{k=1}^{\infty} A_{nk} J_n(\frac{\mu_{nk}}{l}r) = \frac{1}{\pi} \int_0^{2\pi} f(r, \varphi) \cos n\varphi d\varphi,$$

$$b_n(r) = \sum_{k=1}^{\infty} C_{nk} J_n\left(\frac{\mu_{nk}}{l} r\right) = \frac{1}{\pi} \int_0^{2\pi} f(r, \varphi) \sin n\varphi \, d\varphi,$$

for $n = 1, 2, \ldots$. For determining the coefficients A_{nk} and C_{nk} it is sufficient to expand the integrals on the right hand sides into series of Bessel functions. Similarly the coefficients B_{nk} and D_{nk} can be found with the aid of the second initial condition.

At this stage, the problem is completely solved. We recommend comparing this problem with the problem of vibrations of a rectangular membrane in order to understand how radically the geometry of the investigated object influences the choice of the optimal method for solving the corresponding problem.

In conclusion, let us note, without entering into the details, that if the membrane were fixed elastically, then the solution would result in the following changes. Firstly, μ_{nk} would be the roots of the equation $J_n'(\mu) + h\mu J_n(\mu) = 0$, and secondly, the norm of J_n will be certainly changed. Of course, all this requires a detailed justification, which we propose to undertake to the interested reader.

Chapter 7
Methods of Solution of Sturm-Liouville Equations, Direct and Inverse Problems

7.1 Polya's Factorization, Abel's Formula

In the previous chapter we saw that the Sturm-Liouville equation

$$\frac{d}{dx}\left(p(x)\frac{du}{dx}\right) - q(x)u(x) = \lambda r(x)u(x), \quad a < x < b \tag{7.1}$$

naturally arises when solving partial differential equations by the Fourier method of separation of variables. Applications of the Sturm-Liouville equation are not limited to the Fourier method. It is one of the most important mathematical equations. Up to now no formula for its general solution in a closed form is known, unlike the case of a linear differential equation of first order.

It is of primary interest the dependence of the solutions of (7.1) on the spectral parameter λ, and a method for practical solution of (7.1). Below we discuss an approach based on so-called spectral parameter power series (SPPS) representations for solutions of (7.1) which provide convenient formulas and lend themselves to numerical solution of initial, boundary value and spectral problems.

For the sake of simplicity of exposition we will consider Eq. (7.1) under the condition that there exists a particular solution f of the equation

$$\frac{d}{dx}\left(p(x)\frac{du}{dx}\right) - q(x)u(x) = 0 \tag{7.2}$$

such that the functions rf^2 and $1/(pf^2)$ are continuous on $[a, b]$. This condition is obviously fulfilled when p, q, and r are real valued, continuous functions together with p', and $p, r > 0$, that is under the conditions of Sect. 6.4. This is because in such case (7.2) possesses two linearly independent regular real valued solutions f_1 and f_2 whose zeros alternate. For example, f_1 and f_2 can be chosen to satisfy the initial conditions $f_1(a) = 1$, $f_1'(a) = 0$, $f_2(a) = 0$, $f_2'(a) = 1$. Thus, one may choose $f = f_1 + if_2$. Such f will be a convenient solution of (7.2), nonvanishing on $[a, b]$ and guaranteeing the continuity of the functions rf^2 and $1/(pf^2)$. However, the applicability of the SPPS representations goes far beyond this regular situation, see remarks below.

We begin with an auxiliary proposition.

Proposition 72 (Factorization of the Sturm-Liouville Expression) *Let* $f \in C^2[a, b]$ *be a solution of (7.2) such that* $f(x) \neq 0$ *for all* $x \in [a, b]$. *Then for any* $u \in C^2[a, b]$ *the equality is valid*

$$\frac{d}{dx}\left(p\frac{du}{dx}\right) - qu = \frac{1}{f}\frac{d}{dx}\left(pf^2\frac{d}{dx}\left(\frac{u}{f}\right)\right). \tag{7.3}$$

Proof Consider

$$\frac{1}{f}\frac{d}{dx}\left(pf^2\frac{d}{dx}\left(\frac{u}{f}\right)\right) = \frac{1}{f}\frac{d}{dx}\left(pf^2\left(\frac{u'f - uf'}{f^2}\right)\right)$$

$$= \frac{1}{f}\frac{d}{dx}(pu'f - puf')$$

$$= \frac{1}{f}(p'u'f + pu''f - p'uf' - puf'')$$

$$= p'u' + pu'' - \frac{p'uf' + puf''}{f}$$

$$= (pu')' - \frac{p'f' + pf''}{f}u.$$

Since f is a solution of (7.2), we have that

$$q = \frac{(pf')'}{f} = \frac{p'f' + pf''}{f}$$

and hence

$$\frac{1}{f}\frac{d}{dx}\left(pf^2\frac{d}{dx}\left(\frac{u}{f}\right)\right) = (pu')' - qu.$$

∎

The factorization of the Sturm-Liouville expression (7.3) is called the Polya factorization, and is written formally as

$$\frac{d}{dx}p\frac{d}{dx} - q = \frac{1}{f}\frac{d}{dx}pf^2\frac{d}{dx}\frac{1}{f},$$

where all functions are understood as operators of multiplication by the corresponding functions. From this equality it is evident that f is a solution of (7.2), because consecutive application of the operators on the right hand side to f immediately gives zero:

$$\left(\frac{1}{f}\frac{d}{dx}pf^2\frac{d}{dx}\frac{1}{f}\right)[f] = \frac{1}{f}\frac{d}{dx}pf^2\frac{d}{dx}[1] = 0. \qquad (7.4)$$

Moreover, with the aid of the Polya factorization it is easy to derive a formula for the second linearly independent solution of (7.2). Indeed, equality (7.4) shows that zero is obtained due to the annihilating action of the first operator of differentiation on a constant. To make use of the annihilating action of the second operator of differentiation we may consider when the operator $pf^2\frac{d}{dx}\frac{1}{f}$ applied to a function g gives us a constant, e.g., one. Thus, consider the equation

$$pf^2\frac{d}{dx}\frac{1}{f}g = 1$$

or

$$\left(\frac{g}{f}\right)' = \frac{1}{pf^2}.$$

We obtain that the second solution g can be chosen in the form

$$g(x) = f(x)\int_{x_0}^{x}\frac{ds}{f^2(s)p(s)}, \qquad (7.5)$$

where x_0 is an arbitrary point of $[a, b]$. The obtained formula for finding a second linearly independent solution g of (7.2) when a particular solution f is known, it is called Abel's formula. To see that f and g are linearly independent it is sufficient to verify that their Wronskian is different from zero at least at one point, e.g., at x_0. Notice that $g(x_0) = 0$ and

$$g'(x) = f'(x)\int_{x_0}^{x}\frac{ds}{f^2(s)p(s)} + \frac{1}{f(x)p(x)}.$$

Hence

$$g'(x_0) = \frac{1}{f(x_0)p(x_0)}.$$

Thus,

$$W[f, g](x_0) = f(x_0)g'(x_0) - f'(x_0)g(x_0) = \frac{1}{p(x_0)}.$$

Since p does not have zeros on $[a, b]$, we obtain that $W [f, g] (x_0) \neq 0$ and hence the solutions f and g of Eq. (7.2) are linearly independent.

7.2 Spectral Parameter Power Series

Let $f, r, p \in C[a, b]$ be some functions (which can be complex valued), such that $f (x) \neq 0$, $p (x) \neq 0$, for all $x \in [a, b]$, and $x_0 \in [a, b]$. Let us construct the following two sequences of recursive integrals

$$X^{(0)}(x) \equiv 1,$$

$$X^{(n)}(x) = \begin{cases} n \int_{x_0}^{x} X^{(n-1)}(s) \left(f^2(s)r(s) \right) ds, & n \text{ even}, \\ n \int_{x_0}^{x} \frac{X^{(n-1)}(s)}{f^2(s)p(s)} ds, & n \text{ odd}, \end{cases} \qquad (7.6)$$

$$\widetilde{X}^{(0)}(x) \equiv 1,$$

$$\widetilde{X}^{(n)}(x) = \begin{cases} n \int_{x_0}^{x} \widetilde{X}^{(n-1)}(s) \left(f^2(s)r(s) \right) ds, & n \text{ odd}, \\ n \int_{x_0}^{x} \frac{\widetilde{X}^{(n-1)}(s)}{f^2(s)p(s)} ds, & n \text{ even}, \end{cases} \qquad (7.7)$$

for $n = 1, 2, \dots$. Thus, to construct an n-th such integral we multiply the preceding one by the function $f^2(s)r(s)$ or $1/ \left(f^2(s)p(s) \right)$ depending on the parity of n and integrate.

Definition 73 Under the conditions of Proposition 72 the **formal powers** associated with Eq. (7.1) are defined for any $n \in \mathbb{N} \cup \{0\}$ as follows

$$\Phi_n(x) = \begin{cases} f(x)X^{(n)}(x), & n \text{ odd}, \\ f(x)\widetilde{X}^{(n)}(x), & n \text{ even}, \end{cases} \qquad \Psi_n(x) = \begin{cases} \frac{1}{f(x)}X^{(n)}(x), & n \text{ even}, \\ \frac{1}{f(x)}\widetilde{X}^{(n)}(x), & n \text{ odd}. \end{cases}$$

Thus, to construct the sequence of formal powers $\{\Phi_n\}$, we take the odd numbered $X^{(n)}$ and the even numbered $\widetilde{X}^{(n)}$ and multiply them by f. The other "half" of the recursive integrals is used for constructing $\{\Psi_n\}$.

Theorem 74 ([48]) *Assume that on a finite interval $[a, b]$, Eq. (7.2) possesses a particular solution f such that the functions f^2r and $1/(f^2 p)$ are continuous on $[a, b]$. Then the general solution of (7.1) on (a, b) has the form*

$$u = c_1 u_1 + c_2 u_2, \qquad (7.8)$$

where c_1 and c_2 are arbitrary complex constants,

$$u_1(x) = \sum_{k=0}^{\infty} \lambda^k \frac{\Phi_{2k}(x)}{(2k)!} \quad and \quad u_2(x) = \sum_{k=0}^{\infty} \lambda^k \frac{\Phi_{2k+1}(x)}{(2k+1)!}, \tag{7.9}$$

where x_0 is an arbitrary point in $[a, b]$ such that $p(x_0) \neq 0$. Further, both series in (7.9) converge uniformly on $[a, b]$.

Proof The proof consists of the following parts. First, supposing that the series (7.9) converge uniformly with respect to x and, moreover, can be differentiated twice, we prove that u_1 and u_2 defined by (7.9) are indeed solutions of (7.1) for all $\lambda \in \mathbb{C}$. Second, we prove the uniform convergence of the series for u_1, u_2 and for their first and second derivatives. Finally, we prove that u_1 and u_2 are linearly independent and thus, (7.8) is a general solution of (7.1).

Let us make use of Proposition 72. Application of $\frac{1}{r}L$ to u_1 gives

$$\frac{1}{r}Lu_1 = \frac{1}{rf}\frac{d}{dx}\left(pf^2\frac{d}{dx}\sum_{k=0}^{\infty}\lambda^k\frac{\widetilde{X}^{(2k)}}{(2k)!}\right) = \frac{1}{rf}\frac{d}{dx}\sum_{k=1}^{\infty}\lambda^k\frac{\widetilde{X}^{(2k-1)}}{(2k-1)!}$$

$$= f\sum_{k=1}^{\infty}\lambda^k\frac{\widetilde{X}^{(2k-2)}}{(2k-2)!} = \lambda u_1.$$

In a similar way one can check that u_2 satisfies (7.1) as well. In order to give sense to this chain of equalities it is sufficient to prove the uniform convergence of the series involved in u_1 and u_2 as well as of the series obtained by a termwise differentiation. This can be done with the aid of the Weierstrass M-test. Indeed, we have $\left|\widetilde{X}^{(2k)}\right| \leq$ $\left(\max\left|rf^2\right|\right)^k\left(\max\left|\frac{1}{pf^2}\right|\right)^k|b-a|^{2k}$ and the series $\sum_{k=0}^{\infty}\frac{c^k}{(2k)!}$ is convergent where

$$c = |\lambda|\left(\max\left|rf^2\right|\right)\left(\max\left|\frac{1}{pf^2}\right|\right)|b-a|^2. \tag{7.10}$$

The uniform convergence of the series in u_2 as well as of the series of derivatives can be shown similarly.

The last step is to verify that the Wronskian of u_1 and u_2 is different from zero at least at one point (which necessarily implies the linear independence of u_1 and u_2 on the whole segment $[a, b]$). It is easy to see that by definition all the $\widetilde{X}^{(n)}(x_0)$ and $X^{(n)}(x_0)$ vanish except for $\widetilde{X}^{(0)}(x_0)$ and $X^{(0)}(x_0)$ which equal 1. Thus

$$u_1(x_0) = f(x_0), \qquad u_1'(x_0) = f'(x_0), \tag{7.11}$$

$$u_2(x_0) = 0, \qquad u_2'(x_0) = \frac{1}{f(x_0)p(x_0)}, \tag{7.12}$$

and the Wronskian of u_1 and u_2 at x_0 equals $1/p(x_0) \neq 0$.   ∎

Remark 75 In the case $\lambda = 0$, the solution (7.9) becomes $u_1 = f$ and $u_2 =$

$$f \int_{x_0}^{x} \frac{ds}{f^2(s)p(s)}.$$

Remark 76 As it was mentioned in the previous section, in the case of real valued coefficients p, q, and r the existence and construction of the required f presents no difficulty. In this case (7.2) possesses two linearly independent regular solutions v_1 and v_2 whose zeros alternate. Thus one may choose $f = v_1 + i v_2$. Moreover, for the construction of v_1 and v_2 in fact the same SPPS method may be used (see Corollary 82 below).

In general, the existence of a nonvanishing solution for complex-valued coefficients was proved in [48, Remark 5] (see also [14]). Moreover, even when f has zeros the formal powers are still well defined and can be constructed following, e.g., a procedure from [49]. Thus, in what follows we will not restrict ourselves to the requirement $f \neq 0$.

Remark 77 The result of Theorem 74 is valid for infinite intervals as well, the series being uniformly convergent on any finite subinterval.

Remark 78 Theorem 74 points out the fact that considered as functions of λ, the solutions u_1 and u_2 are entire functions (analytic on the whole complex plane).

Remark 79 One of the functions rf^2 or $1/(pf^2)$ may not be continuous on $[a, b]$ and yet u_1 or u_2 may make sense. For example, in the case of the Bessel equation $(xu')' - \frac{1}{x}u = -\lambda x u$, we can choose $f(x) = x/2$. Then $1/(pf^2) \notin C[0, 1]$. Nevertheless all integrals in (7.7) exist and u_1 coincides with the nonsingular $J_1(\sqrt{\lambda}x)$, while u_2 is a singular solution of the Bessel equation.

Remark 80 The procedure for construction of solutions described in Theorem 74 works not only when a solution is available for $\lambda = 0$, but in fact when a solution f of the equation

$$(pf')' - qf = \lambda_0 r f \tag{7.13}$$

is known for some fixed λ_0. The solution (7.9) now takes the form

$$u_1 = \sum_{k=0}^{\infty} (\lambda - \lambda_0)^k \frac{\Phi_{2k}}{(2k)!} \quad \text{and} \quad u_2 = \sum_{k=0}^{\infty} (\lambda - \lambda_0)^k \frac{\Phi_{2k+1}}{(2k+1)!}.$$

This can be easily verified by writing (7.1) as

$$(L - \lambda_0 r) u = (\lambda - \lambda_0) r u.$$

The operator on the left hand side can be factorized exactly as in the proof of the theorem, and the same reasoning carries through.

Remark 81 Other representations of the general solution of (7.1) as a formal power series have been long known (see [21, 73, Theorem 1]) and used for studying qualitative properties of solutions. The complicated manner in which the parameter λ appears in those representations makes that form of a general solution too difficult for quantitative analysis of spectral and boundary value problems. In contrast, the solution (7.8)–(7.9) is a power series with respect to λ, making it quite attractive for numerical solution of spectral, initial value and boundary value problems.

Spectral parameter power series representations for solutions of the Sturm-Liouville equation represent an important mathematical tool. We mention [10, 27, 64, Sect. 10] and [44] among other publications where they appear. To our best knowledge for the first time they were applied for numerical solution of Sturm-Liouville problems in [48]. The reason of this practical underuse of the SPPS lies in the form in which the expansion coefficients were sought. Indeed, in previous works the calculation of coefficients was proposed in terms of successive integrals with the kernels in the form of iterated Green functions (see [10, Sect. 10]). This makes any computation based on such representation difficult, less practical and even proofs of the most basic results like, e.g., the uniform convergence of the spectral parameter power series for any value of $\lambda \in \mathbb{C}$ (established in Theorem 74) are not an easy task. For example, in [10, p. 16] the parameter λ is assumed to be small and no proof of convergence is given.

A special case of Theorem 74, with $q \equiv 0$, $\lambda = 1$, was known to H. Weyl.

Corollary 82 ([85]) *Let $1/p$ and r be continuous on $[a, b]$. The general solution of the equation*

$$(p u')' = r u \tag{7.14}$$

on (a, b) has the form

$$u = c_1 u_1 + c_2 u_2, \tag{7.15}$$

where c_1 and c_2 are arbitrary constants and u_1, u_2 are defined by (7.9) with $f \equiv \lambda = 1$.

That is,

$$u_1 = \sum_{k=0}^{\infty} \frac{\widetilde{X}^{(2k)}}{(2k)!} \quad and \quad u_2 = \sum_{k=0}^{\infty} \frac{X^{(2k+1)}}{(2k+1)!},$$

where

$$X^{(0)}(x) \equiv \widetilde{X}^{(0)}(x) \equiv 1,$$

$$X^{(n)}(x) = \begin{cases} n \int_{x_0}^x X^{(n-1)}(s) r(s) \, ds, & n \text{ even}, \\ n \int_{x_0}^x \frac{X^{(n-1)}(s)}{p(s)} \, ds, & n \text{ odd}, \end{cases}$$

$$\widetilde{X}^{(n)}(x) = \begin{cases} n \int_{x_0}^x \widetilde{X}^{(n-1)}(s) r(s) \, ds, & n \text{ odd}, \\ n \int_{x_0}^x \frac{\widetilde{X}^{(n-1)}(s)}{p(s)} \, ds, & n \text{ even}, \end{cases}$$

for $n = 1, 2, \ldots$.

This corollary enables us to find the particular solution f. Indeed, taking $r = q$ in (7.14) and applying Corollary 82, we obtain a general solution of (7.2).

Remark 83 It is easy to obtain SPPS representations for the derivatives of the solutions u_1 and u_2 from (7.9),

$$u_1' = \frac{f'}{f} u_1 + \frac{1}{p} \sum_{k=1}^{\infty} \lambda^k \frac{\Psi_{2k-1}}{(2k-1)!} = \frac{f'}{f} \sum_{k=0}^{\infty} \lambda^k \frac{\Phi_{2k}}{(2k)!} + \frac{1}{p} \sum_{k=0}^{\infty} \lambda^{k+1} \frac{\Psi_{2k+1}}{(2k+1)!},$$

(7.16)

$$u_2' = \frac{f'}{f} u_2 + \frac{1}{p} \sum_{k=0}^{\infty} \lambda^k \frac{\Psi_{2k}}{(2k)!} = \frac{f'}{f} \sum_{k=0}^{\infty} \lambda^k \frac{\Phi_{2k+1}}{(2k+1)!} + \frac{1}{p} \sum_{k=0}^{\infty} \lambda^k \frac{\Psi_{2k}}{(2k)!} \qquad (7.17)$$

and to prove their uniform convergence.

Example 84 Consider the equation

$$u''(x) = \lambda u(x), \quad 0 < x < b. \tag{7.18}$$

Then the equation for $\lambda = 0$, corresponding to (7.2), becomes

$$u''(x) = 0.$$

Its particular solution, satisfying conditions of Theorem 74 can be chosen as $f \equiv 1$. Then

$$X^{(n)}(x) = \widetilde{X}^{(n)}(x) = x^n,$$

and we obtain two linearly independent solutions

$$u_1(x) = \sum_{k=0}^{\infty} \lambda^k \frac{x^{2k}}{(2k)!} = \cosh\left(\sqrt{\lambda} x\right) \quad \text{and} \quad u_2(x) = \sum_{k=0}^{\infty} \lambda^k \frac{x^{2k+1}}{(2k+1)!} = \frac{\sinh\left(\sqrt{\lambda} x\right)}{\sqrt{\lambda}}.$$

In general, the formal powers cannot be obtained in a closed form. However, the recurrent integration procedure for computing them (formulas (7.6) and (7.7)) is easily realizable on a computer, and usually the numerical computation of a hundred or two formal powers does not present any difficulty and leads to excellent accuracy of the approximate solution, especially for not very large values of λ. When λ is large the spectral shifting procedure from Remark 80 can be used.

7.3 Some Properties of Formal Powers

For simplicity we consider the case when their center x_0 coincides with the origin, $x_0 = 0$, and the seed function f is normalized there, $f(0) = 1$. Moreover, we assume that $p \equiv r \equiv 1$, and thus Eq. (7.1) has the form

$$- u'' + qu = -\lambda u. \tag{7.19}$$

Proposition 85 *Let* $\{\Phi_k\}_{k=0}^{\infty}$ *be a system of formal powers constructed from a solution* $f \in C^2(0, b) \cap C^1[0, b]$ *of the equation*

$$- u'' + qu = 0 \tag{7.20}$$

with $q \in L_2(0, b)$, $f(0) = 1$. *Then for all* $k \in \mathbb{N} \cup \{0\}$ *the asymptotic relation holds*

$$\Phi_k(x) \sim x^k, \quad when \ x \to 0$$

(which means that $\lim_{x \to 0} \frac{\Phi_k(x)}{x^k} = 1$*).*

Thus, asymptotically, near the origin, the formal powers behave like usual powers of the independent variable. We will give a proof of this fact below.

The following proposition establishes the precise dependence of the formal powers on the choice of the particular solution used as a seed function for their construction.

Proposition 86 *Let* f *and* g *be two different solutions of (7.20), both normalized in the origin* $f(0) = 1$, $g(0) = 1$. *Denote* $h_f := f'(0)$ *and* $h_g := g'(0)$. *Let* $\{\Phi_k\}_{k=0}^{\infty}$ *and* $\{\Gamma_k\}_{k=0}^{\infty}$ *be the systems of formal powers constructed from* f *and* g, *respectively. Then*

$$\Gamma_k \equiv \Phi_k, \quad k \ odd \tag{7.21}$$

and

$$\Gamma_k \equiv \Phi_k + (h_g - h_f) \frac{\Phi_{k+1}}{k+1}, \quad k \ even. \tag{7.22}$$

Proof Consider the pair of the solutions $\{u_1, u_2\}$ of (7.19) constructed as it is done in Theorem 74. They are associated with the particular solution f and corresponding formal powers $\{\Phi_k\}_{k=0}^{\infty}$. Analogously, taking g and the corresponding formal powers $\{\Gamma_k\}_{k=0}^{\infty}$ we obtain another pair $\{v_1, v_2\}$ of solutions of (7.19). Since both pairs represent fundamental systems of solutions of the same Eq. (7.19), the solutions from one system can be represented as linear combinations of the solutions from the other system and vice versa. In order to find the coefficients in the linear combinations we recall the initial conditions satisfied by the solutions (7.11) and (7.12). Thus we have

$$u_1(0) = f(0) = 1, \quad u_1'(0) = f'(0) = h_f,$$

$$u_2(0) = 0, \quad u_2'(0) = 1,$$

and

$$v_1(0) = g(0) = 1, \quad v_1'(0) = g'(0) = h_g,$$

$$v_2(0) = 0, \quad v_2'(0) = 1.$$

We notice that u_2 and v_2 satisfy the same initial conditions. Hence, due to the uniqueness of the solution of a Cauchy problem, we have that $u_2 \equiv v_2$ and thus,

$$\sum_{k=0}^{\infty} \lambda^k \frac{\Phi_{2k+1}(x)}{(2k+1)!} \equiv \sum_{k=0}^{\infty} \lambda^k \frac{\Gamma_{2k+1}(x)}{(2k+1)!}$$

for all $\lambda \in \mathbb{C}$. Corresponding coefficients of two equal convergent power series necessarily coincide, and we obtain (7.21).

Turning to the solutions u_1 and v_1 and using their respective initial conditions, we find that

$$v_1 = u_1 + \left(h_g - h_f\right) u_2.$$

Writing down the corresponding spectral parameter power series and equating the corresponding coefficients we arrive at (7.22) that finishes the proof. ∎

7.4 SPPS Method for Solving Sturm-Liouville Problems

The SPPS representation of solutions of the Sturm-Liouville equation offers a simple and easy to implement method for solving Sturm-Liouville problems. First of all, let us mention that the formal powers can be efficiently computed in a computer by implementing the following simple strategy. In order to compute $X^{(n)}$, take $X^{(n-1)}$, multiply it by $f^2 r$ or $1/(f^2 p)$ depending on the parity of n and convert the result into a spline. Integration of a spline is an exact operation, which then gives $X^{(n)}$.

Both operations: the conversion of a vector of values of a function into a spline and the integration of a spline are usually available as built-in routines in modern numerical computing environments, such as Matlab. In practice, to compute accurately two-three hundreds of the formal powers on a laptop presents no difficulty. More on the numerical aspects of calculation of the formal powers and SPPS series can be found, e.g., in [49].

Application of the SPPS representations reduces a Sturm-Liouville problem to a problem of finding zeros of an analytic function given by its Taylor series. Indeed, consider a Sturm-Liouville problem

$$\big(p(x)u'(x)\big)' - q(x)u(x) = \lambda r(x)u(x), \quad 0 < x < l, \tag{7.23}$$

$$\alpha u(0) + \beta u'(0) = 0, \tag{7.24}$$

$$au(l) + bu'(l) = 0. \tag{7.25}$$

Let f be a solution of the equation

$$\big(p(x)f'(x)\big)' - q(x)f(x) = 0, \quad 0 < x < l,$$

such that

$$f(0) = 1.$$

Then the solutions u_1 and u_2 from Theorem 74, constructed with the aid of the formal powers from Definition 73 with $x_0 = 0$, and f as a seed function, satisfy the initial conditions (see equalities (7.11), (7.12))

$$u_1(0) = 1, \quad u_1'(0) = f'(0),$$

$$u_2(0) = 0, \quad u_2'(0) = \frac{1}{p(0)}.$$

Take $u = c_1 u_1 + c_2 u_2$ where the constants c_1 and c_2 are to be chosen such that (7.24) be fulfilled. Consider first the case $\beta \neq 0$. We have

$$\alpha u(0) + \beta u'(0) = \alpha c_1 + \beta \left(c_1 f'(0) + \frac{c_2}{p(0)} \right) = 0,$$

and hence

$$\frac{\beta c_2}{p(0)} = - \left(\alpha + \beta f'(0) \right) c_1,$$

from where we obtain

$$c_2 = -p(0) \left(\frac{\alpha}{\beta} + f'(0) \right) c_1.$$

Thus, u satisfies (7.24) if and only if it has the form

$$u = c_1 \left(u_1 - p(0) \left(\frac{\alpha}{\beta} + f'(0) \right) u_2 \right).$$

If λ is an eigenvalue, then the corresponding eigenfunction necessarily has this form, where c_1 can be chosen to normalize the eigenfunction, for example, by the condition

$$u(0) = 1$$

and thus, $c_1 = 1$. We obtain then a normalized eigenfunction

$$u = u_1 - p(0) \left(\frac{\alpha}{\beta} + f'(0) \right) u_2. \tag{7.26}$$

In the case when $\beta = 0$ we have a homogeneous Dirichlet condition in the origin, and consequently, $u = c u_2$. Choosing $c = p(0)$ we normalize the solution by the condition $u'(0) = 1$. Thus, when $\beta = 0$, a solution satisfying the initial conditions $u(0) = 0$ and $u'(0) = 1$ is given by

$$u(x) = p(0)u_2(x). \tag{7.27}$$

Considering first the case $\beta \neq 0$, we obtain that λ is an eigenvalue of the Sturm-Liouville problem if and only if

$$a \left(u_1(l) - p(0) \left(\frac{\alpha}{\beta} + f'(0) \right) u_2(l) \right) + b \left(u_1'(l) - p(0) \left(\frac{\alpha}{\beta} + f'(0) \right) u_2'(l) \right) = 0.$$

This equality is obtained by substituting (7.26) into (7.25). Now, substitution of the series (7.9), (7.16), and (7.17) leads to the equation

$$a \left(\sum_{k=0}^{\infty} \lambda^k \frac{\Phi_{2k}(l)}{(2k)!} - p(0) \left(\frac{\alpha}{\beta} + f'(0) \right) \sum_{k=0}^{\infty} \lambda^k \frac{\Phi_{2k+1}(l)}{(2k+1)!} \right)$$

$$+ b \left(\frac{f'(l)}{f(l)} \sum_{k=0}^{\infty} \lambda^k \frac{\Phi_{2k}(l)}{(2k)!} + \frac{1}{p(l)} \sum_{k=0}^{\infty} \lambda^{k+1} \frac{\Psi_{2k+1}}{(2k+1)!} \right.$$

$$\left. - p(0) \left(\frac{\alpha}{\beta} + f'(0) \right) \left(\frac{f'(l)}{f(l)} \sum_{k=0}^{\infty} \lambda^k \frac{\Phi_{2k+1}(l)}{(2k+1)!} + \frac{1}{p(l)} \sum_{k=0}^{\infty} \lambda^k \frac{\Psi_{2k}(l)}{(2k)!} \right) \right)$$

$$= 0,$$

which can be written in the form

$$\sum_{k=0}^{\infty} \lambda^k A_k = 0, \tag{7.28}$$

where

$$A_0 = a \left(\Phi_0(l) - p(0) \left(\frac{\alpha}{\beta} + f'(0) \right) \Phi_1(l) \right)$$

$$+ b \left(\frac{f'(l)}{f(l)} \Phi_0(l) - p(0) \left(\frac{\alpha}{\beta} + f'(0) \right) \left(\frac{f'(l)}{f(l)} \Phi_1(l) + \frac{1}{p(l)} \Psi_0(l) \right) \right)$$

and

$$A_k = a \left(\frac{\Phi_{2k}(l)}{(2k)!} - p(0) \left(\frac{\alpha}{\beta} + f'(0) \right) \frac{\Phi_{2k+1}(l)}{(2k+1)!} \right)$$

$$+ b \left(\frac{f'(l)}{f(l)} \frac{\Phi_{2k}(l)}{(2k)!} + \frac{1}{p(l)} \frac{\Psi_{2k-1}}{(2k-1)!} \right.$$

$$\left. - p(0) \left(\frac{\alpha}{\beta} + f'(0) \right) \left(\frac{f'(l)}{f(l)}^k \frac{\Phi_{2k+1}(l)}{(2k+1)!} + \frac{1}{p(l)} \frac{\Psi_{2k}(l)}{(2k)!} \right) \right),$$

for $k = 1, 2, \ldots$.

Thus, λ is an eigenvalue of the Sturm-Liouville problem (7.23), (7.24), (7.25) with $\beta \neq 0$ if and only if λ is a zero of the analytic function (even entire) function (7.28) given by its Taylor series. In order to compute some first eigenvalues of the problem numerically, one may compute zeros of the polynomial $\sum_{k=0}^{N} \lambda^k A_k$ for some N. In practice, several eigenvalues of the Sturm-Liouville problem, which are relatively close to the origin, will be approximated accurately, while for obtaining other eigenvalues a previous spectral parameter shifting from Remark 80 is recommendable.

When in (7.24) $\beta = 0$, the characteristic equation of the Sturm-Liouville problem (7.23), (7.24), (7.25) has the form

$$\sum_{k=0}^{\infty} \lambda^k B_k = 0,$$

where

$$B_k = p(0) \left(a + b \frac{f'(l)}{f(l)} \right) \frac{\Phi_{2k+1}(l)}{(2k+1)!} + b \frac{p(0)}{p(l)} \frac{\Psi_{2k}(l)}{(2k)!}.$$

It is worth mentioning that the SPPS technique is quite universal and can be applied to problems admitting complex eigenvalues (see, e.g., [40, 49]), singular Sturm-Liouville problems [18, 19] and higher order Sturm-Liouville problems [39, 59].

7.5   Transmutation Operators

For simplicity, instead of a general Sturm-Liouville equation (7.23) we will consider an important special case

$$- u'' + q(x)u = \rho^2 u, \tag{7.29}$$

where $\rho \in \mathbb{C}$. This equation is often called the one-dimensional Schrödinger equation. Under quite general conditions on the coefficients the Sturm-Liouville equation (7.23) can be reduced to the form (7.29) with the aid of the Liouville transformation (see [47, 87]).

Let us consider Eq. (7.29) on the interval $(0, l), l > 0$.

Let q be a complex valued function belonging to $L_2(0, l)$. By $\varphi(\rho, x)$ we denote a solution of (7.29) satisfying the initial conditions

$$\varphi(\rho, 0) = 1, \quad \varphi'(\rho, 0) = h, \tag{7.30}$$

where $h \in \mathbb{C}$ and by $s(\rho, x)$ a solution of (7.29) satisfying the initial conditions

$$s(\rho, 0) = 0, \quad s'(\rho, 0) = \rho. \tag{7.31}$$

Let us write these special solutions in terms of SPPS representations. For this we choose the function $f(x) = \varphi(0, x)$ as a seed function for constructing the corresponding formal powers. In Definition 73 take $x_0 = 0$, $p \equiv r \equiv 1$. Comparison of (7.30) and (7.31) with (7.11) and (7.12), respectively, shows that $u_1(x) = \varphi(\rho, x)$

and $u_2(x) = s(\rho, x)/\rho$, where u_1 and u_2 are the solutions from Theorem 74. Hence the SPPS representations for $\varphi(\rho, x)$ and $s(\rho, x)$ have the form

$$\varphi(\rho, x) = \sum_{k=0}^{\infty} (-1)^k \rho^{2k} \frac{\Phi_{2k}(x)}{(2k)!} \quad \text{and} \quad s(\rho, x) = \sum_{k=0}^{\infty} (-1)^k \rho^{2k+1} \frac{\Phi_{2k+1}(x)}{(2k+1)!}.$$
(7.32)

The following important theorem is well known (see, e.g., [66, 67, 69]).

Theorem 87 *There exist functions $G(x, t)$ and $S(x, t)$, defined and continuous in the domain $0 \le t \le x \le l$, such that for all $\rho \in \mathbb{C}$ the equalities hold*

$$\varphi(\rho, x) = \cos \rho x + \int_0^x G(x, t) \cos \rho t \, dt$$
(7.33)

and

$$s(\rho, x) = \sin \rho x + \int_0^x S(x, t) \sin \rho t \, dt.$$
(7.34)

The kernels $G(x, t)$ and $S(x, t)$ satisfy the conditions

$$G(x, x) = h + \frac{1}{2} \int_0^x q(t) \, dt, \quad \left. \frac{\partial}{\partial t} G(x, t) \right|_{t=0} = 0$$
(7.35)

and

$$S(x, x) = \frac{1}{2} \int_0^x q(t) \, dt, \quad S(x, 0) = 0.$$
(7.36)

Of crucial importance is the fact that the integral kernels $G(x, t)$ and $S(x, t)$ are independent of the spectral parameter ρ.

The second kind Volterra integral operators in (7.33) and (7.34) $T_c := I + \int_0^x G(x, t) \cdot dt$ and $T_s := I + \int_0^x S(x, t) \cdot dt$ are called transmutation operators. From Theorem 87 we see that they transmute the elementary functions $\cos \rho x$ and $\sin \rho x$ into solutions of the Schrödinger equation (7.29).

Directly from Theorem 87 and SPPS representations (7.32), the following mapping property of the transmutation operators T_c and T_s can be deduced.

Theorem 88 *Let q be a complex valued function of the independent variable $x \in [0, l]$, $q \in L_2(0, l)$, and let $f(x) = \varphi(0, x)$ be a solution of the Cauchy problem (7.29), (7.30) with $\rho = 0$. Let Φ_k, $k \in \mathbb{N}_0 := \mathbb{N} \cup \{0\}$ be the formal powers defined by Definition 73 where $x_0 = 0$, $p \equiv r \equiv 1$. Then*

$$\Phi_k(x) = T_c\left[x^k\right] = x^k + \int_0^x G(x, t) t^k \, dt, \quad \text{if } k \in \mathbb{N}_0 \text{ is even}$$
(7.37)

and

$$\Phi_k(x) = T_s\left[x^k\right] = x^k + \int_0^x S(x,t)t^k\,dt, \quad if\,k \in \mathbb{N}\,is\,odd. \tag{7.38}$$

Proof Let us prove (7.37), the equality (7.38) can be proved similarly.

Substitution of the SPPS representation of $\varphi(\rho, x)$ from (7.32) into (7.33) together with the Maclaurin series for $\cos \rho x$ gives us the equality

$$\sum_{k=0}^{\infty}(-1)^k\rho^{2k}\frac{\Phi_{2k}(x)}{(2k)!} = \sum_{k=0}^{\infty}(-1)^k\rho^{2k}\frac{T_c\left[x^{2k}\right]}{(2k)!}$$

which in view of the uniform convergence of both series on any compact subset of the complex ρ-plane, yields (7.37).

It is often convenient to work with another transmutation operator which is in fact a combination of the Volterra integral operators defined by (7.33) and (7.34), and which is equally suitable for transmuting odd and even functions. ∎

Theorem 89 *There exists a function $K_h(x, t)$ defined and continuous in the domain $0 \leq |t| \leq |x| \leq l$ such that the operator $\mathbf{T}_h$ defined on $C[-l, l]$ by the equality*

$$\mathbf{T}_h v(x) = v(x) + \int_{-x}^x K_h(x,t)v(t)dt$$

maps any solution v of the equation $v'' + \rho^2 v = 0$ into a solution u of Eq. (7.29) on $(-l, l)$ with the following correspondence of the initial values $u(0) = v(0)$, $u'(0) = v'(0) + hv(0)$ for all $\rho \in \mathbb{C}$. In particular,

$$\mathbf{T}_h[1] = f.$$

The function $K_h(x, t)$ satisfies the following conditions

$$K_h(x,x) = \frac{h}{2} + \frac{1}{2}\int_0^x q(s)\,ds, \qquad K_h(x,-x) = \frac{h}{2}, \qquad 0 \leq x \leq l. \tag{7.39}$$

Moreover, for every $v \in C^2[-l, l]$ the equality holds

$$\left(\frac{d^2}{dx^2} - q(x)\right)\mathbf{T}_h v = \mathbf{T}_h\frac{d^2}{dx^2}v.$$

For the proof of this fact we refer to [68, Theorem 3.1.1]. A slightly different proof and references to earlier publications are given in [50].

Thus, $\mathbf{T}_h$ not only transforms solutions of the elementary equation $u'' + \rho^2 u = 0$ into solutions of (7.29), but also the elementary operator $\frac{d^2}{dx^2}$ itself into the Sturm-

Liouville operator $\frac{d^2}{dx^2} - q(x)$. The relation between the kernels of the transmutation operators has the form

$$K_h(x, t) = \frac{1}{2} \left(G(x, t) + S(x, t) \right).$$

The mapping property of the transmutation operator T_h similar to Theorem 88 has the form

$$\Phi_k(x) = T_h \left[x^k \right], \quad \text{for all } k \in \mathbb{N} \cup \{0\}, \ x \in [-l, l]. \tag{7.40}$$

More on properties of the transmutation operators can be found in [47, 66, 69, 80]. Here we will focus on the possibility to construct the kernels $G(x, t)$ and $S(x, t)$ with the aid of the formal powers, and, as a consequence, to obtain very useful series representations for the solutions $\varphi(\rho, x)$ and $s(\rho, x)$. These series representations turn out to be even more convenient for solving Sturm-Liouville problems than the SPPS representations, especially when one is interested in computing large sets of eigendata.

Let P_n denote the **Legendre polynomial** of order n, $l_{k,n}$ be the corresponding coefficient of x^k, that is $P_n(x) = \sum_{k=0}^{n} l_{k,n} x^k$.

Theorem 90 ([56]) *The integral transmutation kernels $G(x, t)$ and $S(x, t)$ admit the following series representations*

$$G(x, t) = \sum_{n=0}^{\infty} \frac{g_n(x)}{x} P_{2n} \left(\frac{t}{x} \right), \quad 0 < t \leq x \leq l, \tag{7.41}$$

and

$$S(x, t) = \sum_{n=0}^{\infty} \frac{s_n(x)}{x} P_{2n+1} \left(\frac{t}{x} \right), \quad 0 < t \leq x \leq l, \tag{7.42}$$

where

$$g_n(x) = (4n + 1) \left(\sum_{k=0}^{2n} l_{k,2n} \frac{\Phi_k(x)}{x^k} - 1 \right) \tag{7.43}$$

and

$$s_n(x) = (4n + 3) \left(\sum_{k=0}^{2n+1} l_{k,2n+1} \frac{\Phi_k(x)}{x^k} - 1 \right), \tag{7.44}$$

where Φ_k are the formal powers from Definition 73 with $x_0 = 0$, $p \equiv r \equiv 1$ and $f(x) = \varphi(0, x)$ a solution of the Cauchy problem (7.29), (7.30) with $\rho = 0$. For every $x \in (0, l]$ the series converge in the norm of $L_2(0, x)$ and uniformly with respect to $t \in [0, x]$ when $q \in C[0, l]$.

Proof Let us prove (7.41). Equality (7.42) can be proved similarly.

Since for any $x \in (0, l]$ the kernel $G(x, t)$ as a function of t is continuous on $[0, x]$, it admits (see, e.g., [81]) a Fourier-Legendre series of the form $\sum_{n=0}^{\infty} A_n(x) P_{2n}\left(\frac{t}{x}\right)$ that converges in the norm of $L_2(0, x)$ (when $q \in C[0, l]$ the function $G(x, t)$ is continuously differentiable with respect to both arguments and hence its Fourier-Legendre series converges uniformly). For convenience we consider $A_n(x) = \frac{g_n(x)}{x}$. Multiplying (7.41) by $P_{2m}\left(\frac{t}{x}\right)$ and integrating, we obtain

$$\int_0^x G(x, t) P_{2m}\left(\frac{t}{x}\right) dt = \sum_{n=0}^{\infty} \frac{g_n(x)}{x} \int_0^x P_{2n}\left(\frac{t}{x}\right) P_{2m}\left(\frac{t}{x}\right) dt = \frac{g_m(x)}{4m+1}.$$

Hence,

$$g_m(x) = (4m+1) \int_0^x G(x, t) P_{2m}\left(\frac{t}{x}\right) dt$$

$$= (4m+1) \int_0^x G(x, t) \sum_{k=0}^{2m} l_{k,2m} \left(\frac{t}{x}\right)^k dt$$

$$= (4m+1) \sum_{k=0}^{2m} \frac{l_{k,2m}}{x^k} \left(T_c\left[x^k\right] - x^k\right).$$

Using Theorem 88 together with the fact that coefficients at odd powers in even order Legendre polynomials equal zero, we obtain that

$$g_m(x) = (4m+1) \sum_{k=0}^{2m} \frac{l_{k,2m}}{x^k} \left(\Phi_k(x) - x^k\right).$$

This equality gives us (7.43), since

$$P_n(1) = 1 \quad \text{for all } n = 0, 1, \ldots, \tag{7.45}$$

and thus, $\sum_{k=0}^{2m} l_{k,2m} = 1$. $\blacksquare$

Corollary 91 *The following formulas are valid*

$$\sum_{n=0}^{\infty} \frac{g_n(x)}{x} = h + \frac{1}{2} \int_0^x q(t) \, dt \tag{7.46}$$

and

$$\sum_{n=0}^{\infty} \frac{s_n(x)}{x} = \frac{1}{2} \int_0^x q(t)\, dt. \tag{7.47}$$

Proof The proof consists in substituting (7.41) and (7.42) into the first equalities in (7.35) and (7.36) and using the Legendre polynomials property (7.45). ∎

Theorem 90 leads to a very special series representation for the solutions $\varphi(\rho, x)$ and $s(\rho, x)$ belonging to the class of so-called Neumann series of Bessel functions (NSBF). We present it in the next section.

7.6 NSBF for Solutions

Definition 92 (see, e.g., [84, Chapter 16]) Functional series of the form $\sum_{n=0}^{\infty} a_n J_{n+\nu}(x)$ with $\nu \in \mathbb{R}$ are called Neumann series of Bessel functions.

If $f(z)$ is an analytic function inside and on the circle $|z| = R$, and C_R is the contour formed by this circle we have

$$z^{\nu} f(z) = \sum_{n=0}^{\infty} a_n J_{n+\nu}(z),$$

and the relation between the coefficients a_n and the coefficients b_n of the Maclaurin series of the function $f(z)$: $f(z) = \sum_{n=0}^{\infty} b_n z^n$ is given by the formula [84, Sect. 16.13]

$$a_n = (n+\nu) \sum_{m=0}^{\leq n/2} 2^{n+\nu-2m} \frac{\Gamma(n+\nu-m)}{m!} b_{n-2m}.$$

Moreover, due to Pincherle's theorem, both series have identical circles of convergence.

Theorem 93 ([56]) *Let* $q \in L_2(0, l)$ *be a complex valued function and* $\{g_n\}_{n=0}^{\infty}$, $\{s_n\}_{n=0}^{\infty}$ *systems of functions defined by (7.43), (7.44), respectively. Then the solutions* $\varphi(\rho, x)$ *and* $s(\rho, x)$ *of (7.29) admit the series representations*

$$\varphi(\rho, x) = \cos \rho x + \sqrt{\frac{\pi}{2\rho x}} \sum_{n=0}^{\infty} (-1)^n g_n(x) J_{2n+1/2}(\rho x)$$

$$= \cos \rho x + \sum_{n=0}^{\infty} (-1)^n g_n(x) j_{2n}(\rho x) \tag{7.48}$$

and

$$s(\rho, x) = \sin \rho x + \sqrt{\frac{\pi}{2\rho x}} \sum_{n=0}^{\infty} (-1)^n s_n(x) J_{2n+3/2}(\rho x)$$

$$(7.49)$$

$$= \sin \rho x + \sum_{n=0}^{\infty} (-1)^n s_n(x) j_{2n+1}(\rho x),$$

where j_k stands for the spherical Bessel function of order k defined as $j_k(z) :=$ $\sqrt{\frac{\pi}{2z}} J_{k+\frac{1}{2}}(z)$. The series converge pointwise with respect to x on $[0, l]$ and converge uniformly with respect to ρ on any compact subset of the complex ρ-plane.

Proof Formulas (7.48) and (7.49) are obtained by substituting (7.41) and (7.42) into (7.33) and (7.34), respectively. Indeed, for the solutions $\varphi(\rho, x)$ and $s(\rho, x)$ we have

$$\varphi(\rho, x) = \cos \rho x + \sum_{n=0}^{\infty} \frac{g_n(x)}{x} \int_0^x P_{2n}\left(\frac{t}{x}\right) \cos \rho t \, dt$$

and

$$s(\rho, x) = \sin \rho x + \sum_{j=0}^{\infty} \frac{s_n(x)}{x} \int_0^x P_{2n+1}\left(\frac{t}{x}\right) \sin (\rho t) \, dt.$$

Using formula 2.17.7 from [74, p. 433],

$$\int_0^a \left\{ \begin{array}{c} P_{2n+1}\left(\frac{y}{a}\right) \cdot \sin by \\ P_{2n}\left(\frac{y}{a}\right) \cdot \cos by \end{array} \right\} dy = (-1)^n \sqrt{\frac{\pi a}{2b}} J_{2n+\delta+1/2}(ab), \quad \delta = \left\{ \begin{array}{c} 1 \\ 0 \end{array} \right\}, \quad a > 0,$$

we obtain (7.48) and (7.49).

The convergence of the series with respect to ρ can be established by using the fact mentioned above that the circle of convergence of the NSBF of an analytic function coincides with the circle of convergence of its Maclaurin series. Fix $x > 0$ and consider the complex variable $z := \rho x$. Then the Maclaurin series of the function $F(z) := \varphi(\rho, x) - \cos \rho x$, according to (7.32) has the form

$$F(z) = \sum_{k=0}^{\infty} (-1)^k \frac{\Phi_{2k}(x) - x^{2k}}{(2k)! \, x^{2k}} z^{2k}$$

and converges uniformly on any compact subset of the complex z-plane. Hence the function $F(z)$ admits the NSBF representation

$$z^{\frac{1}{2}} F(z) = \sum_{n=0}^{\infty} a_n J_{n+\frac{1}{2}}(z),$$

where the series also converges uniformly on any compact subset of the complex z-plane. Notice that this is precisely equality (7.48) where

$$a_k = \begin{cases} (-1)^n \sqrt{\frac{\pi}{2}} g_n(x) & \text{when } k = 2n, \\ 0 & \text{when } k = 2n + 1. \end{cases}$$

This gives us the uniform convergence of the series in (7.48) with respect to ρ on any compact subset of the ρ-plane. The uniform convergence with respect to ρ of the series (7.49) is proved analogously. ∎

Next theorem reveals an important feature of the representations (7.48) and (7.49) which makes them especially attractive for solving Sturm-Liouville problems. Together with the exact representations (7.48) and (7.49) consider their approximants

$$\varphi_N(\rho, x) = \cos \rho x + \sum_{n=0}^{N} (-1)^n g_n(x) j_{2n}(\rho x) \tag{7.50}$$

and

$$s_N(\rho, x) = \sin \rho x + \sum_{n=0}^{N} (-1)^n s_n(x) j_{2n+1}(\rho x). \tag{7.51}$$

Theorem 94 *Under the conditions of Theorem 93 the following estimates hold*

$$|\varphi(\rho, x) - \varphi_N(\rho, x)| \leq 2|x|\varepsilon_N(x) \quad \text{and} \quad |s(\rho, x) - s_N(\rho, x)| \leq 2|x|\varepsilon_N(x) \tag{7.52}$$

for any $\rho \in \mathbb{R}$, $\rho \neq 0$, and

$$|\varphi(\rho, x) - \varphi_N(\rho, x)| \leq \frac{2\varepsilon_N(x) \sinh(Cx)}{C} \quad \text{and} \quad |s(\rho, x) - s_N(\rho, x)| \leq \frac{2\varepsilon_N(x) \sinh(Cx)}{C} \tag{7.53}$$

for any $\rho \in \mathbb{C}$, $\rho \neq 0$ belonging to the strip $|\text{Im } \rho| \leq C$, $C \geq 0$, where ε_N is a sufficiently small non-negative function tending to zero when $N \to \infty$.

We leave this theorem without proof referring the interested reader to [56] or [47, Sect. 9.2] for it. The importance of these estimates lies in their independence of $\text{Re } \rho$. This is of particular interest when using representations (7.48) and (7.49) for solving spectral problems for (7.29) because they guarantee a uniform approximation of eigendata (see [51, Proposition 7.1]).

Remark 95 Note that the numbers $j_k(z)$ for a fixed z rapidly decrease as $k \to \infty$, see, e.g., [2, (9.1.62)]. Hence, the convergence rate of the series (7.48) and (7.49) for any fixed ρ (and for bounded subsets $\Omega \subset \mathbb{C}$) is, in fact, exponential.

Remark 96 Formulas (7.43) and (7.44) are simple and therefore attractive. They reduce the computation of the coefficients g_k and s_k, $k = 0, 1, \dots$ to the computation of the SPPS formal powers, and the computation of these is easily realizable. However, the direct use of (7.43) and (7.44) encounters an important difficulty. The Legendre polynomial coefficients $l_{k,j}$ grow quite fast and thus a limited number of the coefficients g_k and s_k can be computed in a machine precision arithmetic (numerical examples can be found in [56]). Often quite a reduced number of the coefficients computed is already sufficient for obtaining an accurate solution. However, if one is interested in computing solutions with still higher accuracy, there are available other recurrent formulas for g_k and s_k which are more convenient for computation. We refer the reader to [56] and [47, Sect. 9.2].

Remark 97 NSBF representations for the derivatives with respect to x: $\varphi'(\rho, x)$ and $s'(\rho, x)$, admitting estimates similar to (7.52), (7.53) are also available (see [56] and [47, Sect. 9.2]). They have the form

$$\varphi'(\rho, x) = -\rho \sin \rho x + \left(h + \frac{1}{2} \int_0^x q(s)\,ds \right) \cos \rho x + \sum_{n=0}^{\infty} (-1)^n \gamma_n(x) j_{2n}(\rho x)$$

(7.54)

and

$$s'(\rho, x) = \rho \cos \rho x + \frac{1}{2} \left(\int_0^x q(s)\,ds \right) \sin \rho x + \sum_{n=0}^{\infty} (-1)^n \sigma_n(x) j_{2n+1}(\rho x),$$

(7.55)

where

$$\gamma_n(x) = (4n+1) \left(\sum_{k=0}^{2n} \frac{l_{k,2n} \Phi_k'(x)}{x^k} - \frac{n(2n+1)}{x} - \left(h + \frac{1}{2} \int_0^x q(s)\,ds \right) \right),$$

(7.56)

$$\sigma_n(x) = (4n+3) \left(\sum_{k=0}^{2n+1} \frac{l_{k,2n+1} \Phi_k'(x)}{x^k} - \frac{(n+1)(2n+1)}{x} - \frac{1}{2} \int_0^x q(s)\,ds \right)$$

(7.57)

and

$$\Phi_k'(x) = k \Psi_{k-1}(x) + \frac{f'(x)}{f(x)} \Phi_k(x).$$

Here, as was specified at the beginning of Sect. 7.5, the formal powers Φ_k and Ψ_k are defined by Definition 73 with $f(x) = \varphi(0, x)$, $x_0 = 0$, $p \equiv r \equiv 1$.

7.7 Jost Solutions

The solutions of (7.29) considered in the previous section are tied to conditions at the origin. Often in problems of mathematical physics solutions satisfying certain conditions at infinity are required. In order to introduce the so-called Jost solutions frequently appearing in wave propagation problems we need to impose some additional conditions on the potential at infinity. Consider the one-dimensional Schrödinger equation now on the whole line

$$- u'' + q(x)u = \rho^2 u, \quad x \in (-\infty, \infty) \tag{7.58}$$

with $q(x)$ being a real valued function defined on $(-\infty, \infty)$ and satisfying the condition

$$\int_{-\infty}^{\infty} (1 + |x|) \, |q(x)| \, dx < \infty. \tag{7.59}$$

Roughly speaking this condition means that the potential $q(x)$ decays at $\pm\infty$ sufficiently fast, so that for large $|x|$ the factor ρ^2 (if different from zero) becomes dominant in comparison with q, and one can expect that for large $|x|$ the solutions of (7.58) behave like $e^{\pm i\rho x}$. Potentials satisfying (7.59) are called short-range potentials.

Since in Eq. (7.58) only the square of the parameter ρ appears, it is sufficient to consider it from one half-plane only. So we choose it from the upper half-plane, $\tau := \operatorname{Im} \rho \geq 0$. Denote $\Omega_+ := \{\rho : \tau > 0\}$.

Under the condition (7.59) Eq. (7.58) possesses the unique so-called Jost solutions $u = e(\rho, x)$ and $u = g(\rho, x)$ such that for $\nu = 0, 1$ the following asymptotic relations hold

$$e^{(\nu)}(\rho, x) = (i\rho)^{\nu} e^{i\rho x} (1 + o(1)), \quad x \to +\infty$$

and

$$g^{(\nu)}(\rho, x) = (-i\rho)^{\nu} e^{-i\rho x} (1 + o(1)), \quad x \to -\infty$$

uniformly in $\overline{\Omega_+}$ (see, e.g., [20], [86, Theorem 2.4.1]). For every $\rho \in \Omega_+$ and $\alpha \in \mathbf{R}$: $e(\rho, x) \in L_2(\alpha, \infty)$, $g(\rho, x) \in L_2(-\infty, \alpha)$. The asymptotic relations mean that not only the Jost solutions themselves behave at infinity asymptotically like exponential functions but also their derivatives behave like the derivatives of the exponential functions. Our aim now is to learn how to construct the Jost solutions.

As a starting point we use a representation for Jost solutions called Levin's representation, obtained in [65].

Theorem 98 (Levin's Representation for Jost Solutions) *The functions $e(\rho, x)$ and $g(\rho, x)$ admit the following representations*

$$e(\rho, x) = e^{i\rho x} + \int_x^\infty A(x, t)e^{i\rho t}\,dt \tag{7.60}$$

and

$$g(\rho, x) = e^{-i\rho x} + \int_{-\infty}^x B(x, t)e^{-i\rho t}\,dt, \tag{7.61}$$

where A and B are real valued functions, such that $A(x, \cdot) \in L_2\,(x, \infty)$, $B(x, \cdot) \in L_2\,(-\infty, x)$ for all $x \in \mathbb{R}$,

$$A(x, x) = \frac{1}{2}\int_x^\infty q(t)dt, \tag{7.62}$$

and

$$B(x, x) = \frac{1}{2}\int_{-\infty}^x q(t)dt. \tag{7.63}$$

Formulas (7.60) and (7.61) show that the Jost solutions can be regarded as a result of application of the corresponding transmutation operators to the exponential functions, which are the Jost solutions for the Schrödinger equation with $q \equiv 0$. Similarly to the case of the transmutation operators with conditions at the origin (Theorem 87), here of crucial importance is the fact that the integral kernels $A(x, t)$ and $B(x, t)$ are independent of the spectral parameter ρ.

Again, like in the case of the transmutation operators with conditions at the origin we can construct series representations for the integral kernels $A(x, t)$ and $B(x, t)$ in terms of an appropriate system of orthogonal polynomials. The Laguerre polynomials $L_n(t)$ will be used. Among the notable properties of the Laguerre polynomials we recall that

$$\int_0^\infty e^{-t}L_m(t)L_n(t)dt = \delta_{nm},$$

where δ_{nm} stands for the Kronecker delta, and

$$L_n(0) = 1 \quad \text{for all } n = 0, 1, \ldots. \tag{7.64}$$

The system of the Laguerre polynomials $\{L_n(t)\}_{n=0}^{\infty}$ represents an orthonormal basis in the Hilbert space $L_2\left(0, \infty; e^{-t}\right)$.

Theorem 99 ([26, 46]) *The functions A and B admit the following series representations*

$$A(x, t) = \sum_{n=0}^{\infty} a_n(x) L_n(t - x) e^{\frac{x-t}{2}}$$ (7.65)

and

$$B(x, t) = \sum_{n=0}^{\infty} b_n(x) L_n(x - t) e^{-\frac{x-t}{2}}.$$ (7.66)

For any $x \in \mathbb{R}$ fixed, the series converge in the norm of $L_2(x, \infty)$ and $L_2(-\infty, x)$, respectively.
For the coefficients $a_0(x)$ and $b_0(x)$ the equalities are valid

$$a_0(x) = e\left(\frac{i}{2}, x\right) e^{\frac{x}{2}} - 1$$ (7.67)

and

$$b_0(x) = g\left(\frac{i}{2}, x\right) e^{-\frac{x}{2}} - 1.$$ (7.68)

The coefficients $a_n(x)$, $b_n(x)$, $n = 0, 1, \ldots$ are the unique solutions of the equations

$$L a_0 - a_0' = q,$$ (7.69)

$$L b_0 + b_0' = q,$$ (7.70)

$$L a_n - a_n' = L a_{n-1} + a_{n-1}', \quad n = 1, 2, \ldots,$$ (7.71)

$$L b_n + b_n' = L b_{n-1} - b_{n-1}', \quad n = 1, 2, \ldots,$$ (7.72)

with $L := \frac{d^2}{dx^2} - q(x)$, satisfying the boundary conditions

$$a_n(x) = o(1), \text{ when } x \to +\infty \quad \text{and} \quad b_n(x) = o(1), \text{ when } x \to -\infty.$$ (7.73)

Proof For a complete proof of this theorem we refer the interested reader to [26] or [47]. Here we clarify the particular form of the representations (7.65), (7.66). Consider the kernel A, for the kernel B the reasoning is analogous. Denote

$$a(x, t) := e^{\frac{t}{2}} A(x, x + t).$$

Since $A(x, \cdot) \in L_2(x, \infty)$ we have that $a(x, \cdot)$ belongs to $L_2(0, \infty; e^{-t})$. Indeed,

$$\int_0^\infty e^{-t} a^2(x, t) dt = \int_0^\infty A^2(x, x + t) dt = \int_x^\infty A^2(x, t) dt < \infty.$$

Hence $a(x, t)$ admits the representation (see, e.g., [81])

$$a(x, t) = \sum_{n=0}^\infty a_n(x) L_n(t).$$

For every $x \geq 0$ the series converges in the norm of the space $L_2(0, \infty; e^{-t})$.
Returning to the kernel $A(x, t)$, we obtain the equality

$$A(x, t) = \sum_{n=0}^\infty a_n(x) L_n(t - x) e^{\frac{x-t}{2}}. \tag{7.74}$$

Observe that from here we have that

$$\sum_{n=0}^\infty a_n(x) = A(x, x) = \frac{1}{2} \int_x^\infty q(t) dt. \tag{7.75}$$

This is due to (7.62) and the property (7.64).
On the origin of the system of differential equations satisfied by the coefficients $\{a_n\}$ and $\{b_n\}$ see Remark 102 below. ∎

Below we show that the differential equations (7.69)–(7.72) for the coefficients $\{a_n\}$ and $\{b_n\}$ lead to a simple recurrent integration procedure for their construction. In order to proceed, let us introduce the notation

$$z := \frac{\frac{1}{2} + i\rho}{\frac{1}{2} - i\rho}. \tag{7.76}$$

The map $\rho \mapsto z$ is a Möbius transformation of the upper half-plane of the complex variable ρ onto the unit disc $D = \{z \in \mathbb{C} : |z| \leq 1\}$. It maps the point $\rho = 0$ to $z = 1$, the ray $\rho = i\tau$, $\tau > 0$ to the interval $(-1, 1)$, and when τ runs from 0 to $+\infty$, z runs from 1 to -1. When ρ runs from 0 to $+\infty$ along the real line, z runs from 1 to -1 along the upper unit semicircle. See Fig. 7.1.

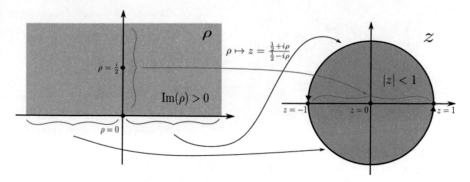

Fig. 7.1 Möbius transformation (7.76)

From (7.76) we have

$$i\rho = \frac{(z-1)}{2(z+1)} \quad \text{and} \quad \frac{1}{2} - i\rho = \frac{1}{z+1}.$$

Theorem 100 ([46]) *The Jost solutions* $e(\rho, x)$ *and* $g(\rho, x)$ *admit the series representations*

$$e(\rho, x) = e^{i\rho x}\left(1 + (z+1)\sum_{n=0}^{\infty}(-1)^n z^n a_n(x)\right) \tag{7.77}$$

and

$$g(\rho, x) = e^{-i\rho x}\left(1 + (z+1)\sum_{n=0}^{\infty}(-1)^n z^n b_n(x)\right), \tag{7.78}$$

where z *is related to* ρ *by (7.76), and* $\{a_n\}$, $\{b_n\}$ *are the coefficients from (7.65), (7.66). For any* $x \in \mathbb{R}$ *fixed, the series in (7.77), (7.78) converge in* D.

Proof Let us prove (7.77), the representation (7.78) is proved analogously. Substitute (7.65) into (7.60). Then

$$e(\rho, x) = e^{i\rho x}\left(1 + \sum_{n=0}^{\infty} a_n(x)\int_0^{\infty} L_n(t)e^{-\left(\frac{1}{2}-i\rho\right)t}\,dt\right).$$

For the integral, use [33, formula 7.414 (2)]. Then

$$\int_0^\infty L_n(t)e^{-\left(\frac{1}{2}-i\rho\right)t}dt = \frac{(-1)^n\left(\frac{1}{2}+i\rho\right)^n}{\left(\frac{1}{2}-i\rho\right)^{n+1}} = (-1)^n z^n(z+1).$$

Thus, (7.77) is obtained.   ∎

Remark 101 For any $x \in \mathbb{R}$ the series $\sum_{n=0}^\infty a_n^2(x)$ and $\sum_{n=0}^\infty b_n^2(x)$ converge, because $\{a_n\}$ and $\{b_n\}$ are the Fourier coefficients with respect to the system of Laguerre polynomials of the corresponding functions $a(x,t) = e^{\frac{t}{2}}A(x,x+t)$ and $b(x,t) = e^{\frac{t}{2}}B(x,x-t)$, both belonging to $L_2(0,\infty; e^{-t})$. Hence for any $x \in \mathbb{R}$ the functions $e(\rho,x)e^{-i\rho x}$ and $g(\rho,x)e^{i\rho x}$ belong to the Hardy space of the unit disc $H^2(D)$ as functions of z (see, e.g., [77, Theorem 17.12]).

Remark 102 The systems of differential equations (7.69)–(7.72) for the coefficients $\{a_n\}$ and $\{b_n\}$ are obtained by substituting the representations (7.77) and (7.78) into Eq. (7.58) and equating terms corresponding to equal powers of z. It is slightly more tricky to prove the asymptotics (7.73). For these proofs we refer the reader to [47, Sect. 10.3].

Remark 103 Substitution of $z = 0$ (that is equivalent to $\rho = i/2$) into (7.77) and (7.78) gives us the formulas for the first coefficients of the series

$$a_0(x) = e\left(\frac{i}{2},x\right)e^{\frac{x}{2}} - 1 \tag{7.79}$$

and

$$b_0(x) = g\left(\frac{i}{2},x\right)e^{-\frac{x}{2}} - 1. \tag{7.80}$$

Thus, to obtain the first coefficients a_0 and b_0 it is sufficient to calculate the Jost solutions of (7.58) corresponding to $\rho = i/2$, that is, to find the Jost solutions of the equation

$$-u'' + \left(q(x) + \frac{1}{4}\right)u = 0. \tag{7.81}$$

Next, we show how the subsequent coefficients $a_n, b_n, n = 1, 2, \ldots$ can be constructed from (7.79) and (7.80) following a recurrent integration procedure. In other words, we show how the Jost solutions of Eq. (7.58) for arbitrary values of the parameter ρ can be calculated from the Jost solutions of (7.81).

Together with the Jost solutions $e(\frac{i}{2},x)$ and $g(\frac{i}{2},x)$ let us consider two other solutions of (7.81), obtained from $e(\frac{i}{2},x)$ and $g(\frac{i}{2},x)$ by Abel's formula as

$$\eta(x) := e\left(\frac{i}{2}, x\right) \int_0^x \frac{dt}{e^2(\frac{i}{2}, t)} \tag{7.82}$$

and

$$\xi(x) := g\left(\frac{i}{2}, x\right) \int_x^0 \frac{dt}{g^2(\frac{i}{2}, t)}. \tag{7.83}$$

It is easy to verify that these solutions of (7.81) satisfy the asymptotic relations $\eta(x) = e^{\frac{x}{2}}(1 + o(1)), x \to \infty$ and $\xi(x) = e^{-\frac{x}{2}}(1 + o(1)), x \to -\infty$, respectively.
The subsequent coefficients $a_n, b_n, n = 1, 2, \ldots$ can be computed as follows

$$a_n(x) = a_0(x) - 2e^{\frac{x}{2}}\left(\eta(x)J_{1,n}(x) - e\left(\frac{i}{2}, x\right)J_{2,n}(x)\right), \tag{7.84}$$

$$b_n(x) = b_0(x) + 2e^{-\frac{x}{2}}\left(\xi(x)I_{1,n}(x) - g(\frac{i}{2}, x)I_{2,n}(x)\right), \tag{7.85}$$

where

$$J_{1,n}(x) = J_{1,n-1}(x) - e^{-\frac{x}{2}}e(\frac{i}{2}, x)a_{n-1}(x) - \int_x^\infty \left(e(\frac{i}{2}, t)e^{-\frac{t}{2}}\right)' a_{n-1}(t)dt, \tag{7.86}$$

$$J_{2,n}(x) = J_{2,n-1}(x) - e^{-\frac{x}{2}}\eta(x)a_{n-1}(x) - \int_x^\infty \left(\eta(t)e^{-\frac{t}{2}}\right)' a_{n-1}(t)dt \tag{7.87}$$

$$I_{1,n}(x) = I_{1,n-1}(x) + e^{\frac{x}{2}}g\left(\frac{i}{2}, x\right)b_{n-1}(x) - \int_{-\infty}^x \left(g\left(\frac{i}{2}, t\right)e^{\frac{t}{2}}\right)' b_{n-1}(t)dt, \tag{7.88}$$

and

$$I_{2,n}(x) = I_{2,n-1}(x) + e^{\frac{x}{2}}\xi(x)b_{n-1}(x) - \int_{-\infty}^x \left(\xi(t)e^{\frac{t}{2}}\right)' b_{n-1}(t)dt \tag{7.89}$$

with $J_{1,0}(x) = J_{2,0}(x) = I_{1,0}(x) = I_{2,0}(x) \equiv 0$. These recurrent integration formulas are obtained by solving the recurrent differential equations (7.69)–(7.72) taking into account the asymptotics (7.73). For the details we refer to [26] and [47].

Thus, the procedure of computation of the coefficients $\{a_n\}$ and $\{b_n\}$ consists of the following steps.

1. Compute the Jost solutions for $\rho = i/2$: the functions $e(\frac{i}{2}, x)$ and $g(\frac{i}{2}, x)$. This can be done in practice by imposing the approximate initial conditions at two points $\alpha < 0$ and $\beta > 0$, located far enough from the origin:

$$e\left(\frac{i}{2},\beta\right) \approx e^{-\frac{\beta}{2}}, \quad e'\left(\frac{i}{2},\beta\right) \approx -\frac{e^{-\frac{\beta}{2}}}{2},$$

$$g\left(\frac{i}{2},\alpha\right) \approx e^{\frac{\alpha}{2}}, \quad g'\left(\frac{i}{2},\alpha\right) \approx \frac{e^{\frac{\alpha}{2}}}{2},$$

and solving the corresponding Cauchy problems by any available numerical method, for example, by the SPPS method. Usually, the numerical methods provide also the first derivative of the solution of the Cauchy problem. Thus, after the first step we obtain the approximate to the functions

$$e\left(\frac{i}{2},x\right), \quad g\left(\frac{i}{2},x\right), \quad \left(e\left(\frac{i}{2},t\right)e^{-\frac{t}{2}}\right)' \text{ and } \left(g\left(\frac{i}{2},t\right)e^{\frac{t}{2}}\right)'.$$

2. Compute η and ξ by (7.82) and (7.83), as well as

$$\left(\eta(x)e^{-\frac{x}{2}}\right)' = \frac{1}{e(\frac{i}{2},x)}\left(\left(e\left(\frac{i}{2},x\right)e^{-\frac{x}{2}}\right)'\eta(x) + e^{-\frac{x}{2}}\right)$$

and

$$\left(\xi(x)e^{\frac{x}{2}}\right)' = \frac{1}{g(\frac{i}{2},x)}\left(\left(g\left(\frac{i}{2},x\right)e^{\frac{x}{2}}\right)'\xi(x) - e^{\frac{x}{2}}\right).$$

These formulas follow from (7.82) and (7.83).

3. Compute a_n, $J_{1,n}$, $J_{2,n}$, b_n, $I_{1,n}$, $I_{2,n}$ by (7.79), (7.80), (7.84)–(7.89).

We emphasize that no numerical differentiation is required at any step, only integration and arithmetic operations.

Similar series representations are available for the derivatives of the Jost solutions $e'(\frac{i}{2},x)$ and $g'(\frac{i}{2},x)$, which also do not require any numerical differentiation for computing the corresponding coefficients (see [26] and [47]). Namely,

$$e'(\rho,x) = e^{i\rho x}\left(\frac{z-1}{2(z+1)} - \frac{1}{2}\int_x^\infty q(t)dt + (z+1)\sum_{n=0}^\infty (-1)^n z^n d_n(x)\right)$$

$$(7.90)$$

$$g'(\rho,x) = e^{-i\rho x}\left(-\frac{z-1}{2(z+1)} + \frac{1}{2}\int_{-\infty}^x q(t)dt + (z+1)\sum_{n=0}^\infty (-1)^n z^n c_n(x)\right),$$

$$(7.91)$$

where the coefficients $\{d_n\}$ and $\{c_n\}$ are obtained from the coefficients $\{a_n\}$ and $\{b_n\}$, respectively, with the aid of the following relations [47, Sect. 10.6]

$$d_0(x) = a_0'(x) - \frac{a_0(x)}{2} + \frac{1}{2} \int_x^\infty q(t)dt,$$

$$c_0(x) = b_0'(x) + \frac{b_0(x)}{2} - \frac{1}{2} \int_{-\infty}^x q(t)dt,$$

$$d_{n+1}(x) = d_n(x) + a_{n+1}'(x) - a_n'(x) - \frac{1}{2}\left(a_{n+1}(x) + a_n(x)\right), \quad n = 0, 1, \ldots,$$

(7.92)

and

$$c_{n+1}(x) = c_n(x) + b_{n+1}'(x) - b_n'(x) + \frac{1}{2}\left(b_{n+1}(x) + b_n(x)\right), \quad n = 0, 1, \ldots.$$

For computing the derivatives $a_n'(x)$ and $b_n'(x)$ the following procedure can be used

$$a_n'(x) = a_0'(x) - 2\left(e^{\frac{x}{2}}\eta(x)\right)' J_{1,n}(x) - 2e^{\frac{x}{2}}\eta(x)J_{1,n}'(x)$$
$$+ 2\left(e^{\frac{x}{2}}e\left(\frac{i}{2}, x\right)\right)' J_{2,n}(x) + 2e^{\frac{x}{2}}e\left(\frac{i}{2}, x\right) J_{2,n}'(x),$$

$$b_n'(x) = b_0'(x) + 2\left(e^{-\frac{x}{2}}\xi(x)\right)' I_{1,n}(x) + 2e^{-\frac{x}{2}}\xi(x)I_{1,n}'(x)$$
$$- 2\left(e^{-\frac{x}{2}}g\left(\frac{i}{2}, x\right)\right)' I_{2,n}(x) - 2e^{-\frac{x}{2}}g\left(\frac{i}{2}, x\right) I_{2,n}'(x),$$

where

$$J_{1,n}'(x) = J_{1,n-1}'(x) - e^{-\frac{x}{2}}e\left(\frac{i}{2}, x\right) a_{n-1}'(x),$$

$$J_{2,n}'(x) = J_{2,n-1}'(x) - e^{-\frac{x}{2}}\eta(x)a_{n-1}'(x),$$

$$I_{1,n}'(x) = I_{1,n-1}'(x) + e^{\frac{x}{2}}g\left(\frac{i}{2}, x\right) b_{n-1}'(x)$$

and

$$I_{2,n}'(x) = I_{2,n-1}'(x) + e^{\frac{x}{2}}\xi(x)b_{n-1}'(x).$$

At first glance, these formulas may seem complicated; however, they are easily realizable on a computer and in fact again they are nothing but a recurrent integration procedure. Below, in Sect. 7.10 we will see how the representations for the Jost solutions and for their derivatives, presented in this section, work in practice for solving problems on infinite intervals, but first we consider the NSBF method and compare it with the SPPS method applied to Sturm-Liouville problems on finite intervals.

7.8 Solving Sturm-Liouville problems with the aid of NSBF representations

Application of the NSBF representations (Theorem 93) to solution of Sturm-Liouville problems follows essentially the same logic as application of the SPPS representations which was explained in Sect. 7.4. The characteristic equation of the Sturm-Liouville problem can be written explicitly in terms of the NSBF representations, and its numerical solution leads to approximate solution of the spectral problem. However, the results are impressively different. As it is prescribed by estimates (7.52), (7.53) a huge amount of the spectral data can be computed with a nondeteriorating accuracy.

Consider the Sturm-Liouville problem

$$-u''(x) + q(x)u(x) = \rho^2 u(x),$$

$$\alpha u(0) + \beta u'(0) = 0,$$

$$a u(l) + b u'(l) = 0,$$

where the coefficients α, β, a, and b are allowed to be not only constants but also entire functions of the parameter ρ, satisfying the conditions $|\alpha| + |\beta| \neq 0$ and $|a| + |b| \neq 0$ for all ρ.

Since we have convenient NSBF representations for the solutions $\varphi(\rho, x)$ and $s(\rho, x)$ which satisfy the initial conditions (7.30) and (7.31), respectively, we can formulate the following algorithm for solving the Sturm-Liouville problem.

1. Find a solution f of the equation $f'' = q(x)f$ that does not vanish on $[0, l]$, normalize it as $f(0) = 1$ and define $h := f'(0)$. The solution f can be constructed by the SPPS method. The choice of a particular f does not affect a lot the final accuracy as long as the functions f and $1/f$ do not take too large values.

2. Compute the functions g_k, s_k and, if necessary, γ_k, σ_k using (7.43), (7.44), (7.56), (7.57) or the alternative formulas from [56], [47, Sect. 9.4].

3. Calculate the approximations $\varphi_N(\rho, l)$ and $s_N(\rho, l)$ of the solutions $\varphi(\rho, l)$ and $s(\rho, l)$ at the point l by (7.50), (7.51). If necessary, calculate the approximations of the derivatives $\varphi'(\rho, l)$ and $s'(\rho, l)$ according to Remark 97.
4. The eigenvalues of the Sturm-Liouville problem coincide with the squares of the zeros of the entire function

$$\Phi(\rho) := \alpha \left(\beta \varphi(\rho, l) - (\alpha + \beta h) \frac{s(\rho, l)}{\rho} \right) + b \left(\beta \varphi'(\rho, l) - (\alpha + \beta h) \frac{s'(\rho, l)}{\rho} \right)$$

and are approximated by squares of zeros of the function

$$\Phi_N(\rho) := \alpha \left(\beta \varphi_N(\rho, l) - (\alpha + \beta h) \frac{s_N(\rho, l)}{\rho} \right) + b \left(\beta \varphi'_N(\rho, l) - (\alpha + \beta h) \frac{s'_N(\rho, l)}{\rho} \right).$$

5. The eigenfunction u_λ corresponding to the eigenvalue $\lambda = \rho^2$ can be taken in the form

$$u_\lambda = \beta \varphi(\rho, x) - (\alpha + \beta h) \frac{s(\rho, x)}{\rho}. \tag{7.93}$$

Hence once the eigenvalues are calculated, the computation of the corresponding eigenfunctions can be performed using formulas (7.50) and (7.51).

Remark 104 The equations (7.46) and (7.47) offer a simple and efficient way for controlling the accuracy of the numerical method and for choosing an optimal N. The differences

$$\varepsilon_{1,N}(x) := \left| \sum_{j=0}^N \frac{g_j(x)}{x} - \left(\frac{h}{2} + \frac{1}{2} \int_0^x q(s)\,ds \right) \right| \quad \text{and} \quad \varepsilon_{2,N}(x) := \left| \sum_{j=0}^N \frac{s_j(x)}{x} - \frac{1}{2} \int_0^x q(t)\,dt \right| \tag{7.94}$$

measure the accuracy of the approximation of the transmutation kernel and hence the accuracy of the approximate solutions (7.50) and (7.51).

7.9 Numerical Example. NSBF vs. SPPS

For illustrating the performance of both methods, the SPPS and the NSBF, we consider the following Sturm-Liouville problem.

Paine Problem A number of spectral problems which have become standard test cases appear in [75]. Consider

$$- u'' + \frac{1}{(x + 0.1)^2} u = \lambda u. \tag{7.95}$$

$$u(0) = 0, \quad u(\pi) = 0. \tag{7.96}$$

The eigenvalues in the following table were calculated via SPPS using integration on 10,000 subintervals for calculating $N = 100$ powers of λ. These eigenvalues were found as roots of a single polynomial (i.e., the shifting of λ as described in Remark 80 was not applied). Due to the sensitivity of the larger roots of the polynomial to errors in the coefficients, 100-digit arithmetic was used.

n	λ_n [75]	λ_n SPPS [48]
0	1.5198658211	1.519865821099
1	4.9433098221	4.943309822144
2	10.284662645	10.28466264509
3	17.559957746	17.55995774633
4	26.782863158	26.78286315899
5	37.964425862	37.96442587941
6	51.113357757	51.11335707578
7	66.236447704	66.23646092491
8	83.338962374	83.33879073183
9	102.42498840	102.4259718823
10	123.49770680	123.512483827

We see that the accuracy of the computed eigenvalues is quite good, though deteriorating for higher index eigenvalues.

Now we compute the first 100 eigenvalues of the same problem (7.95), (7.96) by using the NSBF representation (7.51) with $N = 83$ and in machine precision. Figure 7.2 shows the relative error of the eigenvalues. It can be noticed that it is practically uniform and close to machine precision. Moreover, next several hundreds and even thousands of the eigenvalues are computed with the same relative error. The situation is similar for the computed eigenfunctions of the problem. This is an illustration of the estimate (7.52) which predicts that the approximation of the exact solution by a partial sum of its NSBF series is equally good for both small and large values of the parameter $\rho = \sqrt{\lambda}$.

This makes the NSBF representations from Theorem 93 and Remark 97 especially convenient for solving a variety of problems requiring computing solutions on large intervals with respect to the parameter ρ.

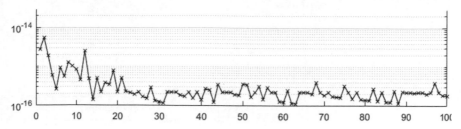

Fig. 7.2 Relative error of the first 100 eigenvalues of problem (7.95), (7.96) computed by using the NSBF representation (7.51) with $N = 83$ in machine precision (in Matlab)

7.10 Scattering Problem on the Line

As an application of the representations for the Jost solutions (Theorem 100), let us consider the following classical scattering problem which is of practical importance in quantum theory, wave propagation theory as well as in the inverse scattering transform method for solving nonlinear partial differential equations of mathematical physics, such as the Korteweg-de Vries equation and the nonlinear Schrödinger equation (see Sect. 7.14).

Consider the one-dimensional Schrödinger equation (7.58):

$$-u'' + q(x)u = \rho^2 u \tag{7.97}$$

on the whole line, $x \in (-\infty, \infty)$ with $q(x)$ a real valued, short range potential (q satisfies the condition (7.59)), Im $\rho \geq 0$. From Sect. 7.7 we know that there exist the unique Jost solutions $e(\rho, x)$ and $g(\rho, x)$ which, roughly speaking, behave asymptotically as $e^{i\rho x}$ when $x \to +\infty$ and $e^{-i\rho x}$ when $x \to -\infty$, respectively.

When $\rho \in \mathbb{R}$, due to the fact that q is real valued, we have that

$$e(-\rho, x) = \overline{e(\rho, x)}, \quad g(-\rho, x) = \overline{g(\rho, x)}$$

and both sets of solutions

$$\left\{ e(\rho, x), \overline{e(\rho, x)} \right\} \quad \text{and} \quad \left\{ g(\rho, x), \overline{g(\rho, x)} \right\}$$

represent fundamental systems of solutions of (7.97) for $\rho \neq 0$. Thus, $g(\rho, x)$ and $\overline{g(\rho, x)}$ can be represented as linear combinations of the fundamental system $\left\{ e(\rho, x), \overline{e(\rho, x)} \right\}$:

$$g(\rho, x) = a(\rho)\overline{e(\rho, x)} + b(\rho)e(\rho, x)$$

and

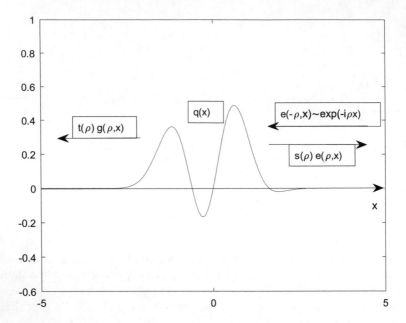

Fig. 7.3 Schematic representation of the scattering model

$$\overline{g(\rho, x)} = \overline{a(\rho)}e(\rho, x) + \overline{b(\rho)e(\rho, x)}.$$

The magnitudes $a(\rho)$ and $b(\rho)$ are called **scattering amplitudes** and can be calculated from the formulas

$$a(\rho) = -\frac{1}{2i\rho} W\left[e(\rho, x), g(\rho, x)\right], \quad \text{Im } \rho \geq 0, \tag{7.98}$$

$$b(\rho) = \frac{1}{2i\rho} W\left[e(\rho, x), g(-\rho, x)\right], \quad \rho \in \mathbb{R}, \tag{7.99}$$

where $W\left[\cdot, \cdot\right]$ stands for the Wronskian of two functions: $W\left[u_1, u_2\right] = u_1 u_2' - u_1' u_2$.

Notice that the expression for $b(\rho)$ is well defined for real values of ρ only, because when $\text{Im } \rho > 0$, a solution of (7.97) behaving as $e^{-i\rho x}$ when $x \to -\infty$ is not unique.

The scattering amplitudes are closely related to the reflection and transmission coefficients. Imagine that a plane wave of a unitary amplitude, asymptotically similar to $e^{-i\rho x}$ when $x \to \infty$, $\rho \in \mathbb{R}$ comes from the right (from the plus infinity) as it is schematically depicted on Fig. 7.3. It interacts with a medium characterized by the potential q. A part of it is transmitted, preserving the same phase but with a new amplitude depending on ρ: $t(\rho)$, while another part is reflected, changing its phase to the opposite and with an amplitude $s(\rho)$. These amplitudes of the transmitted part of a unitary plane wave and of its reflected part are called the *transmission and reflection coefficients*, respectively. From the figure it is seen that

$$\overline{e(\rho, x)} = t(\rho)g(\rho, x) - s(\rho)e(\rho, x).$$

Thus, $t(\rho) = 1/a(\rho)$ and $s(\rho) = b(\rho)/a(\rho)$.

We considered the situation of a wave travelling from the right to the left. Analogously, a wave travelling in the opposite direction can be considered, and hence two reflection coefficients (the right and the left) can be introduced instead. They have the form

$$s^{\pm}(\rho) = \mp \frac{b(\mp\rho)}{a(\rho)}. \tag{7.100}$$

To solve the **scattering problem** means to find a finite set (if it is not empty) of negative eigenvalues $\lambda_n = \rho_n^2 = (i\tau_n)^2$, $0 < \tau_1 < \ldots < \tau_N$, a corresponding set (left or right) of positive norming constants

$$\alpha_n^{+} := \left(\int_{-\infty}^{\infty} e^2(\rho_n, x)dx \right)^{-1} \quad \text{or} \quad \alpha_n^{-} := \left(\int_{-\infty}^{\infty} g^2(\rho_n, x)dx \right)^{-1}$$

and the reflection coefficient (left or right) $s^{+}(\rho)$ or $s^{-}(\rho)$, $\rho \in \mathbb{R}$. The sets

$$J^{\pm} = \left\{ s^{\pm}(\rho), \ \rho \in \mathbb{R}; \ \lambda_n, \alpha_n^{\pm}, \ n = 1, \ldots, N \right\} \tag{7.101}$$

are called the **scattering data**. Thus, the scattering problem consists, given a potential $q(x)$, in finding the associated scattering data J^{+} or J^{-}. Once J^{+} is found, one can also compute J^{-} and vice versa [86]. In practice, the computation of the reflection coefficient is regarded as a computationally challenging problem. However, the representations (7.77), (7.78), (7.90), (7.91) for the Jost solutions and their derivatives lead to a fast and accurate method.

Example 105 As an example, let us consider the potential depicted on Fig. 7.4. It is defined as follows [4]

$$q(x) = \begin{cases} q_1(x), & x < 0 \\ q_2(x), & x > 0, \end{cases} \tag{7.102}$$

where

$$q_1(x) = \frac{16\left(\sqrt{2}+1\right)^2 e^{-2\sqrt{2}x}}{\left(\left(\sqrt{2}+1\right)^2 e^{-2\sqrt{2}x} - 1\right)^2} \tag{7.103}$$

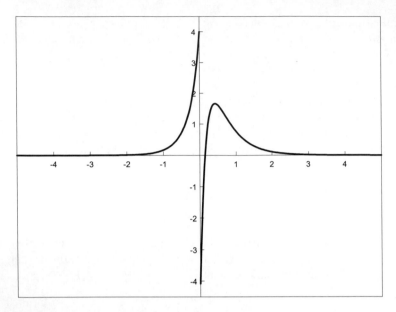

Fig. 7.4 The potential $q(x)$ defined by (7.102)–(7.104)

and

$$q_2(x) = \frac{96e^{2x}\left(81e^{8x} - 144e^{6x} + 54e^{4x} - 9e^{2x} + 1\right)}{\left(36e^{6x} - 27e^{4x} + 12e^{2x} - 1\right)^2}. \tag{7.104}$$

It is known [4] that Eq. (7.97) with this potential does not possess eigenvalues, and the reflection coefficient $s^+(\rho)$ has the form

$$s^+(\rho) = \frac{(\rho + i)(\rho + 2i)(101\rho^2 - 3i\rho - 400)}{(\rho - i)(\rho - 2i)(50\rho^4 + 280i\rho^3 - 609\rho^2 - 653i\rho + 400)}. \tag{7.105}$$

On Fig. 7.5 we show the result of computation of this reflection coefficient by the formulas (7.100), (7.98), (7.99) where the corresponding Jost solutions and their derivatives were computed using the representations (7.77), (7.78), (7.90), (7.91) with 60 terms in corresponding partial sums. The absolute error of the computed reflection coefficient was less than 10^{-12}. Taking into account that the whole computation in Matlab2017 on a usual Laptop took less than a second this is an amazing result.

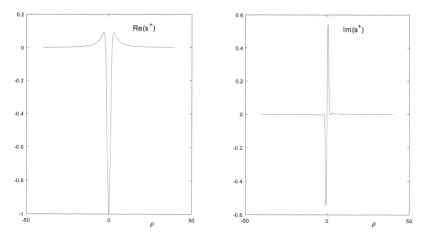

Fig. 7.5 $s^+(\rho)$ computed with the aid of the representations (7.77), (7.78), (7.90), (7.91) with 60 terms in corresponding partial sums. The absolute error of the computed $s^+(\rho)$ was less than 10^{-12}

7.11 Notion of Inverse Spectral Problems

Inverse spectral problems consist in recovering the spectral problem from some given spectral data. For example, the inverse scattering problem on the line consists in recovering the potential $q(x)$ from a given set of scattering data. The inverse Sturm-Liouville problem on a finite interval may be formulated as recovering the potential and the boundary conditions from two sequences of numbers: the eigenvalues $\{\lambda_n\}_{n=0}^{\infty}$ and the so-called norming constants, which are the squares of the L_2-norms of the corresponding normalized eigenfunctions. Below we give a precise statement of the problem.

Inverse spectral problems play a fundamental role in different areas of mathematical physics, and the interest in this topic continuously grows due to a constant emergence of new applications. Main results and approaches of the theory of inverse spectral problems were developed during the second half of the past century. We refer the interested reader to the fundamental books on this subject, such as [20, 30, 36, 47, 66, 69, 73, 76, 86].

Our aim is to present a recently developed approach for efficient solution of inverse spectral problems which allows one to reduce the problem to a system of linear algebraic equations, such that the first component of the solution vector is sufficient for solving the problem. The approach was developed in a series of papers [25, 37, 45, 46, 60] and presented in the book [47].

Here, to explain the approach, we consider two inverse spectral problems: the inverse Sturm-Liouville problem on a finite interval and the inverse scattering problem on the line. The same approach is applicable to a much larger variety of inverse spectral problems.

Let us consider the following direct Sturm-Liouville problem

$$- y'' + q(x)y = \rho^2 y, \quad x \in (0, \pi), \tag{7.106}$$

$$y'(0) - hy(0) = 0, \quad y'(\pi) + Hy(\pi) = 0, \tag{7.107}$$

where q is a real valued function, $q \in L_2(0, \pi)$, h and H are arbitrary real constants. The attentive reader noticed that the boundary conditions (7.107) exclude the possibility of the Dirichlet boundary conditions. Indeed, this is the case; however, it is not difficult to repeat the presented reasoning in the case when one or both boundary conditions are of the Dirichlet type. Then several details change but essentially the same method works.

As earlier (see Sect. 7.5) we denote by $\varphi(\rho, x)$ a solution of (7.106) satisfying the initial conditions

$$\varphi(\rho, 0) = 1, \quad \varphi'(\rho, 0) = h.$$

The direct Sturm-Liouville problem (7.106), (7.107) possesses an infinite sequence of eigenvalues $\{\lambda_n = \rho_n^2\}_{n=0}^{\infty}$, such that $\lambda_n \neq \lambda_m$ when $n \neq m$, associated with the corresponding (real valued) normalized eigenfunctions $\varphi(\rho_n, x)$. The numbers

$$\alpha_n := \int_0^{\pi} \varphi^2(\rho_n, x)\, dx \tag{7.108}$$

are the norming constants of the problem. It is known that the sequences of the real numbers $\{\lambda_n\}_{n=0}^{\infty}$ and $\{\alpha_n\}_{n=0}^{\infty}$ represent spectral data of a Sturm-Liouville problem (7.106), (7.107) if and only if the following relations hold

$$\rho_n := \sqrt{\lambda_n} = n + \frac{\omega}{\pi n} + \frac{k_n}{n}, \quad \alpha_n = \frac{\pi}{2} + \frac{K_n}{n}, \quad \{k_n\}, \{K_n\} \in \ell_2, \tag{7.109}$$

and

$$\alpha_n > 0, \quad \lambda_n \neq \lambda_m \ (n \neq m). \tag{7.110}$$

Here

$$\omega := h + H + \frac{1}{2}\int_0^{\pi} q(t)\, dt. \tag{7.111}$$

The inverse Sturm-Liouville problem is stated as follows. Given two sequences of real numbers $\{\lambda_n\}_{n=0}^{\infty}$ and $\{\alpha_n\}_{n=0}^{\infty}$, such that the asymptotic relations (7.109) hold together with the condition (7.110), find $q(x)$ and the constants h, H, such that $\{\lambda_n\}_{n=0}^{\infty}$ are the eigenvalues of the problem (7.106), (7.107) and $\{\alpha_n\}_{n=0}^{\infty}$ are the corresponding norming constants. This inverse problem is uniquely solvable (see, e.g., [30, 66], [86, Theorem 1.3.2]).

In practice only a limited number of the eigenvalues and norming constants is given, and the method of solution of the inverse Sturm-Liouville problem needs to be able to recover the approximants of $q(x)$, h, H from that finite set of spectral data.

7.12   Direct Method of Solution of the Inverse Sturm-Liouville Problem

The main idea of the method was proposed in [45] (see also [47, Chapter 13]), and in [52] an important refinement of the method was developed. Hence here we mainly follow the exposition from [52] with certain simplifications.

The first key ingredient of the method is a classical result known as a Gelfand-Levitan equation [32]. It establishes that the kernel $G(x, t)$ (see Theorem 87) satisfies the integral equation

$$G(x, t) + F(x, t) + \int_0^x F(t, s)G(x, s)\, ds = 0, \quad 0 < t < x < \pi, \quad (7.112)$$

where the function $F(x, t)$ is constructed from the given spectral data as follows

$$F(x, t) = \sum_{n=0}^{\infty} \left(\frac{\cos \rho_n x \cos \rho_n t}{\alpha_n} - \frac{\cos nx \cos nt}{\alpha_n^0} \right), \quad 0 \le t, x < \pi \quad (7.113)$$

with

$$\alpha_n^0 = \begin{cases} \pi/2, & n > 0, \\ \pi, & n = 0. \end{cases}$$

The Gelfand-Levitan equation apparently leads to a complete solution of the inverse Sturm-Liouville problem. Indeed, for every x fixed, it is nothing but a Fredholm integral equation of the second kind. Having solved it one obtains the kernel $G(x, t)$, and then from (7.35) it follows that

$$q(x) = 2\frac{d}{dx}G(x, x).$$

However, this beautiful idea encounters serious obstacles when attempting to convert it into a practical computational method.

The series in (7.113) is quite tricky. First of all, if trying to sum up each of the two series separately, we face the problem that both are divergent. Second, it turns out that if $\omega \neq 0$, the sum of the series in (7.113) has a jump discontinuity at $t = x = \pi$, though the function $F(x, t)$ itself is continuous on $0 \leq t, x \leq \pi$, including the point $t = x = \pi$. This discrepancy explains, why in the equality (7.113) the point $t = x = \pi$ is not included.

From the theoretical viewpoint this is not a big deal: the function $F(x, t)$ defined by the series (7.113) is extended onto the whole domain $0 \leq t, x \leq \pi$ by continuity, thus eliminating the jump discontinuity. However, from the practical viewpoint this discontinuity is a serious obstacle for applying the Gelfand-Levitan equation. First of all it slows down the convergence of the series and, second, the accuracy of results of any computation made on base of Eq. (7.112) considerably deteriorates in the vicinity of π.

An important improvement in representing the function $F(x, t)$ was achieved in [41] where it was shown that

$$F(x,t) = \sum_{n=1}^{\infty} \left(\frac{\cos \rho_n x \, \cos \rho_n t}{\alpha_n} - \frac{\cos nx \, \cos nt}{\alpha_n^0} + \frac{2\omega}{\pi^2 n} \left(x \sin nx \, \cos nt + t \sin nt \, \cos nx \right) \right)$$

$$+ \frac{\cos \rho_0 x \, \cos \rho_0 t}{\alpha_0} - \frac{1}{\pi} - \frac{\omega}{\pi^2} \left(\pi \max\{x, t\} - x^2 - t^2 \right), \tag{7.114}$$

and the series converges absolutely and uniformly. To the difference from representation (7.113), in (7.114) there appears the parameter ω. Thus, in order to make use of the improved series representation (7.114) we need first to find a way to compute ω.

Now we are in a position to formulate the method for solving the inverse Sturm-Liouville problem.

In a first step we change the problem so that $\rho_0 = 0$. That is, the first eigenvalue of the problem is zero. It is always possible to achieve by a simple shift of the potential. If originally $\rho_0 \neq 0$, then we consider a new potential $\widetilde{q}(x) := q(x) - \rho_0^2$. Obviously, the eigenvalues $\{\lambda_n\}_{n=0}^{\infty}$ shift by the same amount, while the numbers h, H and $\{\alpha_n\}_{n=0}^{\infty}$ do not change. After recovering $\widetilde{q}(x)$ from the shifted eigenvalues, one gets the original potential $q(x)$ by adding ρ_0^2 back.

Thus, from now on and without loss of generality, we suppose that

$$\rho_0 = 0 \tag{7.115}$$

and preserve the same notation $q(x)$ for the potential.

Next, we recover the parameter ω as follows. We know that for all eigenvalues ρ_k at the point $x = \pi$ the equality holds

$$\varphi'(\rho_k, \pi) + H\varphi(\rho_k, \pi) = 0, \tag{7.116}$$

where H is still unknown. With the aid of the NSBF representations for $\varphi(\rho, x)$ and $\varphi'(\rho, x)$ (formulas (7.48), (7.54)) we can write (7.116) in the form

$$-\rho_k \sin \rho_k \pi + \omega \cos \rho_k \pi + \sum_{n=0}^{\infty} (-1)^n \left(\gamma_n(\pi) + H g_n(\pi) \right) j_{2n}(\rho_k \pi) = 0.$$

Denote

$$h_n := \gamma_n(\pi) + H g_n(\pi), \qquad n \geq 0.$$

Then

$$\sum_{n=0}^{\infty} (-1)^n h_n j_{2n}(\rho_k \pi) = \rho_k \sin \rho_k \pi - \omega \cos \rho_k \pi, \quad k = 0, 1, \ldots. \tag{7.117}$$

Notice that due to the supposition (7.115) we have that

$$\varphi'(0, \pi) + H \varphi(0, \pi) = 0. \tag{7.118}$$

From (7.43) we find that

$$g_0(x) = \varphi(0, x) - 1, \tag{7.119}$$

and from (7.56) that

$$\gamma_0(x) = \varphi'(0, x) - h - \frac{1}{2} \int_0^x q(s) \, ds. \tag{7.120}$$

Hence (7.118) can be written as

$$\gamma_0(\pi) + h + \frac{1}{2} \int_0^\pi q(s) \, ds + H \left(g_0(\pi) + 1 \right) = 0,$$

or equivalently,

$$h_0 + \omega = 0.$$

Thus,

$$\omega = -h_0. \tag{7.121}$$

Notice that this equality coincides with (7.117) when $k = 0$, while for $k = 1, 2, \ldots$ equality (7.117) can be written in the form

$$h_0 \cos \rho_k \pi - \sum_{n=0}^{\infty} (-1)^n h_n j_{2n}(\rho_k \pi) = -\rho_k \sin \rho_k \pi, \quad k = 1, 2, \ldots.$$

Hence the numbers h_n including $h_0 = -\omega$ can be approximated by solving the system of linear algebraic equations

$$h_0 \cos \rho_k \pi - \sum_{n=0}^{N_1-1} (-1)^n h_n j_{2n}(\rho_k \pi) = -\rho_k \sin \rho_k \pi, \quad k = 1, \ldots, N_1. \quad (7.122)$$

Moreover, the knowledge of the numbers $\{h_n\}_{n=0}^{N_1-1}$ means the knowledge of the (approximate) characteristic function $\varphi'(\rho, \pi) + H\varphi(\rho, \pi)$ of the problem (7.106), (7.107) at all points $\rho \in \mathbb{R}$, that allows us to compute more of the eigenvalues ρ_k^2 as zeros of the characteristic function and even more norming constants (though this last possibility we leave outside this presentation and refer the interested reader to [53] for related details). This again is possible due to the remarkable estimate (7.52).

On the next step we make use of the Gelfand-Levitan equation, in which the function $F(x, t)$ is substituted in the form (7.114) while the kernel $G(x, t)$ in the form (7.41). In this way we arrive at a system of linear algebraic equations for the coefficients $\{g_n\}_{n=0}^{\infty}$. We formulate this result in the form of the theorem.

Theorem 106 ([52]) *The coefficients $g_m(x)$ satisfy the system of linear algebraic equations*

$$\frac{g_k(x)}{(4k+1)x} + \sum_{m=0}^{\infty} g_m(x)\widetilde{C}_{km}(x) = \widetilde{d}_k(x), \quad k = 0, 1, \ldots \quad (7.123)$$

where

$$\widetilde{c}_{km}(x) = -\frac{\omega x}{8\pi} \left(\frac{\delta_{m,k-1}}{(2k-3/2)_3} - \frac{2\delta_{m,k}}{(2k-1/2)_3} + \frac{\delta_{m,k+1}}{(2k+1/2)_3} \right)$$

$$+ (-1)^{k+m} \sum_{n=1}^{\infty} \left[\frac{j_{2k}(\rho_n x) j_{2m}(\rho_n x)}{\alpha_n} - \frac{2 j_{2k}(nx) j_{2m}(nx)}{\pi} \right. \quad (7.124)$$

$$\left. + \frac{2\omega}{\pi^2 n} \left(x j_{2k}(nx) j_{2m+1}(nx) + x j_{2k+1}(nx) j_{2m}(nx) - \frac{2(k+m) j_{2k}(nx) j_{2m}(nx)}{n} \right) \right],$$

$$\widetilde{C}_{0m}(x) = \left(\frac{1}{\alpha_0} - \frac{1}{\pi} + \frac{2\omega x^2}{3\pi^2} \right) \delta_{0m} + \frac{2\omega x^2}{15\pi^2} \delta_{1m} + \widetilde{c}_{0m}(x), \quad (7.125)$$

$$\widetilde{C}_{1m}(x) = \frac{2\omega x^2}{15\pi^2} \delta_{0m} + \widetilde{c}_{1m}(x), \quad (7.126)$$

$$\widetilde{C}_{km}(x) = \widetilde{c}_{km}(x), \quad k = 2, 3, \ldots, \ m \in \mathbb{N}_0, \quad (7.127)$$

and

$$\tilde{d}_k(x) = -\left(\frac{1}{\alpha_0} - \frac{1}{\pi} + \frac{4\omega x^2}{3\pi^2} - \frac{\omega x}{\pi}\right)\delta_{k0} - \frac{2\omega x^2}{15\pi^2}\delta_{k1}$$

$$- (-1)^k \sum_{n=1}^{\infty}\left[\frac{\cos\rho_n x}{\alpha_n}j_{2k}(\rho_n x) - \frac{2\cos nx}{\pi}j_{2k}(nx)\right.$$

$$\left. + \frac{2\omega}{\pi^2 n}\left(x\sin nx\, j_{2k}(nx) + x\cos nx\, j_{2k+1}(nx) - \frac{2k}{n}\cos nx\, j_{2k}(nx)\right)\right],$$

$$\tag{7.128}$$

where $\delta_{k,m}$ stands for the Kronecker delta and $(k)_m$ denotes the Pochhammer symbol, $(k)_m = k\,(k+1)\cdots(k+m-1)$.

For all $x \in [0, \pi]$ and $k = 0, 1, \ldots$ the series in (7.123) converges.

It is of crucial importance that from system (7.123) we need to find the first coefficient $g_0(x)$ only, and then taking into account equality (7.119) we recover $q(x)$ from the equality

$$q(x) = \frac{\varphi''(0, x)}{\varphi(0, x)} = \frac{g_0''(x)}{g_0(x) + 1}.\tag{7.129}$$

Notice that $g_0(x) + 1 \neq 0$ when $x \in [0, \pi]$. This is due to the fact that $\rho_0 = 0$ and hence $\varphi(0, x)$ is the first eigenfunction of the problem (7.106), (7.107), and consequently does not have zeros on $[0, \pi]$ (see, e.g., [6, Theorem 8.4.5]).

Thus, the method for solving the inverse Sturm-Liouville problem consists of the following steps.[1]

1. Solve system (7.122) to compute ω.
2. Compute the coefficients of system (7.123).
3. Solve system (7.123) to compute $g_0(x)$.
4. Compute $q(x)$ from (7.129).

This simple and direct method can be enriched by some additional techniques which may lead to somewhat more accurate results (see [52]) but even in its "basic configuration" (steps 1–4) it is accurate and efficient.

7.12.1   Numerical Realization

The numerical realization of the method in a modern numerical computing environment such as Matlab, Maple, or Mathematica does not present any difficulty and mainly involves standard built-in routines. In particular, we used Matlab 2017. The differentiation of $g_0(x)$ at step 4 was performed by converting $g_0(x)$ into a spline with the aid of the routine "*spapi*" with a subsequent application of the

---

[1]Here we assume already that $\rho_0 = 0$, see p. 172.

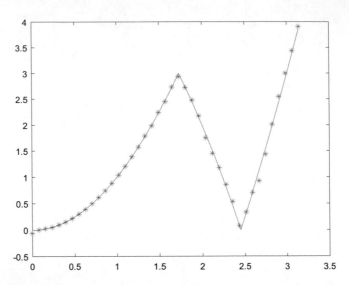

Fig. 7.6 The potential $q(x) = \left|3 - \left|x^2 - 3\right|\right|$ recovered from 200 eigenvalues, with 6 equations

Fig. 7.7 The potential
(7.130) recovered from 200
eigenvalues

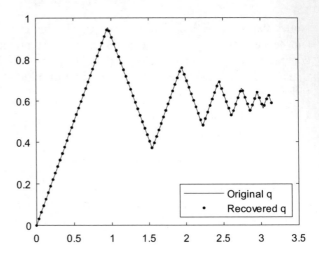

routine "*fnder.*" Without much detail we illustrate the performance of the method by two figures. On Fig. 7.6 the potential $q(x) = \left|3 - \left|x^2 - 3\right|\right|$ with $h = H = 0$ is recovered from 200 eigenvalues. Here only 6 equations of the truncated system (7.123) were used.

On Fig. 7.7 the "sawtooth" potential

$$q(x) = \int_0^x \text{sign}\left(\sin\left(\frac{10t}{4 - t}\right)\right) dt \qquad (7.130)$$

with $h = 1$, $H = 2$ is recovered from 200 eigenvalues [52]. The values of the recovered constants h and H were $h_{rec} = 1.0000031$ and $H_{rec} = 1.999980$, respectively.

Additional details regarding the implementation of the method the reader can find in [52].

7.13   Direct Method of Solution of the Inverse Scattering Problem

Interestingly enough, essentially the same approach is applicable to a large variety of inverse spectral problems both on finite and infinite intervals. As an example, we consider here the inverse scattering problem on the line. It is stated as follows. Given a set of the scattering data J^+ (or J^-), find the potential $q(x)$ of Eq. (7.97). We refer to Sect. 7.10 for the definition of the scattering data. For definiteness we assume that J^+ is given.

Again, two key ingredients of the method are the corresponding Gelfand-Levitan equation, which in this case is often called the Gelfand-Levitan-Marchenko equation, and the series expansion of the transmutation integral kernel (7.65) from which the first coefficient $a_0(x)$ is sufficient for recovering the potential $q(x)$. Indeed, from the equality (7.67), taking into account that $e(\frac{i}{2}, x)$ is a solution of (7.81), we obtain that

$$q = \frac{a_0'' - a_0'}{a_0 + 1}. \tag{7.131}$$

The transmutation operator kernel A satisfies the corresponding Gelfand-Levitan-Marchenko equation (see, e.g., [20, 66, 86])

$$F(x + y) + A(x, y) + \int_x^\infty A(x, t)F(t + y)dt = 0, \quad y > x, \tag{7.132}$$

where

$$F(x) = \sum_{k=1}^N \alpha_k^+ e^{-\tau_k x} + \frac{1}{2\pi} \int_{-\infty}^\infty s^+(\rho) e^{i\rho x} d\rho.$$

Equation (7.132) is uniquely solvable. However, the method we explain here does not involve its solution. We use Eq. (7.132) in order to obtain a system of linear algebraic equations for the coefficients $\{a_n\}_{n=0}^\infty$, which allows us to compute a_0 and hence to recover q by (7.131).

Let us introduce the following notations

$$A_{mn}(x) := (-1)^{m+n} \left(\sum_{k=1}^{N} \alpha_k^+ e^{-2\tau_k x} \frac{\left(\frac{1}{2} - \tau_k \right)^{m+n}}{\left(\frac{1}{2} + \tau_k \right)^{m+n+2}} + \frac{1}{2\pi} \int_{-\infty}^{\infty} s^+(\rho) e^{2i\rho x} \frac{\left(\frac{1}{2} + i\rho \right)^{m+n}}{\left(\frac{1}{2} - i\rho \right)^{m+n+2}} d\rho \right),$$

(7.133)

$$r_m(x) := (-1)^{m+1} \left(\sum_{k=1}^{N} \alpha_k^+ e^{-2\tau_k x} \frac{\left(\frac{1}{2} - \tau_k \right)^{m}}{\left(\frac{1}{2} + \tau_k \right)^{m+1}} + \frac{1}{2\pi} \int_{-\infty}^{\infty} s^+(\rho) e^{2i\rho x} \frac{\left(\frac{1}{2} + i\rho \right)^{m}}{\left(\frac{1}{2} - i\rho \right)^{m+1}} d\rho \right).$$

(7.134)

Theorem 107 *The coefficients $\{a_n\}_{n=0}^{\infty}$ satisfy the following system of linear algebraic equations*

$$a_m(x) + \sum_{n=0}^{\infty} a_n(x) A_{mn}(x) = r_m(x), \quad m = 0, 1, \dots$$

(7.135)

For every m the series in (7.135) converges pointwise.

The proof consists in substitution of (7.65) into (7.132) and evaluation of the corresponding integrals.

7.13.1 Numerical Realization

Based on Theorem 107, thus, a numerical method for solving the inverse scattering problem can be formulated as follows.

1. Given a set of scattering data J^+. Choose a number of equations N_s, so that the truncated system

$$a_m(x) + \sum_{n=0}^{N_s} a_n(x) A_{mn}(x) = r_m(x), \quad m = 0, \dots, N_s$$

(7.136)

   is to be solved.
2. Compute $r_m(x)$ and $A_{mn}(x)$ according to the formulas from above.
3. Solve the system (7.136) to find $a_0(x)$.
4. Compute q with the aid of (7.131).

Let us discuss some relevant aspects of this numerical approach.

First, consider the question regarding an appropriate way for computing the integrals in (7.133), (7.134). Notice that they are Fourier transforms of the reflection coefficients multiplied by fractions of a special form. Here it would be interesting to apply some of the available techniques for numerical computation of the Fourier transform. However, the special form of the fractional factors suggests another possibility for computing the integrals, which proved to provide good results.

The integral we are interested in computing has the form

$$I(x) = \int_{-\infty}^{\infty} s(\rho) e^{2i\rho x} \frac{\left(\frac{1}{2} + i\rho\right)^m}{\left(\frac{1}{2} - i\rho\right)^n} d\rho,$$

where $n - m \geq 1$. Let $z := \frac{\frac{1}{2} + i\rho}{\frac{1}{2} - i\rho}$. Then, when ρ runs along the real axis $(-\infty, \infty)$, the variable z runs along the unitary circle, so that $z = e^{i\theta}$ with $\theta \in (-\pi, \pi)$. Thus, the change of the integration variable has the form $e^{i\theta} = \frac{\frac{1}{2} + i\rho}{\frac{1}{2} - i\rho}$. Then $e^{i\theta} d\theta = \frac{d\rho}{\left(\frac{1}{2} - i\rho\right)^2}$, $2i\rho = \frac{e^{i\theta} - 1}{e^{i\theta} + 1}$ and $\frac{1}{\frac{1}{2} - i\rho} = e^{i\theta} + 1$. Hence the integral $I(x)$ can be written as

$$I(x) = \int_{-\pi}^{\pi} s\left(\frac{i(1 - e^{i\theta})}{2(1 + e^{i\theta})}\right) \exp\left(\frac{e^{i\theta} - 1}{e^{i\theta} + 1} x\right) e^{i\theta(m+1)} \left(e^{i\theta} + 1\right)^{n-m-2} d\theta.$$

$$(7.137)$$

All the integrals involved were computed using formula (7.137) with the aid of the Matlab routine "trapz," evaluating the integrand in $N_i = 10^4$ points, uniformly distributed on the interval $(-\pi, \pi)$.

Example 108 Consider Example 105. This time we are given the reflection coefficient (7.105) and need to recover from it the potential $q(x)$, which is defined by (7.102)–(7.104). In the numerical examples presented below we computed separately the recovered potential for $x < 0$ and $x > 0$ considering the truncated system (7.136) for the intervals $-\pi < x < 0$ and $0 < x < \pi$, respectively, with 201 points $\{x_l\}$ uniformly distributed in both cases. The differentiation of $a_0(x)$ at step 4 was performed by fitting the computed $a_0(x)$ with a partial sum of a Fourier series containing eight elements (the maximum allowed by the Matlab routine "fit") and applying the Matlab command "differentiate."

Already with one equation in the truncated system we obtain a reasonably good approximation of the potential $q_2(x)$ presented on Fig. 7.8.

With five equations the recovered potential $q_2(x)$ practically coincides with the exact one (Fig. 7.9).

For an accurate recovery of $q_1(x)$ more equations were necessary. On Fig. 7.10 the result obtained with 15 equations in the truncated system is presented. The maximum absolute error in this case was 0.378 attained at the origin.

With 25 equations in the truncated system the recovered potential is presented on Fig. 7.11. The maximum absolute error in this case was 0.097 attained at the origin.

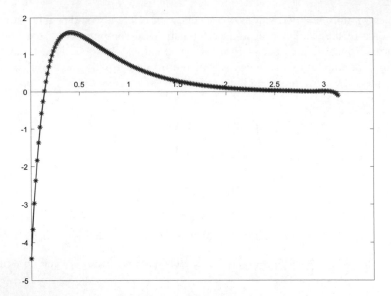

Fig. 7.8 With one equation only the recovered potential approximates the exact $q_2(x)$ with the maximal absolute error 0.344 attained at the origin

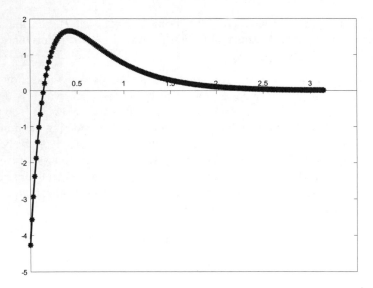

Fig. 7.9 With five equations the recovered potential $q_2(x)$ practically coincides with the exact one

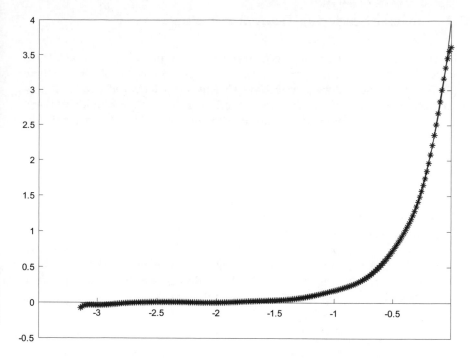

Fig. 7.10 The recovered potential $q_1(x)$ with 15 equations in the truncated system

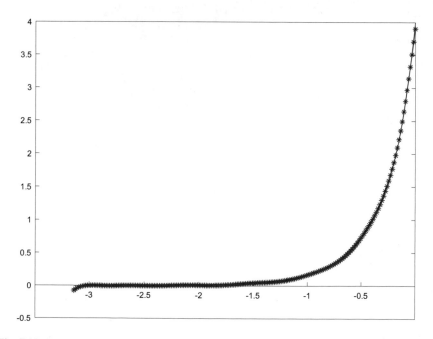

Fig. 7.11 The recovered potential $q_1(x)$ with 25 equations in the truncated system

7.14   Inverse Scattering Transform Method

The inverse scattering transform method is one of the most important and spectacular discoveries of mathematical physics of the second half of the twentieth century. The method allows one to solve Cauchy's problems for a wide class of important nonlinear physical equations by reducing them to the solution of direct and inverse scattering problems for linear differential equations. For example, the Cauchy problem for the nonlinear partial differential equation

$$u_t + u_{xxx} - 6uu_x = 0, \qquad (7.138)$$

which arises in modeling wave propagation on shallow water surfaces and is known as the **Korteweg–de Vries equation**, can be solved by reducing it to the direct and inverse scattering problems considered in Sects. 7.10 and 7.13, respectively. The inverse scattering transform method is applicable to other important partial differential equations, such as the **nonlinear Schrödinger equation** whose principal applications are to the propagation of light in nonlinear optical fibers (see, e.g., [79])

$$iu_t + u_{xx} \pm 2ku \, |u|^2 = 0.$$

Schematically, for the Korteweg–de Vries equation the inverse scattering transform method can be presented as follows. Given $q(x)$. It is required to solve the Korteweg–de Vries equation (7.138) with the initial condition

$$u(x, 0) = q(x). \qquad (7.139)$$

The first step is to solve the direct scattering problem for $q(x)$, formulated in Sect. 7.10. This produces the scattering data corresponding to $t = 0$ (see the first line of Diagram 1). Then a simple linear evolution law is applied to the scattering data which gives us the scattering data for all $t > 0$. Namely,

$$J^{\pm}(t) = \left\{ s^{\pm}(t, \rho), \ \rho \in \mathbb{R}; \ \lambda_n(t), \alpha_n^{\pm}(t), \ n = 1, \ldots, N \right\},$$

where

$$s^{\pm}(t, \rho) = s^{\pm}(0, \rho)e^{\pm 8i\rho^3 t}, \quad \lambda_n(t) = \lambda_n(0), \quad \alpha_n^{\pm}(t) = \alpha_n^{\pm}(0)e^{\pm 8\tau_n^3 t} \quad (\lambda_n = -\tau_n^2).$$

Finally, for every instant $t > 0$ for which the solution $u(x, t)$ of the Cauchy problem (7.138), (7.139) is to be computed, an inverse scattering problem (see Sect. 7.13) should be solved for a corresponding set of the scattering data $J^+(t)$ or $J^-(t)$. Then the solution of this inverse problem is precisely the sought for solution $u(x, t)$.

Diagram 1   Schematic representation of the IST

$$u(x, 0) = q(x) \quad \begin{array}{c} -y''+q(x)y=\lambda y \\ ----\!-\!-\!- \longrightarrow \\ \text{direct scattering problem} \end{array} \quad J^+ or\, J^-(t=0)$$

$$\begin{array}{c} \mid \\ \text{linear evolution law} \\ \downarrow \end{array}$$

$$u(x, t) \quad \begin{array}{c} \text{inverse scattering problem} \\ \longleftarrow ----\!-\!- \end{array} \quad J^+ or\, J^-(t)$$

Often this procedure is referred to as a direct and inverse **nonlinear Fourier transform** [1].

The method of solution of the direct and inverse scattering problems presented in Sects. 7.10 and 7.13, thus allows us to solve such nonlinear equations as the Korteweg–de Vries equation (7.138) among others by the inverse scattering transform method.

We conclude this part of the book by commenting that the approach presented here for solving inverse spectral problems is direct, simple in its numerical realization and allows one to recover the potential from the spectral or scattering data quickly and accurately. It was proposed first in [45] for recovering the Sturm-Liouville problem from its spectral density function on a finite interval and in [46] for solving the inverse scattering problem on the line and developed further in [25] for solving the inverse Sturm-Liouville problem on the half-line, in [37] for solving the inverse scattering problem on the half-line, in [60] for solving the inverse quantum scattering problem for the perturbed Bessel equation and in [52] for a variety of the inverse Sturm-Liouville problems on a finite interval. Inverse spectral problems, in general, are computationally challenging. Hence, the mere existence of a universal, simple, and direct approach for their solution is surprising. However, this advancement is based in good measure on a well-developed and beautiful theory and thus stands on the shoulders of giants, such as Ch. Sturm, J. Liouville, H. Weyl, D. Hilbert, V. A. Steklov, J. Delsarte, G. Borg, E. Ch. Titchmarsh, A. Ya. Povzner, I. M. Gelfand, B. M. Levitan, V. A. Marchenko, B. Ya. Levin and many other brilliant mathematicians and physicists who contributed to its creation.

Chapter 8
Boundary Value Problems for the Heat Equation

One of the aims of this chapter is to prove the uniqueness theorems for some fundamental boundary value problems for the heat equation. Since their proof is based on the maximum (minimum) principle for solutions of the heat equation, we begin by deducing this principle.

8.1 Maximum and Minimum Principle for the Heat Equation

For the heat equation

$$\frac{\partial u}{\partial t} = a^2 \left(\frac{\partial^2 u}{\partial x^2} + \frac{\partial^2 u}{\partial y^2} + \frac{\partial^2 u}{\partial z^2} \right) \tag{8.1}$$

a maximum principle analogous to that for analytic functions is valid. Let us agree on some notations. Let Ω be a bounded simply connected domain in $\mathbb{R}^3$, $(x, y, z) \in \Omega$, and the variable t takes values from 0 to T. Consider a cylinder C_T in the space of four variables (x, y, z, t), whose base is the domain Ω, and the generating lines are parallel to the t-axis. Thus,

$$C_T = \{(x, y, z, t) : (x, y, z) \in \Omega, 0 < t < T\}.$$

The corresponding closed cylinder is denoted by $\overline{C}_T$,

$$\overline{C}_T = \{(x, y, z, t) : (x, y, z) \in \overline{\Omega}, 0 \le t \le T\},$$

where $\overline{\Omega} = \Omega \cup S$, and S is the boundary of Ω in the three-dimensional space (see Fig. 8.1). Finally, let Γ denote the lower base of the cylinder together with its side surface ("the glass"),

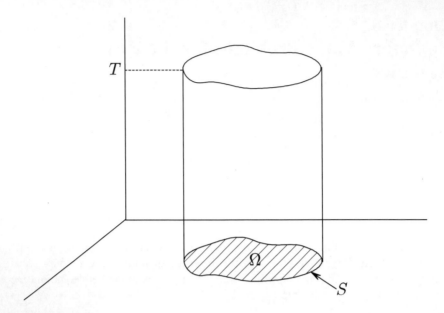

Fig. 8.1 Cylinder C_T

$$\Gamma = \{(x, y, z, t) : (x, y, z) \in \overline{\Omega},\, t = 0\} \cup \{(x, y, z, t) : (x, y, z) \in S,\, 0 \leq t \leq T\}.$$

The following theorem is valid.

Theorem 109 (The Maximum and Minimum Principle for the Heat Equation)
If the function $u(x, y, z, t)$ satisfies the heat equation (8.1) in the cylinder C_T and is continuous up to the boundary of the cylinder, then u attains its maximal and minimal values on Γ, that is on the base of the cylinder ($t = 0$) or on its side surface ($(x, y, z) \in S$).

Proof Let M denote the maximum value of the function u in $\overline{C}_T$ and m its maximum value on Γ,

$$M = \max_{(x,y,z,t) \in \overline{C}_T} u, \quad m = \max_{(x,y,z,t) \in \Gamma} u.$$

Clearly, $M \geq m$. We want to prove that in fact $M = m$. Suppose the opposite, that $M > m$, and let $P_0 = (x_0, y_0, z_0, t_0)$ be the point where this maximum is attained. Obviously, $(x_0, y_0, z_0) \in \Omega, 0 < t \leq T$.

Let us introduce an auxiliary function

$$v(x, y, z, t) = u(x, y, z, t) + \frac{M - m}{6d^2}\left((x - x_0)^2 + (y - y_0)^2 + (z - z_0)^2\right),$$

where d is the diameter of the domain Ω. We have

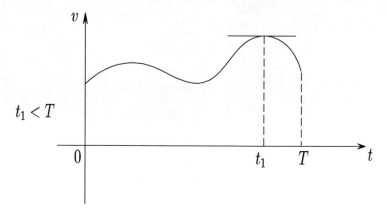

Fig. 8.2 Maximum when $t_1 < T$

$$v(x, y, z, t)|_\Gamma \leq m + \frac{M - m}{6} = \frac{M}{6} + \frac{5}{6}m < M.$$

Here we took into account that $(x - x_0)^2 + (y - y_0)^2 + (z - z_0)^2 \leq d^2$. Thus,

$$v(x, y, z, t)|_\Gamma < M.$$

On the other hand,

$$v(x_0, y_0, z_0, t_0) = M$$

and hence the function v does not attain its maximum on Γ but at some other point $P_1 = (x_1, y_1, z_1, t_1)$, such that $(x_1, y_1, z_1) \in \Omega$, $0 < t_1 \leq T$. The following necessary conditions of a maximum must be fulfilled at P_1:

$$\frac{\partial^2 v}{\partial x^2}, \frac{\partial^2 v}{\partial y^2}, \frac{\partial^2 v}{\partial z^2} \leq 0, \quad \frac{\partial v}{\partial t} \geq 0$$

and, moreover, $\frac{\partial v}{\partial t}\big|_{P_1} = 0$ if $t_1 < T$ (see Fig. 8.2) and $\frac{\partial v}{\partial t}\big|_{P_1} > 0$ if $t_1 = T$ (see Fig. 8.3).

Consequently,

$$\frac{\partial v}{\partial t} - a^2 \left(\frac{\partial^2 v}{\partial x^2} + \frac{\partial^2 v}{\partial y^2} + \frac{\partial^2 v}{\partial z^2} \right) \geq 0$$

at the point P_1. At the same time,

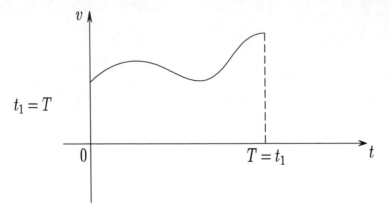

Fig. 8.3 Maximum when $t_1 = T$

$$\left(\frac{\partial v}{\partial t} - a^2 \Delta v\right)\Bigg|_{P_1} = \left(\frac{\partial u}{\partial t} - a^2 \Delta u\right)\Bigg|_{P_1} - a^2 \frac{M - m}{d^2} = -a^2 \frac{M - m}{d^2} < 0$$

which is a contradiction.

The second part of the theorem, concerning the minimum value of the solution u can be proved analogously, because, if u specified by the conditions of the theorem attains its minimum at some point, then the function $-u$, which also satisfies the conditions of the theorem attains there its maximum.   ∎

The maximum principle for the heat equation agrees with the second law of thermodynamics which establishes that spontaneous heat transfer always occurs from a region of higher temperature to a region of lower temperature. Hence in absence of sources of heat in a volume the temperature inside it can be increased only by providing a higher temperature on its surface.

Corollary 110 *The solution of the first boundary value problem for the heat equation is unique.*

Proof Consider the first boundary value problem for the heat equation

$$\begin{cases} \frac{\partial u}{\partial t} - a^2 \Delta u = 0, \\ u|_{t=0} = \varphi(x, y, z), \\ u|_S = \psi(x, y, z, t), \quad (x, y, z) \in S. \end{cases}$$

Suppose the opposite, that there exist two different solutions of the problem u_1 and u_2. Then their difference $u = u_1 - u_2$ satisfies the heat equation as well as homogeneous boundary conditions

$$\begin{cases} \frac{\partial u}{\partial t} - a^2 \Delta u = 0, \\ u|_{t=0} = 0, \\ u|_S = 0. \end{cases}$$

Thus, for all $T > 0$ the solution u is identically zero on the surface Γ of the cylinder C_T, and due to the maximum and minimum principle, we obtain that

$$u \equiv 0$$

for all points of C_T. Since T is arbitrary, $u \equiv 0$ for all $(x, y, z) \in \Omega$ and $0 < t < \infty$, that proves the corollary, that is $u_1 \equiv u_2$. ∎

Corollary 111 *The solution of the first boundary value problem for the heat equation depends continuously on the initial and boundary conditions.*

Proof Consider two problems

$$\begin{cases} \frac{\partial u}{\partial t} - a^2 \Delta u = 0, \\ u|_{t=0} = \varphi_1, \\ u|_S = \psi_1, \end{cases} \qquad \begin{cases} \frac{\partial u}{\partial t} - a^2 \Delta u = 0, \\ u|_{t=0} = \varphi_2, \\ u|_S = \psi_2. \end{cases}$$

Let u_1 and u_2 be the solutions of these problems, respectively. Then the function $u = u_1 - u_2$ is a solution of the problem

$$\begin{cases} \frac{\partial u}{\partial t} - a^2 \Delta u = 0, \\ u|_{t=0} = \varphi_1 - \varphi_2, \\ u|_S = \psi_1 - \psi_2. \end{cases}$$

Due to the maximum and minimum principle we have the estimate

$$\max |u| \leq \max \{|\varphi_1 - \varphi_2|, |\psi_1 - \psi_2|\},$$

that is,

$$\max |u_1 - u_2| \leq \max \{|\varphi_1 - \varphi_2|, |\psi_1 - \psi_2|\}.$$

Thus, if $\max |\varphi_1 - \varphi_2| < \varepsilon$ and $\max |\psi_1 - \psi_2| < \varepsilon$, then $\max |u_1 - u_2| < \varepsilon$ as well. ∎

Thus, we proved that the first boundary value problem for the heat equation is well posed.

8.2   Uniqueness Theorem for Unbounded Domains (on the Example of a Line)

Theorem 112 *The solution of the Cauchy problem for the heat equation in the whole space is unique.*

Proof Since an unbounded domain is considered, additionally to the Cauchy problem

$$\begin{cases} \frac{\partial u}{\partial t} - a^2 \Delta u = 0, \\ u|_{t=0} = \varphi(x, y, z) \end{cases}$$

we need to assume the boundedness of the solution in the whole space. That is, $|u(x, y, z, t)| < M$ for all $-\infty < x, y, z < \infty, 0 \le t < \infty$. For simplicity, let us give the proof of the theorem in the case of a line.

$$\frac{\partial u}{\partial t} - a^2 \frac{\partial^2 u}{\partial x^2} = 0, \quad u(x, 0) = \varphi(x).$$

Let u_1 and u_2 be two different solutions of this Cauchy problem. Denote $u = u_1 - u_2$. Then u is a solution of the problem

$$\frac{\partial u}{\partial t} - a^2 \frac{\partial^2 u}{\partial x^2} = 0, \quad u(x, 0) = 0.$$

Since $|u_1(x, t)| < M_1$ and $|u_2(x, t)| < M_2$, we have that $|u| < M = M_1 + M_2$.

The maximum principle is not applicable directly because the domain under consideration is infinite. In order to apply the maximum principle we introduce the finite domain $|x| \le L, 0 \le t \le T$, in which we consider the function

$$v(x, t) = \frac{2(M_1 + M_2)}{L^2} \left(\frac{x^2}{2} + a^2 t \right).$$

Obviously,

$$\frac{\partial v}{\partial t} - a^2 \frac{\partial^2 v}{\partial x^2} = \frac{2(M_1 + M_2)}{L^2} \left(a^2 - a^2 \right) = 0$$

and thus $v(x, t)$ is a solution of the heat equation. Moreover,

$$v(x, 0) \ge u(x, 0) = 0,$$

$$v(\pm L, t) \ge M_1 + M_2 > |u(\pm L, t)|.$$

Applying the maximum principle to the difference of the functions $v(x, t)$ and $\pm u(x, t)$ in the domain under consideration we obtain

$$v(x, t) - u(x, t) \geq 0$$

and

$$v(x, t) + u(x, t) \geq 0$$

for $|x| \leq L$ and $0 \leq t \leq T$. Hence for all (x, t) in this domain the inequality holds

$$|u(x, t)| \leq v(x, t) = \frac{2(M_1 + M_2)}{L^2} \left(\frac{x^2}{2} + a^2 t \right).$$

Now fix the point (x, t) and tend L to infinity. Then $|u(x, t)| \leq 0$, that is, $u(x, t) = 0$ and hence $u_1(x, t) = u_2(x, t)$ for all (x, t) from the half-plane $-\infty < x < \infty$, $0 \leq t < \infty$.   ∎

8.3   Heat Transfer in an Infinite Rod

Having the uniqueness theorem, let us now construct a solution of the one-dimensional problem

$$\begin{cases} \frac{\partial u}{\partial t} - a^2 \frac{\partial^2 u}{\partial x^2} = 0, \\ u(x, 0) = \varphi(x), \\ |u(x, t)| < \infty. \end{cases} \tag{8.2}$$

Let us apply the Fourier method of separation of variables. We look for a solution of the heat equation in the form $u(x, t) = X(x)T(t)$. Then $T'X = a^2 T X''$ or

$$\frac{T'}{a^2 T} = \frac{X''}{X} = -\lambda^2.$$

Solving the equations

$$X'' + \lambda^2 X = 0$$

and

$$T' + (\lambda a)^2 T = 0,$$

we obtain

$$X(x) = \widetilde{A}(\lambda) \cos \lambda x + \widetilde{B}(\lambda) \sin \lambda x$$

and

$$T(t) = C(\lambda)e^{-a^2\lambda^2 t}.$$

Since no boundary conditions are specified, the value of the parameter λ remains arbitrary, and for all $\lambda \in \mathbb{R}$ the function

$$u_\lambda(x, t) = X(x)T(t) = e^{-a^2\lambda^2 t} \left(A(\lambda) \cos \lambda x + B(\lambda) \sin \lambda x \right)$$

is a solution of the heat equation. Due to the arbitrariness of λ and according to the superposition principle, it is natural to write a general solution of the heat equation in the form

$$u(x, t) = \int_{-\infty}^{\infty} u_\lambda(x, t)\, d\lambda = \int_{-\infty}^{\infty} e^{-a^2\lambda^2 t} \left(A(\lambda) \cos \lambda x + B(\lambda) \sin \lambda x \right) d\lambda.$$

$$(8.3)$$

It remains to choose $A(\lambda)$ and $B(\lambda)$ such that the initial condition $u(x, 0) = \varphi(x)$ be fulfilled. Thus,

$$\int_{-\infty}^{\infty} \left(A(\lambda) \cos \lambda x + B(\lambda) \sin \lambda x \right) d\lambda = \varphi(x).$$

To obtain $A(\lambda)$ and $B(\lambda)$ in terms of the initial data, let us apply the Fourier formula for $\varphi(x)$. We have

$$\varphi(x) = \frac{1}{2\pi} \int_{-\infty}^{\infty} d\lambda \int_{-\infty}^{\infty} \varphi(\xi) \cos \lambda(\xi - x)\, d\xi.$$

Then

$$\int_{-\infty}^{\infty} \left(A(\lambda) \cos \lambda x + B(\lambda) \sin \lambda x \right) d\lambda = \frac{1}{2\pi} \int_{-\infty}^{\infty} \left(\int_{-\infty}^{\infty} (\cos \lambda \xi)\, \varphi(\xi) d\xi \right) \cos \lambda x\, d\lambda$$

$$+ \frac{1}{2\pi} \int_{-\infty}^{\infty} \left(\int_{-\infty}^{\infty} (\sin \lambda \xi)\, \varphi(\xi) d\xi \right) \sin \lambda x\, d\lambda.$$

Thus,

$$\begin{cases} A(\lambda) = \frac{1}{2\pi} \int_{-\infty}^{\infty} (\cos \lambda \xi)\, \varphi(\xi) d\xi, \\ B(\lambda) = \frac{1}{2\pi} \int_{-\infty}^{\infty} (\sin \lambda \xi)\, \varphi(\xi) d\xi. \end{cases} \qquad (8.4)$$

Substituting (8.4) into (8.3) we obtain

$$u(x,t) = \frac{1}{2\pi} \int_{-\infty}^{\infty} e^{-a^2\lambda^2 t} d\lambda \int_{-\infty}^{\infty} (\cos \lambda x \cos \lambda \xi + \sin \lambda x \sin \lambda \xi)\, \varphi(\xi) d\xi$$

$$= \frac{1}{2\pi} \int_{-\infty}^{\infty} e^{-a^2\lambda^2 t} d\lambda \int_{-\infty}^{\infty} \cos \lambda(x - \xi)\varphi(\xi) d\xi = \frac{1}{2\pi} \int_{-\infty}^{\infty} \varphi(\xi) d\xi \int_{-\infty}^{\infty} \cos \lambda(x - \xi) e^{-a^2\lambda^2 t} d\lambda.$$

Let us calculate the inner integral

$$\mathcal{I} = \frac{1}{2\pi} \int_{-\infty}^{\infty} \cos \lambda(x - \xi) e^{-a^2\lambda^2 t} d\lambda = \frac{1}{\pi} \int_{0}^{\infty} \cos \lambda(x - \xi) e^{-a^2\lambda^2 t} d\lambda.$$

Let $a^2\lambda^2 t =: z^2$, that is, $\lambda = \frac{z}{a\sqrt{t}}$. Then

$$\mathcal{I} = \frac{1}{\pi} \int_{0}^{\infty} e^{-z^2} \cos\left(z \frac{x - \xi}{a\sqrt{t}}\right) \frac{dz}{a\sqrt{t}}.$$

Denote $\frac{x-\xi}{a\sqrt{t}} = \mu$ and consider the integral

$$\mathcal{I}_0(\mu) = a\sqrt{t}\,\mathcal{I} = \frac{1}{\pi} \int_{0}^{\infty} e^{-z^2} \cos \mu z\, dz.$$

We have

$$\mathcal{I}_0'(\mu) = -\frac{1}{\pi} \int_{0}^{\infty} z e^{-z^2} \sin \mu z\, dz.$$

Differentiation under the integral sign is justified because the integrand and its derivative with respect to μ are both continuous with respect to μ and z in the region $-\infty < \mu < \infty$, $z \geq 0$, the integral $\mathcal{I}_0(\mu)$ converges, and $\mathcal{I}_0'(\mu)$ converges uniformly due to the Weierstrass criterion:

$$\left| \int_{0}^{\infty} z e^{-z^2} \sin \mu z\, dz \right| < \int_{0}^{\infty} z e^{-z^2}\, dz < \infty.$$

Integrating $\mathcal{I}_0'(\mu)$ by parts we obtain

$$\mathcal{I}_0'(\mu) = \frac{1}{2\pi} \int_{0}^{\infty} \sin \mu z\, d\left(e^{-z^2}\right) = \frac{1}{2\pi} \sin \mu z\, e^{-z^2} \Big|_{0}^{\infty} - \frac{\mu}{2\pi} \int_{0}^{\infty} e^{-z^2} \cos \mu z\, dz = -\frac{\mu}{2}\mathcal{I}_0(\mu).$$

Hence the function $\mathcal{I}_0(\mu)$ satisfies the differential equation

$$\mathcal{I}_0'(\mu) + \frac{\mu}{2}\mathcal{I}_0(\mu) = 0.$$

Its general solution has the form

$$\mathcal{I}_0(\mu) = Ce^{-\frac{\mu^2}{4}}.$$

Note that

$$\mathcal{I}_0(0) = \frac{1}{\pi} \int_0^\infty e^{-z^2}\, dz = \frac{1}{\pi}\frac{\sqrt{\pi}}{2} = \frac{1}{2\sqrt{\pi}}.$$

Thus,

$$\mathcal{I}_0(\mu) = \frac{1}{2\sqrt{\pi}} e^{-\frac{\mu^2}{4}}.$$

Finally, for the integral $\mathcal{I}$ we obtain

$$\mathcal{I} = \frac{1}{a\sqrt{t}}\mathcal{I}_0(\mu) = \frac{1}{2a\sqrt{\pi t}} e^{-\frac{(x-\xi)^2}{4a^2 t}}.$$

Hence $u(x, t)$ has the form

$$u(x, t) = \frac{1}{2a\sqrt{\pi t}} \int_{-\infty}^{\infty} \varphi(\xi) e^{-\frac{(x-\xi)^2}{4a^2 t}}\, d\xi. \tag{8.5}$$

This representation is known as the **Poisson integral** or **Poisson formula**.
   Denote

$$G(x, \xi, t) = \frac{1}{2a\sqrt{\pi t}} e^{-\frac{(x-\xi)^2}{4a^2 t}}.$$

Then

$$u(x, t) = \int_{-\infty}^{\infty} G(x, \xi, t)\varphi(\xi)d\xi.$$

The function $G(x, \xi, t)$ is called the **fundamental solution of the heat equation**. It is easy to verify that for any ξ fixed the function $G(x, \xi, t)$ satisfies the heat equation. Indeed,

$$\frac{\partial G}{\partial x} = -\frac{2(x-\xi)}{\sqrt{\pi}\,(4a^2 t)^{3/2}} e^{-\frac{(x-\xi)^2}{4a^2 t}},$$

$$\frac{\partial^2 G}{\partial x^2} = \frac{2}{\sqrt{\pi}}\left(-\frac{1}{(4a^2 t)^{3/2}} + \frac{2(x-\xi)^2}{(4a^2 t)^{5/2}}\right) e^{-\frac{(x-\xi)^2}{4a^2 t}},$$

$$\frac{\partial G}{\partial t} = \frac{1}{2a\sqrt{\pi}}\left(-\frac{1}{2t^{3/2}} + \frac{(x-\xi)^2}{4a^2 t^2\sqrt{t}}\right) e^{-\frac{(x-\xi)^2}{4a^2 t}}.$$

From here it is easy to see that $\frac{\partial G}{\partial t} = a^2 \frac{\partial^2 G}{\partial x^2}$.

Let us justify the Poisson formula. For this purpose we need to prove that the function (8.5) is indeed a solution of the Cauchy problem (8.2).

The Poisson integral (8.5) can be differentiated under the integral sign due to the uniform convergence of the involved integrals. Thus,

$$\left(\frac{\partial}{\partial t} - a^2 \frac{\partial^2}{\partial x^2}\right) u = \int_{-\infty}^{\infty} \left(\frac{\partial}{\partial t} - a^2 \frac{\partial^2}{\partial x^2}\right) (G(x,\xi,t)\varphi(\xi)) \, d\xi$$

$$= \int_{-\infty}^{\infty} \varphi(\xi) \left(\frac{\partial G}{\partial t} - a^2 \frac{\partial^2 G}{\partial x^2}\right) d\xi = 0.$$

Let us show that if $|\varphi(x)| \le M < \infty$, then $|u(x,t)| \le M$ as well. Indeed,

$$|u(x,t)| \le \int_{-\infty}^{\infty} |G(x,\xi,t)| \, |\varphi(\xi)| \, d\xi \le M \frac{1}{2a\sqrt{\pi t}} \int_{-\infty}^{\infty} e^{-\frac{(x-\xi)^2}{4a^2 t}} \, d\xi = \frac{M}{\sqrt{\pi}} \int_{-\infty}^{\infty} e^{-z^2} dz = M.$$

Finally, we prove that the solution $u(x,t)$ satisfies the initial condition. Since, apparently the value $t = 0$ cannot be simply substituted into the expression (8.5) the proof is somewhat cumbersome. We begin by showing that for small enough t, close to zero, the inequality

$$|u(x,t) - \varphi(x)| < \varepsilon$$

holds for small $\varepsilon > 0$.

Notice that

$$u(x,t) = \frac{1}{\sqrt{\pi}} \int_{-\infty}^{\infty} \varphi(x + 2az\sqrt{t}) e^{-z^2} dz$$

with

$$z = -\frac{(x-\xi)}{2a\sqrt{t}},$$

and

$$\varphi(x) = \varphi(x) \cdot 1 = \varphi(x) \cdot \frac{1}{\sqrt{\pi}} \int_{-\infty}^{\infty} e^{-z^2} dz = \frac{1}{\sqrt{\pi}} \int_{-\infty}^{\infty} \varphi(x) e^{-z^2} dz.$$

Hence

$$|u(x,t) - \varphi(x)| = \frac{1}{\sqrt{\pi}} \left| \int_{-\infty}^{\infty} e^{-z^2} \left(\varphi(x + 2az\sqrt{t}) - \varphi(x)\right) dz \right|$$

$$\leq \frac{1}{\sqrt{\pi}} \int_{-\infty}^{\infty} e^{-z^2} \left| \varphi(x + 2az\sqrt{t}) - \varphi(x) \right| dz$$

$$= \frac{1}{\sqrt{\pi}} \left(\int_{-\infty}^{-N} + \int_{-N}^{N} + \int_{N}^{\infty} \right).$$

Obviously,

$$\frac{1}{\sqrt{\pi}} \int_{-\infty}^{-N} e^{-z^2} \left| \varphi(x + 2az\sqrt{t}) - \varphi(x) \right| dz \leq \frac{2M}{\sqrt{\pi}} \int_{-\infty}^{-N} e^{-z^2} dz \leq \frac{\varepsilon}{3},$$

$$\frac{1}{\sqrt{\pi}} \int_{N}^{\infty} e^{-z^2} \left| \varphi(x + 2az\sqrt{t}) - \varphi(x) \right| dz \leq \frac{2M}{\sqrt{\pi}} \int_{N}^{\infty} e^{-z^2} dz \leq \frac{\varepsilon}{3},$$

where N is chosen large enough, so that

$$\frac{1}{\sqrt{\pi}} \int_{-\infty}^{-N} e^{-z^2} dz < \frac{\varepsilon}{6M}, \qquad \frac{1}{\sqrt{\pi}} \int_{N}^{\infty} e^{-z^2} dz < \frac{\varepsilon}{6M}.$$

It remains to estimate the integral

$$\frac{1}{\sqrt{\pi}} \int_{-N}^{N} e^{-z^2} \left| \varphi(x + 2az\sqrt{t}) - \varphi(x) \right| dz.$$

Since the function $\varphi(x)$ is continuous, for $|z| \leq N$ and for all t sufficiently close to zero we have

$$\left| \varphi(x + 2az\sqrt{t}) - \varphi(x) \right| < \frac{\varepsilon}{3}.$$

Then

$$\frac{1}{\sqrt{\pi}} \int_{-N}^{N} e^{-z^2} \left| \varphi(x + 2az\sqrt{t}) - \varphi(x) \right| dz < \frac{\varepsilon}{3} \frac{1}{\sqrt{\pi}} \int_{-N}^{N} e^{-z^2} dz < \frac{\varepsilon}{3} \frac{1}{\sqrt{\pi}} \int_{-\infty}^{\infty} e^{-z^2} dz = \frac{\varepsilon}{3}.$$

Thus,

$$|u(x, t) - \varphi(x)| < \varepsilon$$

for all t sufficiently close to zero. Since ε is arbitrary, it follows that

$$\lim_{t \to 0} u(x, t) = \varphi(x).$$

This finishes the proof of the fact that the function (8.5) is the unique solution of the Cauchy problem (8.2).

8.4   Corollaries from the Poisson Formula

Corollary 113 *Heat flows along the rod instantaneously. Indeed, if initially the rod temperature equals zero everywhere except for an interval (α, β) where the temperature is given by some function $\varphi(x) > 0$, then, by Poisson's formula the temperature distribution for subsequent times has the form*

$$u(x, t) = \frac{1}{2a\sqrt{\pi t}} \int_\alpha^\beta \varphi(\xi) e^{-\frac{(x-\xi)^2}{4a^2 t}} \, d\xi. \tag{8.6}$$

From here it can be seen that $u(x, t) > 0$ at any arbitrarily distant point x and after an arbitrarily short time t. This infinite velocity of the heat flux is explained by inexactness of the physical premises that lie at the basis of the theory of heat flow. The Poisson formula gives an approximation to the real physical phenomenon; however, its imprecision is relatively small since the value of the integral is small for large x and small t. Thus, although the heat comes instantaneously to any distant point x, for a small t it is negligible.

Corollary 114 *The solution of the problem of heat flow in an infinite rod is a continuously differentiable function with respect to x and t any number of times, independently of the differentiability of the initial data. This smoothness of the solution independent of the smoothness of the initial data differs the heat equation from, e.g., the wave equation.*

Corollary 115 (Physical Meaning of the Fundamental Solution) *Let x_0 be an arbitrary point of the infinite rod and $(x_0 - h, x_0 + h)$ its neighborhood. Let an initial temperature distribution have the form*

$$\varphi(x) = \begin{cases} u_0 = \text{Const}, & x \in (x_0 - h, x_0 + h), \\ 0, & x \notin (x_0 - h, x_0 + h). \end{cases}$$

Physically this means that at an initial instant an amount of heat $Q = 2hc\rho u_0$ is transferred to the element of the rod $(x_0 - h, x_0 + h)$, where c is the thermal capacity and ρ is the density of the rod. The temperature distribution has then the form

$$u(x, t) = \int_{x_0-h}^{x_0+h} G(x, \xi, t) u_0 d\xi = \frac{Q}{2hc\rho} \int_{x_0-h}^{x_0+h} G(x, \xi, t) d\xi.$$

Let tend h to zero, while assuming Q to remain constant. This means that the same amount of heat is transferred to an ever smaller segment. In the limit we obtain a point source of heat. Due to the mean value theorem we have

$$u(x, t) = \lim_{h \to 0} \frac{Q}{2hc\rho} 2hG(x, \overline{\xi}, t),$$

where $x_0 - h < \overline{\xi} < x_0 + h$, and thus

$$u(x, t) = \frac{Q}{c\rho} G(x, x_0, t).$$

Hence

$$u(x, t) = \frac{Q}{2ac\rho\sqrt{\pi t}} e^{-\frac{(x-x_0)^2}{4a^2 t}}$$

is the temperature at the point x and at the instant t generated by a point source of heat located at the instant $t = 0$ at the point x_0 and of capacity Q. The general solution

$$u(x, t) = \int_{-\infty}^{\infty} G(x, \xi, t)\varphi(\xi)d\xi$$

can be interpreted as a result of summation of solutions generated by corresponding point sources.

Remark 116 The Cauchy problem of heat flow on the plane or in the space can be studied similarly to the case of the rod (i.e., to the one-dimensional case). Let

$$\frac{\partial u}{\partial t} - a^2 \left(\frac{\partial^2 u}{\partial x_1^2} + \frac{\partial^2 u}{\partial x_2^2} + \ldots + \frac{\partial^2 u}{\partial x_n^2} \right) = 0,$$

$$u|_{t=0} = \varphi(M), \quad M = (x_1, x_2, \ldots, x_n).$$

Then it can be shown that

$$u(x_1, x_2, \ldots, x_n, t) = \int_{-\infty}^{\infty} \int_{-\infty}^{\infty} \ldots \int_{-\infty}^{\infty} \varphi(P)G(M, P, t)dV_P,$$

where the fundamental solution $G(M, P, t)$ has the form

$$G(M, P, t) = \frac{1}{(2a\sqrt{\pi t})^n} e^{-\frac{r_{PM}^2}{4a^2 t}},$$

where $P = (\xi_1, \xi_2, \ldots, \xi_n)$, $M = (x_1, x_2, \ldots, x_n)$ and

$$r_{PM} = \sqrt{(x_1 - \xi_1)^2 + (x_2 - \xi_2)^2 + \ldots + (x_n - \xi_n)^2}.$$

8.5   Heat Flow in a Semi-Infinite Rod

Consider the problem of heat flow in a semi-infinite rod

$$u_t = a^2 u_{xx}, \quad 0 \leq x < \infty.$$

In this case besides the initial temperature distribution $u(x, 0) = \varphi(x)$ a boundary condition at the endpoint $x = 0$ needs to be imposed. Let us consider first the situation when at the endpoint of the rod a zero temperature is maintained and thus the problem has the form

$$\begin{cases} \frac{\partial u}{\partial t} - a^2 \frac{\partial^2 u}{\partial x^2} = 0, & x > 0, \\ u(0, t) = 0, \\ u(x, 0) = \varphi(x). \end{cases} \tag{8.7}$$

Let us extend the function $\varphi(x)$ defined for $x > 0$ onto the negative half-line $x < 0$ in an arbitrary way. Then we can write down the Poisson formula

$$u(x, t) = \frac{1}{2a\sqrt{\pi t}} \int_{-\infty}^{\infty} \varphi(\xi) e^{-\frac{(x-\xi)^2}{4a^2 t}} \, d\xi = \frac{1}{2a\sqrt{\pi t}} \int_{0}^{\infty} \left(\varphi(\xi) e^{-\frac{(x-\xi)^2}{4a^2 t}} + \varphi(-\xi) e^{-\frac{(x+\xi)^2}{4a^2 t}} \right) d\xi. \tag{8.8}$$

This function satisfies the Cauchy problem

$$\frac{\partial u}{\partial t} - a^2 \frac{\partial^2 u}{\partial x^2} = 0, \tag{8.9}$$

$$u(x, 0) = \varphi(x) \tag{8.10}$$

for all x: $-\infty < x < \infty$ and hence the initial condition is fulfilled for $x > 0$. It remains to fulfill the homogeneous boundary condition:

$$u(0, t) = \frac{1}{2a\sqrt{\pi t}} \int_{0}^{\infty} (\varphi(\xi) + \varphi(-\xi)) e^{-\frac{\xi^2}{4a^2 t}} \, d\xi = 0.$$

This can be achieved by extending $\varphi(x)$ onto the negative half-line in a special way. Indeed, extending $\varphi(x)$ as an odd function, $\varphi(x) = -\varphi(-x)$, we obtain that the solution of the problem (8.7) is given by the formula

$$u(x, t) = \frac{1}{2a\sqrt{\pi t}} \int_{0}^{\infty} \varphi(\xi) \left(e^{-\frac{(x-\xi)^2}{4a^2 t}} - e^{-\frac{(x+\xi)^2}{4a^2 t}} \right) d\xi. \tag{8.11}$$

Next, consider the problem in the case when the end of the rod is thermally insulated

$$\begin{cases} \frac{\partial u}{\partial t} - a^2 \frac{\partial^2 u}{\partial x^2} = 0, & x > 0, \\ \frac{\partial u(0,t)}{\partial x} = 0, \\ u(x,0) = \varphi(x). \end{cases} \qquad (8.12)$$

The function (8.8) still satisfies the Cauchy problem (8.9), (8.10), so it remains to satisfy the boundary condition:

$$\frac{\partial u(0,t)}{\partial x} = \frac{1}{2a\sqrt{\pi t}} \int_0^\infty \frac{\xi}{2a^2 t} \left(\varphi(\xi) - \varphi(-\xi) \right) e^{-\frac{\xi^2}{4a^2 t}} d\xi = 0.$$

This is valid if $\varphi(x)$ is extended onto the negative half-line as an even function, i.e., $\varphi(\xi) = \varphi(-\xi)$. Thus,

$$u(x,t) = \frac{1}{2a\sqrt{\pi t}} \int_0^\infty \varphi(\xi) \left(e^{-\frac{(x-\xi)^2}{4a^2 t}} + e^{-\frac{(x+\xi)^2}{4a^2 t}} \right) d\xi$$

is a solution of the problem (8.12).

Consider a more general model with a given thermal regime maintained at the endpoint of the rod

$$\begin{cases} \frac{\partial u}{\partial t} - a^2 \frac{\partial^2 u}{\partial x^2} = 0, & x > 0, \\ u(0,t) = \gamma(t), \\ u(x,0) = \varphi(x). \end{cases} \qquad (8.13)$$

This problem can be split into two problems

$$\begin{cases} \frac{\partial w}{\partial t} - a^2 \frac{\partial^2 w}{\partial x^2} = 0, & x > 0, \\ w(0,t) = 0, \\ w(x,0) = \varphi(x); \end{cases} \qquad \begin{cases} \frac{\partial v}{\partial t} - a^2 \frac{\partial^2 v}{\partial x^2} = 0, & x > 0, \\ v(0,t) = \gamma(t), \\ v(x,0) = 0, \end{cases} \qquad (8.14)$$

and

$$u = v + w.$$

The function w is defined by the formula (8.11), thus, it remains to find a way to solve the second problem in (8.14).

Lemma 117 *If $\gamma(t) = \gamma_0 = $ Const, then the solution of the second problem in (8.14) is the function*

$$v(x,t) = \gamma_0 \left(1 - \mathrm{erf}\left(\frac{x}{2a\sqrt{t}} \right) \right),$$

where erf *z is the Gauss error function defined by the formula*

$$\operatorname{erf} z = \frac{2}{\sqrt{\pi}} \int_0^z e^{-x^2} dx = \Phi(z).$$

Proof It is easy to verify that $v(x, t)$ satisfies the initial and boundary conditions. It remains to show that it satisfies the heat equation. It is sufficient to prove that the function $\operatorname{erf} z = \operatorname{erf} \left(\frac{x}{2a\sqrt{t}} \right)$ is a solution of the heat equation. The corresponding problem has the form

$$\begin{cases} \frac{\partial u}{\partial t} - a^2 \frac{\partial^2 u}{\partial x^2} = 0, & x > 0, \\ u(0, t) = 0, \\ u(x, 0) = 1. \end{cases} \tag{8.15}$$

The last two equalities are satisfied by $\operatorname{erf} z$. Indeed,

$$\operatorname{erf} \left(\frac{x}{2a\sqrt{t}} \right)\Big|_{x=0} = 0, \quad \operatorname{erf} \left(\frac{x}{2a\sqrt{t}} \right)\Big|_{t=0} = 1.$$

A solution of the problem (8.15) can be found by the formula (8.11) with $\varphi(\xi) \equiv 1$. We have

$$u(x, t) = \frac{1}{2a\sqrt{\pi t}} \left(\int_0^\infty e^{-\frac{(x-\xi)^2}{4a^2 t}} d\xi - \int_0^\infty e^{-\frac{(x+\xi)^2}{4a^2 t}} d\xi \right).$$

Making the change of variable $\frac{\xi - x}{2a\sqrt{t}} = s$ in the first integral and $\frac{\xi + x}{2a\sqrt{t}} = s$ in the second, we obtain that

$$u(x, t) = \frac{1}{\sqrt{\pi}} \left(\int_{-\frac{x}{2a\sqrt{t}}}^\infty e^{-s^2} ds - \int_{\frac{x}{2a\sqrt{t}}}^\infty e^{-s^2} ds \right)$$

$$= \frac{1}{\sqrt{\pi}} \int_{-\frac{x}{2a\sqrt{t}}}^{\frac{x}{2a\sqrt{t}}} e^{-s^2} ds = \frac{2}{\sqrt{\pi}} \int_0^{\frac{x}{2a\sqrt{t}}} e^{-s^2} ds$$

$$= \operatorname{erf} \left(\frac{x}{2a\sqrt{t}} \right) = \Phi \left(\frac{x}{2a\sqrt{t}} \right).$$

■

Thus, the second problem in (8.14) is solved for $\gamma(t) = \gamma_0$. Similarly, if

$$\gamma(t) = \begin{cases} 0, & t < \tau, \\ 1, & t > \tau, \end{cases}$$

(see Fig. 8.4) then the function

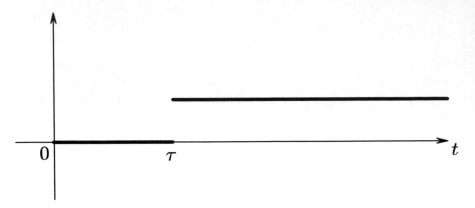

Fig. 8.4 γ equals zero until $t = \tau$, and one afterwards

$$v_\tau(x, t) = \begin{cases} 0, & 0 \le t \le \tau, \\ 1 - \mathrm{erf}\left(\frac{x}{2a\sqrt{t-\tau}}\right), & t \ge \tau \end{cases}$$

is a solution of the second problem in (8.14).

Furthermore, if $\gamma(t)$ equals one during a certain period of time and the rest of time is zero, i.e.,

$$\gamma(t) = \begin{cases} 0, & t < \tau \text{ or } t > \tau + \Delta\tau, \\ 1, & \tau < t < \tau + \Delta\tau, \end{cases}$$

(see Fig. 8.5) then the solution of the same problem is the function $v_\tau(x, t) - v_{\tau+\Delta\tau}(x, t)$. Indeed,

$$v_\tau(x, t) - v_{\tau+\Delta\tau}(x, t)|_{t<\tau,\, x=0} = 0,$$

$$v_\tau(x, t) - v_{\tau+\Delta\tau}(x, t)|_{t>\tau+\Delta\tau,\, x=0} = 0,$$

$$v_\tau(x, t) - v_{\tau+\Delta\tau}(x, t)|_{\tau<t<\tau+\Delta\tau,\, x=0} = 1,$$

and hence $v_\tau(0, t) - v_{\tau+\Delta\tau}(0, t) = \gamma(t)$. On the other hand

$$v_\tau(x, t) - v_{\tau+\Delta\tau}(x, t) \simeq -\frac{\partial v_\tau(x, t)}{\partial \tau} d\tau,$$

and thus for an arbitrary $\gamma(t)$ the solution has the form

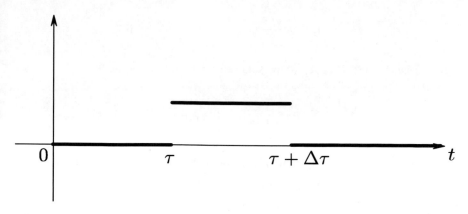

Fig. 8.5 γ equals one on a finite interval, outside of which it is zero

$$v(x, t) = -\int_0^t \gamma(t)\frac{\partial v_\tau(x, t)}{\partial \tau}d\tau. \tag{8.16}$$

Let us calculate the kernel $-\frac{\partial v_\tau(x,t)}{\partial \tau}$ for $\tau < t$. We have

$$-\frac{\partial v_\tau(x, t)}{\partial \tau} = \frac{\partial}{\partial \tau}\,\mathrm{erf}\left(\frac{x}{2a\sqrt{t - \tau}}\right) = \frac{\partial}{\partial \tau}\int_0^{\frac{x}{2a\sqrt{t-\tau}}}e^{-s^2}ds$$

$$= \frac{2}{\sqrt{\pi}}e^{-\frac{x^2}{4a^2(t-\tau)}}\frac{\partial}{\partial \tau}\left(\frac{x}{2a\sqrt{t - \tau}}\right)$$

$$= \frac{x}{2a\sqrt{\pi}\,(t - \tau)^{\frac{3}{2}}}e^{-\frac{x^2}{4a^2(t-\tau)}}.$$

Thus,

$$v(x, t) = \frac{x}{2a\sqrt{\pi}}\int_0^t \frac{\gamma(t)}{(t - \tau)^{\frac{3}{2}}}e^{-\frac{x^2}{4a^2(t-\tau)}}d\tau. \tag{8.17}$$

After the substitution $\xi = \frac{x}{2a\sqrt{t-\tau}}$ we obtain

$$v(x, t) = \frac{2}{\sqrt{\pi}}\int_{\frac{x}{2a\sqrt{t}}}^{\infty}\gamma(t - \frac{x^2}{4a^2\xi^2})e^{-\xi^2}d\xi. \tag{8.18}$$

Notice that the function $\frac{\partial v_\tau(x,t)}{\partial \tau}$ is related to the fundamental solution $G(x, \xi, t)$ by the equality

$$\frac{\partial v_\tau(x,t)}{\partial \tau} = -2a^2 \frac{\partial G(x,0,t-\tau)}{\partial x} = 2a^2 \left. \frac{\partial G}{\partial \xi} \right|_{\xi=0}.$$

Hence, instead of (8.17) or (8.18) the solution can be written in the form

$$v(x,t) = 2a^2 \int_0^t \gamma(t) \frac{\partial G(x,0,t-\tau)}{\partial x} d\tau. \tag{8.19}$$

Obviously, combining this solution of the second problem in (8.14) with the solution of the first problem we finally arrive at the solution of the problem (8.13).

Remark 118 Solution of the second problem in (8.14) was done as follows. First, a solution was obtained for a constant γ. Then with its aid the problem was solved for γ which is a nonzero constant on an interval $(\tau, \tau + \Delta\tau)$ and zero outside it. For an arbitrary γ the solution was obtained by a passage to the limit, dividing the half-axis into small intervals and considering γ a constant on each of them. So that for a variable $\gamma(t)$ the solution took the form (8.19). This procedure is known as **Duhamel's principle** and is often used when solving boundary value problems.

8.6   Heat Flow on a Finite Interval

Consider the problem of heat flow on a segment at the endpoints of which a zero temperature is maintained

$$\begin{cases} \frac{\partial u}{\partial t} - a^2 \frac{\partial^2 u}{\partial x^2} = 0, & 0 < x < l, \\ u(x,0) = \varphi(x), \\ u(0,t) = u(l,t) = 0, \end{cases} \tag{8.20}$$

where $\varphi(x)$ is supposed to be a continuous function possessing a piecewise continuous derivative. For the reason of compatibility of the initial and boundary conditions we need to assume additionally that $\varphi(0) = \varphi(l) = 0$. Applying the Fourier method: $u(x,t) = X(x)T(t)$ we obtain the equalities

$$\frac{T'}{a^2 T} = \frac{X''}{X} = -\lambda^2,$$

where λ is a constant, and hence the solution $u(x,t)$ in the form of the series

$$u(x,t) = \sum_{k=1}^{\infty} A_k e^{-\left(\frac{k\pi}{l}\right)^2 a^2 t} \sin\left(\frac{k\pi}{l}x\right). \tag{8.21}$$

The coefficients A_k are found as the coefficients of the series expansion of the function $\varphi(x)$ in terms of the sine functions on the interval $(0, l)$:

$$A_k = \frac{2}{l} \int_0^l \varphi(x) \sin\left(\frac{k\pi}{l} x\right) dx.$$

The series in (8.21) converges absolutely and uniformly, and hence the function $u(x, t)$ defined by the series is continuous in the domain $0 \le x \le l, t \ge 0$. For $0 < x < l, t > 0$ the function $u(x, t)$ satisfies the heat equation under the assumption that the series resulting from differentiating once with respect to t and twice with respect to x are again absolutely and uniformly convergent, but this is true due to the exponential factors in (8.21). The series resulting after differentiation can be majorized by convergent numerical series and thus are subject to the Weierstrass criterion. Moreover, in spite of $\varphi(x)$ being just continuous, the solution $u(x, t)$ results to be infinitely many times continuously differentiable. This significantly distinguishes the heat equation from the wave equation.

Consider the problem of heat flow in a rod at the ends of which a free heat exchange with the environment takes place

$$\begin{cases} \frac{\partial u}{\partial t} - a^2 \frac{\partial^2 u}{\partial x^2} = 0, & 0 < x < l, \\ u(x, 0) = \varphi(x), & \\ \left(\frac{\partial u}{\partial x} - hu\right)_{x=0} = 0, & \left(\frac{\partial u}{\partial x} + hu\right)_{x=l} = 0, \end{cases} \tag{8.22}$$

where h is a positive constant. Application of the Fourier method leads to the Sturm-Liouville problem

$$\begin{cases} X''(x) + \lambda^2 X(x) = 0, & 0 < x < l, \\ X'(0) - hX(0) = 0, & \\ X'(l) + hX(l) = 0. \end{cases} \tag{8.23}$$

Solving the differential equation and satisfying the boundary conditions we find that a nontrivial solution $X(x)$ exists under the condition

$$\begin{vmatrix} h & -\lambda \\ h\cos\lambda l - \lambda\sin\lambda l & h\sin\lambda l + \lambda\cos\lambda l \end{vmatrix} = 0.$$

Introducing the notations $\lambda l = \mu$, $hl = p > 0$, we obtain the characteristic equation of the problem

$$2\cot\mu = \frac{\mu}{p} - \frac{p}{\mu}.$$

This equation possesses an infinite number of real roots. Denote the positive roots as $\mu_1, \mu_2, \ldots, \mu_n, \ldots$ Hence the eigenvalues of the problem (8.23) are of the form

$$\lambda_n^2 = \left(\frac{\mu_n}{l}\right)^2, \quad n = 1, 2, \ldots.$$

According to the general Sturm-Liouville theory, the multiplicity of every eigen-value is one and hence to each λ_n there is associated one linearly independent eigenfunction

$$X_n(x) = \cos\left(\frac{\mu_n}{l}x\right) + \frac{p}{\mu_n}\sin\left(\frac{\mu_n}{l}x\right).$$

Thus, the solution of the problem (8.22) takes the form

$$u(x,t) = \sum_{n=1}^{\infty} A_n e^{-\left(\frac{\mu_n a}{l}\right)^2 t}\left(\cos\left(\frac{\mu_n}{l}x\right) + \frac{p}{\mu_n}\sin\left(\frac{\mu_n}{l}x\right)\right). \tag{8.24}$$

The coefficients A_n can be found by expanding $\varphi(x)$ into a series in terms of the eigenfunctions and using the orthogonality of the eigenfunctions belonging to different eigenvalues,

$$\int_0^l X_m(x)X_n(x)dx = 0, \quad m \neq n,$$

as well as the equality for the norms of the eigenfunctions

$$\|X_n(x)\|^2 = \int_0^l \left(\cos\left(\frac{\mu_n}{l}x\right) + \frac{p}{\mu_n}\sin\left(\frac{\mu_n}{l}x\right)\right)^2 dx = \frac{l}{2}\frac{p(p+2)+\mu_n^2}{\mu_n^2}.$$

We have

$$u(x,0) = \varphi(x) = \sum_{n=1}^{\infty} A_n\left(\cos\left(\frac{\mu_n}{l}x\right) + \frac{p}{\mu_n}\sin\left(\frac{\mu_n}{l}x\right)\right) = \sum_{n=1}^{\infty} A_n X_n(x),$$

from where the formula for the coefficients A_n is obtained

$$A_n = \frac{2}{l}\frac{\mu_n^2}{p(p+2)+\mu_n^2}\int_0^l \varphi(x)\left(\cos\left(\frac{\mu_n}{l}x\right) + \frac{p}{\mu_n}\sin\left(\frac{\mu_n}{l}x\right)\right)dx. \tag{8.25}$$

Thus, the solution of the problem (8.22) has the form (8.24) with the coefficients A_n calculated by (8.25).

8.7   Heat Source Function

Let V denote a volume (or an area in the two-dimensional case). Consider the problem of heat flow in V. Let $\varphi(x, y, z)$ (or $\varphi(x, y)$) be the initial temperature distribution. If there exists a function $G(x, y, z, \xi, \eta, \zeta, t)$ such that the temperature distribution in the volume at any instant $t > 0$ can be written in the form

$$u(x, y, z, t) = \int G(x, y, z, \xi, \eta, \zeta, t)\varphi(\xi, \eta, \zeta)d\xi d\eta d\zeta,$$

then G is called the **source function**. Let us show that G corresponds to the temperature distribution generated by a point heat source. For simplicity consider the one-dimensional situation, $V = [0, l]$. Let $x_0 \in (0, l)$. Suppose that the initial temperature distribution has the form

$$\varphi_\varepsilon(x) = \begin{cases} \varphi(x), & x \in (x_0 - \varepsilon, x_0 + \varepsilon) \subset (0, l), \\ 0, & x \notin (x_0 - \varepsilon, x_0 + \varepsilon), \end{cases}$$

and $u(0, t) = u(l, t) = 0$. For simplicity we assume that the rod is being heated, and $\varphi(x) = \frac{Q}{2c\rho\varepsilon} > 0$. We have

$$u_\varepsilon(x, t) = \int_0^l G(x, \xi, t)\varphi_\varepsilon(\xi)d\xi = \int_{x_0-\varepsilon}^{x_0+\varepsilon} G(x, \xi, t)\frac{Q}{2c\rho\varepsilon}d\xi$$

and hence

$$u_\varepsilon(x, t) = \frac{Q}{c\rho}G(x, \overline{\xi}, t),$$

where $x_0 - \varepsilon < \overline{\xi} < x_0 + \varepsilon$. Tending ε to zero we obtain

$$\lim_{\varepsilon \to 0} u_\varepsilon(x, t) = \frac{Q}{c\rho}G(x, x_0, t).$$

Thus, the source function $G(x, x_0, t)$ considered as a function of x represents the temperature distribution in the rod at the time t, if the temperature in the time preceding $t = 0$ was zero, and at the initial instant $t = 0$ at the point x_0 an amount of heat $Q = c\rho$ is released instantaneously.

Observe that $G(x, x_0, t) \geq 0$ for all x, x_0 and $t > 0$. Indeed, since $\varphi_\varepsilon(x) \geq 0$, due to the minimum principle $u_\varepsilon(x, t) \geq 0$. Then $G(x, \overline{\xi}, t) \geq 0$ for $\overline{\xi}$ sufficiently close to x_0. Passing to the limit we find that $G(x, x_0, t) \geq 0$.

Example 119 Consider the problem (8.20):

$$\begin{cases} \frac{\partial u}{\partial t} - a^2 \frac{\partial^2 u}{\partial x^2} = 0, & 0 < x < l, \\ u(x, 0) = \varphi(x), \\ u(0, t) = u(l, t) = 0. \end{cases}$$

We have

$$u(x, t) = \sum_{n=1}^{\infty} \left(\frac{2}{l} \int_0^l \varphi(\xi) \sin\left(\frac{n\pi}{l}\xi\right) d\xi \right) e^{-\left(\frac{n\pi}{l}\right)^2 a^2 t} \sin\left(\frac{n\pi}{l}x\right)$$

$$= \int_0^l G(x, \xi, t) \varphi(\xi) d\xi,$$

where

$$G(x, \xi, t) = \frac{2}{l} \sum_{n=1}^{\infty} e^{-\left(\frac{n\pi}{l}\right)^2 a^2 t} \sin\left(\frac{n\pi}{l}x\right) \sin\left(\frac{n\pi}{l}\xi\right).$$

Here the order of summation and integration can be changed because the series converges uniformly.

Chapter 9
Harmonic Functions and Their Properties

9.1 Definition of Harmonic Function, Fundamental Solution of the Laplace Equation

Consider the Laplace equation

$$\Delta u \equiv \frac{\partial^2 u}{\partial x_1^2} + \frac{\partial^2 u}{\partial x_2^2} + \ldots + \frac{\partial^2 u}{\partial x_n^2} = 0$$

in an n-dimensional Euclidean space $\mathbb{R}^n$. Let $V \subset \mathbb{R}^n$ be a domain in this space, $x = (x_1, x_2, \ldots, x_n) \in V$ a point in the domain V.

Definition 120 A function $u(x)$, $x \in V$ is called harmonic in a bounded domain V, if it is twice continuously differentiable in V and satisfies the Laplace equation there.

If the domain V is infinite, a function $u(x)$ is called harmonic in V, if it is twice continuously differentiable and satisfies the Laplace equation in V, and for large $|x| = \sqrt{x_1^2 + x_2^2 + \ldots + x_n^2}$, the inequality holds

$$|u(x)| \leq \frac{C}{|x|^{n-2}}.$$

Notice that for $n = 2$, according to this inequality, the function must be bounded for large $|x|$, and for $n \geq 3$, the function u tends to zero,

$$\lim_{|x| \to \infty} u(x) = 0$$

uniformly in all directions.

© The Author(s), under exclusive license to Springer Nature Switzerland AG 2022
A. N. Karapetyants, V. V. Kravchenko, *Methods of Mathematical Physics*,
https://doi.org/10.1007/978-3-031-17845-0_9

Let $M\left(x_1^0, x_2^0, \ldots, x_n^0\right)$ be a fixed point and $P\left(x_1, x_2, \ldots, x_n\right)$ an arbitrary point of the space $\mathbb{R}^n$. Denote by

$$r = r_{PM} = \sqrt{\left(x_1 - x_1^0\right)^2 + \left(x_2 - x_2^0\right)^2 + \ldots + \left(x_n - x_n^0\right)^2}$$

the distance between them.

Let us show that the number m can be chosen such that the function r^m will be harmonic everywhere in $\mathbb{R}^n$ except for the point M. We have

$$\frac{\partial}{\partial x_j}\left(r^m\right) = mr^{m-1}\frac{2\left(x_j - x_j^0\right)}{2r} = mr^{m-2}\left(x_j - x_j^0\right),$$

and

$$\frac{\partial^2}{\partial x_j^2}\left(r^m\right) = mr^{m-2} + m\left(x_j - x_j^0\right)(m-2)r^{m-3}\frac{x_j - x_j^0}{r} = mr^{m-2}\left(1 + (m-2)\frac{\left(x_j - x_j^0\right)^2}{r^2}\right).$$

Thus,

$$\Delta\left(r^m\right) = \sum_{j=1}^{n}\frac{\partial^2}{\partial x_j^2}\left(r^m\right) = mr^{m-2}\sum_{j=1}^{n}\left(1 + (m-2)\frac{\left(x_j - x_j^0\right)^2}{r^2}\right)$$

$$= mr^{m-2}\left(n + (m-2)\right).$$

From here it follows that the function r^m satisfies the Laplace equation iff $m = 2-n$. Hence the function

$$u(P) = \frac{1}{r_{PM}^{n-2}}$$

is twice continuously differentiable and satisfies the Laplace equation everywhere except for the point $P = M$. Moreover, $u(P)$ fulfills the decay condition at infinity when $n \geq 3$. When $n = 2$, we obtain a constant which is of little interest as a solution of the Laplace equation. In this case we choose another function

$$u(P) = \ln\frac{1}{r_{PM}} = \ln\frac{1}{\sqrt{\left(x_1 - x_1^0\right)^2 + \left(x_2 - x_2^0\right)^2}}.$$

Again, a straightforward calculation shows that $\ln \frac{1}{r_{PM}}$ satisfies the Laplace equation everywhere except for the point $P = M$ and logarithmically grows at infinity.

The function

$$
u(P) = \begin{cases} \frac{1}{r_{PM}^{n-2}}, & n \geq 3, \\[2mm] \ln \frac{1}{r_{PM}}, & n = 2 \end{cases}
$$

is called the **fundamental solution of the Laplace equation**.

As one can observe, there is a profound difference between the two-dimensional ($n = 2$) and the multidimensional ($n \geq 3$) situations. This difference radicates in the fact that the fundamental solution of the Laplace operator in a two-dimensional case has two singular points: zero and infinity, which is not the case for higher dimensional cases.

Finally, it is worth mentioning the Laplace operator in the following specific forms.

In polar coordinates ($n = 2$),

$$
\Delta_{r,\theta} u = \frac{\partial^2 u}{\partial r^2} + \frac{1}{r} \frac{\partial u}{\partial r} \frac{1}{r^2} \frac{\partial^2 u}{\partial \theta^2}.
$$

In spherical coordinates ($n = 3$),

$$
\Delta_{r,\theta,\varphi} u = \frac{1}{r^2} \frac{\partial}{\partial r} \left(r^2 \frac{\partial u}{\partial r} \right) + \frac{1}{r^2 \sin \theta} \frac{\partial}{\partial \theta} \left(\sin \theta \frac{\partial u}{\partial \theta} \right) + \frac{1}{r^2 \sin^2 \theta} \frac{\partial^2 u}{\partial \varphi^2} = 0.
$$

It is a good exercise for a reader to verify these formulas by performing standard calculus of change of variables in differential operators.

Remark 121 Let M coincide with the origin, for simplicity, so $r = r_{PM} = |P|$. The functions

$$
\ln \frac{1}{r} \quad \text{and} \quad r^{2-n}
$$

up to multiplier constants are the only nontrivial spherically symmetric functions, which satisfy Laplace equation in the cases $n = 2$ and $n \geq 3$, respectively.

Proof For the case $n = 2$ and $n = 3$, this statement follows directly by solving the Laplace equation in polar and spherical coordinates, see formulas above. The case $n > 3$ is similar. We leave details to the interested reader. ∎

9.2   Green's Function and Integral Representation of Functions from the Class $C^2(V)$

Let $\overline{A}(P) = \{A_1(P), A_2(P), \ldots, A_n(P)\}$ be a vector-valued function in the n-dimensional space. The Gauss–Ostrogradski theorem establishes the equality

$$\int_V \operatorname{div} \overline{A}\, dV = \int_\Sigma A_\nu d\sigma, \qquad (9.1)$$

where V is a bounded domain of $\mathbb{R}^n$, Σ is its boundary, and ν is the exterior normal to Σ. Let us choose the functions $A_1(P), A_2(P), \ldots, A_n(P)$ in the form

$$A_1(P) = \varphi(P)\frac{\partial \psi}{\partial x_1}, \quad A_2(P) = \varphi(P)\frac{\partial \psi}{\partial x_2}, \ldots, A_n(P) = \varphi(P)\frac{\partial \psi}{\partial x_n},$$

i.e.,

$$\overline{A} = \varphi \operatorname{grad} \psi.$$

Then

$$\operatorname{div} \overline{A} = \operatorname{div} (\varphi \operatorname{grad} \psi) = \varphi \Delta \psi + (\operatorname{grad} \varphi, \operatorname{grad} \psi)$$

and

$$A_\nu = \varphi \frac{\partial \psi}{\partial \nu}.$$

Hence the Gauss–Ostrogradski formula (9.1) for the function $\overline{A}$ reads as

$$\int_V \varphi \Delta \psi \, dV + \int_V (\operatorname{grad} \varphi, \operatorname{grad} \psi) \, dV = \int_\Sigma \varphi \frac{\partial \psi}{\partial \nu} d\sigma. \qquad (9.2)$$

This equality is known as the **first Green identity**. Interchanging in this equality φ and ψ, we obtain

$$\int_V \psi \Delta \varphi \, dV + \int_V (\operatorname{grad} \varphi, \operatorname{grad} \psi) \, dV = \int_\Sigma \psi \frac{\partial \varphi}{\partial \nu} d\sigma.$$

Subtracting this equality from (9.2), we arrive at the **second Green identity**

$$\int_V (\varphi \Delta \psi - \psi \Delta \varphi) \, dV = \int_\Sigma \left(\varphi \frac{\partial \psi}{\partial \nu} - \psi \frac{\partial \varphi}{\partial \nu} \right) d\sigma.$$

Notice that the first Green identity is valid for $\varphi \in C^1(V) \cap C(\overline{V})$, $\psi \in C^2(V) \cap C^1(\overline{V})$, while the second Green identity is valid for functions $\varphi, \psi \in C^2(V) \cap C^1(\overline{V})$.

Let us obtain an integral representation for functions from $C^2(V) \cap C^1(\overline{V})$. Let $M = (x_1^0, x_2^0, \ldots, x_n^0)$ be a fixed point of the domain V. Consider the fundamental solution $\frac{1}{r_{PM}^{n-2}} = r_{PM}^{2-n}$ in M, $P \in V$.

Choose a spherical neighborhood $B_\varepsilon(M) = \{P : r_{PM} < \varepsilon\}$ of radius ε centered at M. By $\partial B_\varepsilon(M) = \{P : r_{PM} = \varepsilon\}$, we denote the boundary of the ball $B_\varepsilon(M)$. In the domain $V \setminus \overline{B}_\varepsilon(M)$, the function $\frac{1}{r_{PM}^{n-2}}$ is twice continuously differentiable up to the boundary. Let φ be an arbitrary function from the class $C^2(V) \cap C^1(\overline{V})$. Then, for the functions φ and $\psi = \frac{1}{r_{PM}^{n-2}}$ (for simplicity, we write $\frac{1}{r^{n-2}} = \frac{1}{r_{PM}^{n-2}}$), the second Green identity in the domain $V \setminus \overline{B}_\varepsilon(M)$ is valid

$$\int_{V \setminus \overline{B}_\varepsilon(M)} \varphi \Delta \left(\frac{1}{r^{n-2}} \right) dV - \int_{V \setminus \overline{B}_\varepsilon(M)} \Delta \varphi \, \frac{1}{r^{n-2}} dV$$

$$= \int_{\Sigma + \partial B_\varepsilon(M)} \varphi \frac{\partial}{\partial v} \left(\frac{1}{r^{n-2}} \right) d\sigma - \int_{\Sigma + \partial B_\varepsilon(M)} \frac{1}{r^{n-2}} \frac{\partial \varphi}{\partial v} d\sigma.$$

We have

$$\int_\Sigma \left(\varphi \frac{\partial}{\partial v} \left(\frac{1}{r^{n-2}} \right) - \frac{1}{r^{n-2}} \frac{\partial \varphi}{\partial v} \right) d\sigma$$

$$= \lim_{\varepsilon \to 0} \left\{ -\int_{V \setminus \overline{B}_\varepsilon(M)} \Delta \varphi \, \frac{1}{r^{n-2}} dV - \int_{\partial B_\varepsilon(M)} \varphi \frac{\partial}{\partial v} \left(\frac{1}{r^{n-2}} \right) d\sigma + \int_{\partial B_\varepsilon(M)} \frac{1}{r^{n-2}} \frac{\partial \varphi}{\partial v} d\sigma \right\}.$$

Note that v is an exterior normal to $\Sigma + \partial B_\varepsilon(M)$. Hence, at the points of the sphere $\partial B_\varepsilon(M)$, the vector v is directed to the interior of the sphere along the radius and thus $\frac{\partial}{\partial v} = -\frac{\partial}{\partial r}$. We have

$$\int_\Sigma \left(\varphi \frac{\partial}{\partial v} \left(\frac{1}{r^{n-2}} \right) - \frac{1}{r^{n-2}} \frac{\partial \varphi}{\partial v} \right) d\sigma$$

$$= -\int_V \Delta \varphi \, \frac{1}{r^{n-2}} dV + \lim_{\varepsilon \to 0} \int_{\partial B_\varepsilon(M)} \left(\varphi \frac{2-n}{r^{n-1}} + \frac{1}{r^{n-2}} \frac{\partial \varphi}{\partial v} \right) d\sigma$$

$$= -\int_V \Delta \varphi \, \frac{1}{r^{n-2}} dV + \lim_{\varepsilon \to 0} \left\{ \frac{2-n}{\varepsilon^{n-1}} \int_{\partial B_\varepsilon(M)} \varphi \, d\sigma + \frac{1}{\varepsilon^{n-2}} \int_{\partial B_\varepsilon(M)} \frac{\partial \varphi}{\partial v} d\sigma \right\}.$$

Application of the mean value theorem to the last two integrals gives us the equality

$$\lim_{\varepsilon \to 0} \left\{ \frac{2-n}{\varepsilon^{n-1}} \int_{\partial B_\varepsilon(M)} \varphi \, d\sigma + \frac{1}{\varepsilon^{n-2}} \int_{\partial B_\varepsilon(M)} \frac{\partial \varphi}{\partial v} d\sigma \right\}$$

$$= \lim_{\varepsilon \to 0} \left\{ -\frac{n-2}{\varepsilon^{n-1}} \varphi\left(M_1\right) \left|\partial B_\varepsilon(M)\right| + \frac{\partial \varphi\left(M_2\right)}{\partial v} \frac{\left|\partial B_\varepsilon(M)\right|}{\varepsilon^{n-2}} \right\},$$

where by $\left|\partial B_\varepsilon(M)\right|$ we denoted the area of the surface $\partial B_\varepsilon(M)$. Since $\left|\partial B_\varepsilon(M)\right| = \varepsilon^{n-1} \frac{2\pi^{n/2}}{\Gamma(n/2)}$ and when $\varepsilon \to 0$ we have that $M_1 \to M$, $M_2 \to M$, we obtain

$$\int_\Sigma \left(\varphi \frac{\partial}{\partial v} \left(\frac{1}{r^{n-2}} \right) - \frac{1}{r^{n-2}} \frac{\partial \varphi}{\partial v} \right) d\sigma = -\frac{(n-2)\,2\pi^{n/2}}{\Gamma(n/2)} \varphi(M) - \int_V \Delta\varphi \frac{1}{r^{n-2}} dV.$$

Thus,

$$\int_\Sigma \left(\frac{1}{r^{n-2}} \frac{\partial \varphi}{\partial v} - \varphi \frac{\partial}{\partial v} \left(\frac{1}{r^{n-2}} \right) \right) d\sigma - \int_V \Delta\varphi \frac{1}{r^{n-2}} dV = \frac{(n-2)\,2\pi^{n/2}}{\Gamma(n/2)} \varphi(M)$$

or which is the same

$$\varphi(M) = C_n \int_\Sigma \left(\frac{1}{r^{n-2}} \frac{\partial \varphi}{\partial v} - \varphi \frac{\partial}{\partial v} \left(\frac{1}{r^{n-2}} \right) \right) d\sigma - C_n \int_V \Delta\varphi \frac{1}{r^{n-2}} dV, \quad r = r_{PM},$$

$$(9.3)$$

where $C_n = \frac{\Gamma(n/2)}{(n-2)2\pi^{n/2}}$. In particular, when $n = 3$, we obtain

$$\varphi(x, y, z) = \frac{1}{4\pi} \int_\Sigma \left(\frac{1}{r} \frac{\partial \varphi}{\partial v} - \varphi \frac{\partial}{\partial v} \left(\frac{1}{r} \right) \right) d\sigma - \frac{1}{4\pi} \int_V \Delta\varphi \frac{1}{r} dV.$$

Analogously, in the case $n = 2$, one can deduce the formula

$$\varphi(x, y) = \frac{1}{2\pi} \int_l \left(\ln\left(\frac{1}{r}\right) \frac{\partial \varphi}{\partial v} - \varphi \frac{\partial}{\partial v} \left(\ln\left(\frac{1}{r}\right) \right) \right) dl - \frac{1}{2\pi} \int_S \Delta\varphi \ln\left(\frac{1}{r}\right) ds.$$

9.3   Properties of Harmonic Functions

Property 1 (Property of the Normal Derivative) If the function $u(M)$ is harmonic in V, then

$$\int_S \frac{\partial u}{\partial v} d\sigma = 0,$$

where S is any closed surface entirely lying in the domain V.

Proof Application of the first Green identity with $\varphi = 1$ and $\psi = u$ gives us the result.   ∎

Property 2 (Integral Representation) Let u be harmonic in V and continuously differentiable up to the boundary $\Sigma = \partial V$. Then, at any point $M \in V$, the value of the function u can be computed from the values of u and of its normal derivative $\frac{\partial u}{\partial v}$ on the boundary Σ according to the formula

$$u(M) = C_n \int_\Sigma \left(\frac{1}{r^{n-2}} \frac{\partial u}{\partial v} - u \frac{\partial}{\partial v} \left(\frac{1}{r^{n-2}} \right) \right) d\sigma,$$

where $C_n = \frac{\Gamma(n/2)}{(n-2)2\pi^{n/2}}$ and $r = r_{PM}$ (P runs through all points of the surface Σ).

Proof The formula follows directly from the integral representation (9.3). ∎

Property 3 (Differentiability of Harmonic Functions) Any function u which is harmonic in a domain V possesses continuous derivatives of any order in V.

Proof Let $M(x_1, \ldots, x_n)$ be an arbitrary point in V. Choose a closed surface S enclosing M and entirely lying in the domain V. Then due to Property 2,

$$u(M) = C_n \int_\Sigma \left(\frac{1}{r^{n-2}} \frac{\partial u}{\partial v} - u \frac{\partial}{\partial v} \left(\frac{1}{r^{n-2}} \right) \right) d\sigma,$$

where the integrand possesses continuous partial derivatives with respect to the variables $x_1, \ldots, x_n$ of any order. Hence the integral can be differentiated any number of times as well. ∎

Property 4 (The Mean Value Property) Let u be harmonic in a ball and continuous up to its boundary. Then the value of u at the center of the ball is given by the average value of u on the surface of the ball.

Proof Let $B_R(M)$ be a ball of radius R, centered at a point M. By $B_{R_0}(M)$ we denote a ball centered at the same point but of a smaller radius $R_0 < R$. Let $\partial B_R(M)$ and $\partial B_{R_0}(M)$ denote the boundaries of the balls, respectively. Let u be harmonic in $B_R(M)$. Then, obviously, in the smaller ball $B_{R_0}(M)$, it is harmonic and twice continuously differentiable up to the boundary. Hence, due to Property 2, we have

$$u(M) = C_n \int_{\partial B_{R_0}(M)} \left(\frac{1}{r^{n-2}} \frac{\partial u}{\partial v} - u \frac{\partial}{\partial v} \left(\frac{1}{r^{n-2}} \right) \right) d\sigma$$

$$= \frac{C_n}{R_0^{n-2}} \int_{\partial B_{R_0}(M)} \frac{\partial u}{\partial v} d\sigma + \frac{(n-2) C_n}{R_0^{n-1}} \int_{\partial B_{R_0}(M)} u d\sigma.$$

Due to Property 1, the first integral equals zero, while the second one gives us the equality

$$u(M) = \frac{\int_{\partial B_{R_0}(M)} u(P) d\sigma_P}{\left| \partial B_{R_0}(M) \right|},$$

where $|\partial B_{R_0}(M)|$ stands for the area of the sphere $\partial B_{R_0}(M)$.

Consider the limit of this equality when $R_0 \to R$. Since $u(P)$ is continuous in the closed ball $\overline{B}_R(M)$, in the limit, we obtain

$$u(M) = \frac{\int_{\partial B_R(M)} u(P)d\sigma_P}{|\partial B_R(M)|},$$

which is the mean value property.   ∎

Property 5 (Maximum and Minimum Principle) Let V be a simply connected domain in $\mathbb{R}^n$. A non-constant harmonic function in V cannot attain its maximum (or minimum) value at an interior point of V.

Proof First, let us note that in the case of an unbounded domain the maximum principle is valid when limiting the consideration to finite interior points of the domain, i.e., assigning the infinitely distant point to the boundary.

Now suppose the opposite that the function u attains its maximum at an interior point $M \in V$. Let $B_R(M)$ be a ball of radius R, centered at M and together with its boundary belonging to V. Then, due to the mean value property,

$$u(M) = \frac{\int_{\partial B_R(M)} u(P)d\sigma_P}{|\partial B_R(M)|}.$$

Let us show that the supposition implies that $u(P) = u(M)$ for all $P \in \partial B_R(M)$. Since the maximum of u is supposed to be attained at M, we have that

$$u(P) \le u(M).$$

Suppose that at some point $P_1 \in \partial B_R(M)$, the strict inequality holds $u(P_1) < u(M)$. Since u is continuous, this inequality is valid also for some part $S_{R'}$ of the sphere $\partial B_R(M)$: $u(P) < u(M)$, $P \in S_{R'}$. The mean value property then gives us the equality

$$u(M) = \frac{1}{|\partial B_R(M)|} \left(\int_{S_{R'}} u(P)d\sigma_P + \int_{\partial B_R(M) \backslash S_{R'}} u(P)d\sigma_P \right).$$

On the other hand,

$$\int_{S_{R'}} u(P)d\sigma_P < u(M)|S_{R'}|$$

and

$$\int_{\partial B_R(M) \backslash S_{R'}} u(P)d\sigma_P \le u(M)|\partial B_R(M) \backslash S_{R'}|.$$

Thus,

$$u(M) < u(M)\frac{|S_{R'}| + |\partial B_R(M) \setminus S_{R'}|}{|\partial B_R(M)|} = u(M),$$

which is a contradiction. Hence

$$u(P) = u(M) \quad \text{for all} \quad P \in \partial B_R(M).$$

Since R is arbitrary, this equality holds not only on the sphere $\partial B_R(M)$ but also in the whole ball $B_R(M)$. Thus,

$$u(P) = u(M) = \text{Const}$$

in the whole ball $B_R(M)$ centered at M. It remains to prove that the equality is valid on the whole domain V. Let P be an arbitrary point of the domain. Let us join the points M and P by a continuous line γ, and let d be the minimal distance from the line to the boundary of V. Let M_1 be the point of intersection of the sphere $\partial B_R(M)$ with the line γ. At this point $u(M_1) = u(M)$. Consider a ball of radius d centered at M_1. By just proved, the function u is constant everywhere in this ball, so $u(M_2) = u(M_1) = u(M)$, where M_2 is another point of intersection of the sphere of radius d centered at M_1 with the line γ. Repeating this reasoning a finite number of times, we will arrive at the ball containing the point P and hence

$$u(P) = u(M_n) = u(M_{n-1}) = \ldots = u(M_1) = u(M).$$

The minimum principle is proved by replacing $u(M)$ with $-u(M)$. ∎

Corollary 122 *If two functions u and U are continuous in a closed domain $\overline{V}$ and harmonic in V, then the inequality*

$$u \leq U \quad on \ \Sigma = \partial V$$

implies the inequality

$$u \leq U \quad everywhere \ in \ V.$$

Proof The function $U - u$ is continuous in $\overline{V}$, harmonic in V, and $U - u \geq 0$ on Σ. Due to the maximum and minimum principle, we have that $U - u \geq 0$ everywhere in V, from which the validity of the statement follows. ∎

Corollary 123 *If two functions u and U are continuous in a closed domain $\overline{V}$ and harmonic in V, then the inequality*

$$|u| \leq U \quad on \ \Sigma = \partial V$$

implies the inequality

$$|u| \leq U \quad everywhere \ in \ V. \tag{9.4}$$

Proof We have that the three functions $-U$, u, and U satisfy the inequalities

$$-U \leq u \leq U \quad on \ \Sigma.$$

Applying Corollary 122 twice, we obtain that

$$-U \leq u \leq U \quad everywhere \ in \ V$$

or which is the same as the inequality (9.4).   ∎

Property 6 (Lemma on Removable Singularities) Let the function u be harmonic in V everywhere except, possibly, for a point $M \in V$. If at this point

$$u(P) = o\left(\frac{1}{r_{MP}^{n-2}}\right), \quad P \to M, \quad n \geq 3,$$

$$u(P) = o\left(\ln \frac{1}{r_{MP}}\right), \quad P \to M, \quad n = 2,$$

then u can be redefined so that it will be harmonic at M as well.

Proof For simplicity let us prove this statement in the case $n = 3$. The proof for other values of n is similar. Let us choose $R > 0$ sufficiently small, such that the ball $B_R(M) \subset V$. It can be proved without using this lemma that it is always possible to construct a harmonic function in the ball $B_R(M)$ coinciding on its surface $\partial B_R(M)$ with an arbitrary continuous function. Denote by v that function which is harmonic in $B_R(M)$ and coincides with u on $\partial B_R(M)$. Consider the function

$$w(P) = u(P) - v(P) - \varepsilon\left(\frac{1}{r_{MP}} - \frac{1}{R}\right),$$

where $\varepsilon > 0$ is small. The function w is harmonic in $B_R(M) \setminus \overline{B}_\varepsilon(M)$ and $w(P) = 0$ on $\partial B_R(M)$. For sufficiently small ε, on $\partial B_\varepsilon(M)$, the term $-\varepsilon\left(\frac{1}{r_{MP}} - \frac{1}{R}\right)$ becomes dominant. Thus, $w \leq 0$ on $\partial B_\delta(M)$ for a small enough δ. Hence, by the maximum principle, $w \leq 0$ in $B_R(M) \setminus \overline{B}_\delta(M)$. That is,

$$u \leq v + \varepsilon\left(\frac{1}{r_{MP}} - \frac{1}{R}\right) \quad in \ B_R(M) \setminus B_\delta(M).$$

Thus, by letting $\varepsilon \to 0$, we get

$$u \leq v \quad \text{in } B_R(M) \setminus B_\delta(M).$$

This is true for any sufficiently small δ, so it is true for all $P \in B_R(M) \setminus \{M\}$.
By reverting u and v in the definition of w, we can get

$$v \leq u \quad \text{in } B_R(M) \setminus \{M\}.$$

Thus, defining $u(M) = v(M)$, we extend u to be a harmonic function in $B_R(M)$ and hence in V. ∎

Remark 124 It follows from Property 6 that if a harmonic function $u(P)$ has a non-removable singularity at some point M, then when approximating to this point the function grows not slower than $\ln \frac{1}{r_{PM}}$ when $n = 2$ and $\frac{1}{r_{PM}^{n-2}}$ when $n > 2$. The functions $\ln \frac{1}{r_{PM}}$ when $n = 2$ and $\frac{1}{r_{PM}^{n-2}}$ when $n > 2$ give us examples of harmonic functions having a non-removable singularity at M.

Property 7 (Kelvin's Transform and Behavior of Harmonic Functions at Infinity)
Let us introduce a concept of symmetry with respect to a sphere in n-dimensional space. Let $x = (x_1, x_2, \ldots, x_n)$ be an arbitrary point in $\mathbb{R}^n$. Let us put in correspondence to x the point

$$x^* = \frac{R^2}{|x|^2} x, \tag{9.5}$$

where $R > 0$ and $|x| = \sqrt{x_1^2 + x_2^2 + \ldots + x_n^2}$. Let us express x in terms of x^*:

$$x = \frac{|x|^2}{R^2} x^*.$$

Obviously, $|x^*| = \frac{R^2}{|x|^2} |x| = \frac{R^2}{|x|}$ and $|x| = \frac{R^2}{|x^*|}$. Hence

$$x = \left(\frac{R^2}{|x^*|} \right)^2 \frac{1}{R^2} x^* = \frac{R^2}{|x^*|^2} x^*. \tag{9.6}$$

Note another important property of this symmetry

$$|x| \, |x^*| = R^2. \tag{9.7}$$

The transform (9.5) is called the **inversion** with respect to a sphere of radius R (centered at the origin). Due to (9.6), the inverse transform is also the inversion. The points x and x^* are called conjugate or **symmetric** with respect to a sphere of radius R. Sometimes the notation is used

$$x^* = \text{In } V_R \, x$$

and, due to (9.6),

$$x = \text{In } V_R \, x^*.$$

Obviously, $(x^*)^* \equiv x$ or which is the same $\text{In } V_R (\text{In } V_R \, x) = x$. Equality (9.7) means that the product of distances of symmetric points to the center of the sphere equals the square of the radius. Hence the points P_1 and P_2 are symmetric with respect to the sphere of radius R if and only if:

1. They are located on the same ray whose initial point is the origin.
2. $O P_1 \cdot O P_2 = R^2$.

Let a function u be harmonic at a point M. Let us study the question whether it can be transformed into a harmonic function at the symmetric point or not.

Definition 125 Let

$$v(x) = \frac{R^{n-2}}{|x|^{n-2}} u \left(\frac{R^2}{|x|^2} x \right) \tag{9.8}$$

or equivalently

$$v(x) = \frac{R^{n-2}}{|x|^{n-2}} u \left(x^* \right). \tag{9.9}$$

Transformation of a function $u(x)$ into a function $v(x)$ according to this rule is called **Kelvin's transform**.

Note that (9.9) can be written in the form

$$v(\text{In } V_R \, x^*) = \frac{R^{n-2}}{|x|^{n-2}} u \left(x^* \right) = \frac{R^{n-2}}{\left| \frac{R^2}{|x^*|} \right|^{n-2}} u \left(x^* \right) = \frac{|x^*|^{n-2}}{R^{n-2}} u \left(x^* \right).$$

Hence

$$u \left(x^* \right) = \frac{R^{n-2}}{|x^*|^{n-2}} v(\text{In } V_R \, x^*) = \frac{R^{n-2}}{|x^*|^{n-2}} v(x),$$

and thus, if v is a Kelvin transform of u, then u is a Kelvin transform of v.

Theorem 126 *If the function u is harmonic inside the ball $B_R(0)$ centered at the origin, then its Kelvin transform is harmonic outside $\overline{B}_R(0)$.*

Proof First, let us check the behavior of a transformed function at infinity. Let $n = 2$. In this case the Kelvin transform of a function u has the form $v(x) = u(x^*)$.

When $x \to \infty$, we have that $x^* \to 0$. Since u is bounded at the origin, we obtain that $|v(x)| \le$ Const when $x \to \infty$, and hence v fulfills the condition of boundedness at infinity (see Definition 120). Let $n \ge 3$. Then, from (9.9), we obtain

$$|v(x)| = \frac{R^{n-2}}{|x|^{n-2}} \left| u \left(x^* \right) \right|.$$

Since u is bounded at the origin, we have that

$$|v(x)| \le \frac{\text{Const}}{|x|^{n-2}} \to 0 \quad \text{when } x \to \infty.$$

Thus, it remains to show that everywhere outside $\overline{B}_R(0)$ we have $\Delta v = 0$. For simplicity let us consider the case $n = 3$. For other values of n, the proof is similar.
   Consider the spherical coordinates

$$x = (r, \theta, \varphi).$$

Obviously, the symmetric point then has the form

$$x^* = \left(\frac{R^2}{r}, \theta, \varphi \right),$$

and the Kelvin transform is written as

$$v(x) = v(r, \theta, \varphi) = \frac{R}{r} u \left(\frac{R^2}{r}, \theta, \varphi \right).$$

Denote $\rho = \frac{R^2}{r}$. Thus,

$$v(r, \theta, \varphi) = \frac{\rho}{R} u(\rho, \theta, \varphi).$$

We know that $\Delta_\rho u = 0$, $\rho < R$ and need to prove that $\Delta_r v = 0$ for $r > R$. Let us write the Laplacian in spherical coordinates

$$\Delta_\rho u = \frac{1}{\rho^2} \frac{\partial}{\partial \rho} \left(\rho^2 \frac{\partial u}{\partial \rho} \right) + \frac{1}{\rho^2 \sin \theta} \frac{\partial}{\partial \theta} \left(\sin \theta \frac{\partial u}{\partial \theta} \right) + \frac{1}{\rho^2 \sin^2 \theta} \frac{\partial^2 u}{\partial \varphi^2} = 0.$$

We have

$$\Delta_r v = \frac{1}{R} \left(\frac{1}{r^2} \frac{\partial}{\partial r} \left(r^2 \frac{\partial (\rho u)}{\partial r} \right) + \frac{1}{r^2 \sin \theta} \frac{\partial}{\partial \theta} \left(\sin \theta \frac{\partial (\rho u)}{\partial \theta} \right) + \frac{1}{r^2 \sin^2 \theta} \frac{\partial^2 (\rho u)}{\partial \varphi^2} \right).$$

Let us substitute r by ρ taking into account that

$$\frac{\partial}{\partial r} = \frac{\partial}{\partial \rho}\frac{\partial \rho}{\partial r} = -\frac{R^2}{r^2}\frac{\partial}{\partial \rho} = -\frac{\rho^2}{R^2}\frac{\partial}{\partial \rho}.$$

Thus,

$$
\begin{aligned}
\Delta_r v &= \frac{1}{R}\left\{\frac{\rho^2}{R^4}\frac{\rho^2}{R^2}\frac{\partial}{\partial \rho}\left(\frac{R^4}{\rho^2}\frac{\rho^2}{R^2}\frac{\partial (\rho u)}{\partial \rho}\right) + \frac{\rho^3}{R^4 \sin \theta}\frac{\partial}{\partial \theta}\left(\sin \theta \frac{\partial u}{\partial \theta}\right) + \frac{\rho^3}{R^4 \sin^2 \theta}\frac{\partial^2 u}{\partial \varphi^2}\right\} \\
&= \frac{\rho^5}{R^5}\left\{\frac{1}{\rho}\frac{\partial^2 (\rho u)}{\partial \rho^2} + \frac{1}{\rho^2 \sin \theta}\frac{\partial}{\partial \theta}\left(\sin \theta \frac{\partial u}{\partial \theta}\right) + \frac{1}{\rho^2 \sin^2 \theta}\frac{\partial^2 u}{\partial \varphi^2}\right\} \\
&= \frac{\rho^5}{R^5}\left\{\frac{1}{\rho}\frac{\partial}{\partial \rho}\left(u + \rho \frac{\partial u}{\partial \rho}\right) + \Delta_\rho u - \frac{1}{\rho^2}\frac{\partial}{\partial \rho}\left(\rho^2 \frac{\partial u}{\partial \rho}\right)\right\} \\
&= \frac{\rho^5}{R^5}\left\{\frac{1}{\rho}\frac{\partial u}{\partial \rho} + \frac{1}{\rho}\frac{\partial u}{\partial \rho} + \frac{\partial^2 u}{\partial \rho^2} + \Delta_\rho u - \frac{\partial^2 u}{\partial \rho^2} - \frac{2\rho}{\rho^2}\frac{\partial u}{\partial \rho}\right\} \\
&= \frac{\rho^5}{R^5}\Delta_\rho u.
\end{aligned}
$$

Hence

$$\Delta_r v = \frac{\rho^5}{R^5}\Delta_\rho u, \tag{9.10}$$

and if $\Delta_\rho u = 0$ for $\rho < R$, then $\Delta_r v = 0$ for $r > R$. ∎

Theorem 127 (*Inverse Theorem*) *Let u be harmonic outside* $\overline{B}_R(0)$. *Then its Kelvin transform is harmonic in* $B_R(0)$.

Proof Let us write (9.10) in the form

$$\Delta_r v = \frac{R^5}{\rho^5}\Delta_\rho u.$$

Since $\Delta_\rho u = 0$ for all $\rho > R$, then obviously, $\Delta_r v = 0$ for all points of V_R, except possibly for its center where $r = 0$. Let us show that v is harmonic at the origin as well. When $n = 2$, the function $v(x) = u(x^*)$ is bounded when $x \to 0$ and hence can be redefined there so that it will be harmonic in the whole $B_R(0)$. Let $n \geq 3$. We have

$$v(x) = \frac{R^{n-2}}{r^{n-2}}u\left(x^*\right).$$

Let $x \to 0$. Then $x^* \to \infty$. Since u is harmonic at infinity, $\lim_{x^* \to \infty} u\left(x^*\right) = 0$, and hence

$$v(x) = o\left(\frac{1}{r^{n-2}}\right), \quad r \to 0.$$

Then the theorem on the removable singularity is applicable which finishes the proof. ∎

9.4 Behavior of Harmonic Functions at Infinity

Theorem 128 *Let u be a harmonic function outside a closed ball $\overline{B}_R(0)$ of radius R centered at the origin. Then there exist such number $A > 0$ and $r_0 > R$ that for all $r = |P| \geq r_0$ the estimates are valid*

$$|u(P)| < \frac{A}{r^{n-2}}, \quad \left|\frac{\partial u(P)}{\partial x_i}\right| < \frac{A}{r^{n-1}}, \quad i = 1,\ldots,n \quad \text{when } n \geq 3.$$

When $n = 2$, there exists a finite limit at infinity $\lim_{r \to +\infty} u(P)$ and $\left|\frac{\partial u(P)}{\partial x_i}\right| < \frac{A}{r^2}$, $i = 1, 2$.

Proof Under the condition of the theorem when $n \geq 3$, the function u tends to zero uniformly in all directions when $r \to +\infty$. Thus, there exists such function $\alpha(r) > 0$ that

$$|u(P)| < \alpha(r)$$

for all $r = |P| \geq r_0 > R$, and $\alpha(r) \to 0$ when $r \to +\infty$. When $n = 2$, the function u is bounded at infinity, i.e., $u(P) = O(1)$ when $r \to +\infty$.

Due to Kelvin's theorem (Theorem 127), the function

$$v(\xi) = \left(\frac{R}{\rho}\right)^{n-2} u\left(\frac{R^2}{\rho^2}\xi\right) = \left(\frac{R}{\rho}\right)^{n-2} u(P) \tag{9.11}$$

is harmonic in $B_R(0) \setminus \{\rho = 0\}$ and satisfies the following condition when $\rho = |\xi| \to 0$:

$$v(\xi) = \begin{cases} O(1), & \text{when } n = 2, \\ \frac{\tilde{\alpha}(\rho)}{\rho^{n-2}}, & \text{when } n > 2, \end{cases}$$

where $\tilde{\alpha}(\rho) = R^{n-2}\alpha\left(\frac{R^2}{\rho^2}\right) \to 0$ when $\rho \to 0$. Then, due to the removable singularity property, the function v is harmonic at the point $\xi = 0$ and hence everywhere in $B_R(0)$. Then, from (9.11) for all $r = |P| \geq r_0 = \frac{1}{\rho_0}$, we have

$$|u(P)| = \frac{R^{n-2}}{r^{n-2}}|v(\xi)| < \frac{A}{r^{n-2}},$$

where $A \geq R^{n-2}A_1$, $A_1 = \sup|v(\xi)|$ for $|\xi| < \rho_0$.

Furthermore, using (9.11), we calculate the derivative

$$\frac{\partial u}{\partial x_i} = R^{n-2}(2-n)r^{-n}x_i v(\xi) - 2R^{n-2}r^{-n-2}x_i \sum_{k=1}^{n} \frac{\partial v(\xi)}{\partial \xi_k} x_k$$

$$+ R^{n-2}r^{-n}\frac{\partial v(\xi)}{\partial \xi_i}. \tag{9.12}$$

From here, using the estimates

$$\frac{|x_i|}{r} \leq 1, \quad |v(\xi)| \leq A_1, \quad \left|\frac{\partial v(\xi)}{\partial \xi_k}\right| \leq A_2,$$

we obtain

$$\left|\frac{\partial u}{\partial x_i}\right| \leq \frac{R^{n-2}(n-2)A_1}{r^{n-1}} + \frac{2nR^{n-2}A_2}{r^n} + \frac{R^{n-2}A_2}{r^n}$$

$$= \frac{1}{r^{n-1}}\left(R^{n-2}(n-2)A_1 + (2n+1)\frac{R^{n-2}A_2}{r}\right) \leq \frac{A}{r^{n-1}}.$$

Let $n = 2$. Since $v(\xi)$ is harmonic at the point $\xi = 0$, it is continuous there. Hence there exists a finite limit

$$\lim_{r \to +\infty} u(P) = \lim_{\rho \to 0} v(\xi) = v(0).$$

Now, setting $n = 2$ in the equality (9.12), we obtain the estimate

$$\left|\frac{\partial u}{\partial x_i}\right| \leq \frac{A}{r^2}, \quad i = 1, 2.$$

∎

Chapter 10
Boundary Value Problems for the Laplace Equation

10.1 Statement of Interior and Exterior Boundary Value Problems

Let Σ be a closed surface in $\mathbb{R}^n$, V_+ a domain enclosed by Σ, and V_- a domain exterior with respect to Σ (see Fig. 10.1). For the Laplace equation $\Delta u = 0$, we will study the following boundary value problems.

1. **Interior Dirichlet problem** (Problem D_+)
   Given a continuous function φ defined on Σ, find a harmonic function u in V_+, such that

$$u(P) = \varphi(P) \quad \text{for all } P \in \Sigma.$$

2. **Interior Neumann problem** (Problem N_+)
   Given a continuous function f defined on Σ, find a harmonic function u in V_+, such that

$$\frac{\partial u(P)}{\partial v} = f(P) \quad \text{for all } P \in \Sigma.$$

3. **Exterior Dirichlet problem** (Problem D_-)
   Given a continuous function φ defined on Σ, find a harmonic function u in V_-, such that

$$u(P) = \varphi(P) \quad \text{for all } P \in \Sigma.$$

4. **Exterior Neumann problem** (Problem N_-)
   Given a continuous function f defined on Σ, find a harmonic function u in V_-, such that

© The Author(s), under exclusive license to Springer Nature Switzerland AG 2022
A. N. Karapetyants, V. V. Kravchenko, *Methods of Mathematical Physics*,
https://doi.org/10.1007/978-3-031-17845-0_10

Fig. 10.1 Interior and
exterior domains

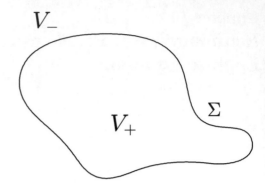

$$\frac{\partial u(P)}{\partial v} = f(P) \quad \text{for all } P \in \Sigma.$$

The Dirichlet and Neumann problems are also known as the first and second
boundary value problems for the Laplace equation, respectively. Often the third
(or mixed, or Robin) boundary value problem is considered, when the boundary
condition has the form

$$\frac{\partial u(P)}{\partial v} + \beta(P)u(P) = f(P) \quad \text{for all } P \in \Sigma.$$

We restrict ourselves to the study of the first four problems.

10.2   Uniqueness Theorems for Dirichlet and Neumann Problems

Theorem 129 *A solution of Problem D_+ is unique.*

Proof Suppose that there exist two different solutions u_1 and u_2. Consider their
difference $u = u_1 - u_2$. Then

$$\begin{cases} \Delta u = 0 & \text{in } V_+, \\ u|_\Sigma = 0. \end{cases}$$

Due to the maximum and minimum principle theorem, $u \equiv 0$ in V_+, and hence
$u_1 \equiv u_2$. ∎

Theorem 130 *A solution of Problem N_+ is unique up to an arbitrary additive
constant.*

Proof Let u_1 and u_2 be two different solutions of Problem N_+. Then their difference $u = u_1 - u_2$ is a solution of the problem

$$\begin{cases} \Delta u = 0 & \text{in } V_+, \\ \frac{\partial u}{\partial v}\big|_\Sigma = 0. \end{cases}$$

Let us apply the first Green identity, choosing $\varphi = \psi = u$:

$$\int_{V_+} u \, \Delta u \, dV + \int_{V_+} (\operatorname{grad} u, \operatorname{grad} u) \, dV = \int_\Sigma u \frac{\partial u}{\partial v} \, d\sigma.$$

From here, we obtain

$$\int_{V_+} (\operatorname{grad} u, \operatorname{grad} u) \, dV = 0$$

or which is the same as

$$\int_{V_+} \left(\left(\frac{\partial u}{\partial x_1}\right)^2 + \left(\frac{\partial u}{\partial x_2}\right)^2 + \ldots + \left(\frac{\partial u}{\partial x_n}\right)^2 \right) dV = 0,$$

which is possible if only

$$\frac{\partial u}{\partial x_1} = \frac{\partial u}{\partial x_2} = \ldots = \frac{\partial u}{\partial x_n} = 0.$$

This implies that $u \equiv \text{Const}$ and thus $u_1 = u_2 + \text{Const}$. ■

Remark 131 Problem N_+ can have no solution. Indeed, according to Property 1 of harmonic functions,

$$\int_\Sigma \frac{\partial u}{\partial v} \, d\sigma = 0.$$

Hence the function f from the boundary condition must satisfy the equality

$$\int_\Sigma f(P) \, d\sigma_P = 0. \tag{10.1}$$

Thus, condition (10.1) is necessary for the existence of a solution of Problem N_+. Later on its sufficiency will be shown.

Theorem 132 *A solution of Problem D_- is unique.*

Proof Let first $n \geq 3$. Consider a ball $B_R(0)$ of a sufficiently large radius R such that $\Sigma \subset B_R(0)$ and a domain T enclosed between the sphere $\partial B_R(0)$ and the surface Σ. Obviously, $T = B_R(0)\backslash\overline{V}_+$. Let $u = u_1 - u_2$, where u_1 and u_2 are two different hypothetically existing solutions of Problem D_-. We have $u|_\Sigma = 0$, and moreover, choosing R sufficiently large, we obtain $\|u\|_{\partial B_R(0)} < \varepsilon$. Hence $|u| < \varepsilon$ on the boundary $\Sigma + \partial B_R(0)$ of the domain T. Due to the maximum principle,

$$|u(M)| < \varepsilon \quad \text{for all } M \in T.$$

Fixing the point M and letting $R \to \infty$, we observe that $\varepsilon \to 0$ and hence $u(M) = 0$. Since the point M is arbitrary, we obtain that $u \equiv 0$, which finishes the proof.

Let now $n = 2$. Let V_- be the exterior domain on the plane with the boundary Σ. Let u_1 and u_2 be two different solutions of Problem D_-. Hence they are bounded, so

$$|u_1| \le N_1 \quad \text{and} \quad |u_2| \le N_2,$$

and consequently

$$|u| = |u_1 - u_2| \le N_1 + N_2 = N.$$

Choose an arbitrary point $M_0 \in V_+$ and a circle $C_R(M_0) \subset V_+$ of radius R centered at M_0. The harmonic function $\ln\left(1/r_{MM_0}\right)$ has no singularity in the exterior domain V_- and the function $\ln\left(r_{MM_0}/R\right)$ is positive on the whole V_- including Σ. Choose $R_1 > 0$ large enough, so that the circle $C_{R_1}(M_0)$ of radius R_1 centered at M_0 contains Σ. Denote by T the domain bounded by Σ and $C_{R_1}(M_0)$. The function

$$v(M) = N \frac{\ln\left(r_{MM_0}/R\right)}{\ln\left(R_1/R\right)}$$

is harmonic in T, equal to N on the circle $C_{R_1}(M_0)$ and positive on Σ. We have

$$0 = \|u\|_\Sigma < v|_\Sigma,$$

$$\|u\|_{C_{R_1}(M_0)} \le N = v|_{C_{R_1}(M_0)}.$$

Hence, due to Corollary 123,

$$|u(M)| \le N \frac{\ln\left(r_{MM_0}/R\right)}{\ln\left(R_1/R\right)} \quad \text{for all } M \in T.$$

Fixing the point M and letting $R_1 \to \infty$, we obtain that $v(M) \to 0$ and hence $u(M) = 0$. Since M is arbitrary, $u \equiv 0$ in V_-. ∎

Theorem 133 *A solution of Problem* N_- *is unique when* $n \geq 3$ *and unique up to an arbitrary additive constant when* $n = 2$.

Proof Consider the closed surface Σ dividing $\mathbb{R}^n$ into the bounded domain V_+ and the unbounded complement V_-. Let $R > 0$ be sufficiently large, so that the ball $B_R(0)$ contains V_+. Consider the domain $T = B_R(0) \setminus \overline{V}_+$ enclosed between Σ and $\partial B_R(0)$. Suppose there exist two different solutions of the problem u_1 and u_2. Consider $u = u_1 - u_2$. Apply the first Green identity choosing $\varphi = \psi = u$:

$$\int_T u \, \Delta u \, dV + \int_T (\operatorname{grad} u, \operatorname{grad} u) \, dV = \int_\Sigma u \frac{\partial u}{\partial \nu} d\sigma + \int_{\partial B_R(0)} u \frac{\partial u}{\partial \nu} d\sigma.$$

Since $\frac{\partial u}{\partial \nu}\big|_\Sigma = 0$, we obtain

$$\int_T (\operatorname{grad} u, \operatorname{grad} u) \, dV = \int_{\partial B_R(0)} u \frac{\partial u}{\partial \nu} d\sigma.$$

Consider first $n \geq 3$. Due to Theorem 128 on the behavior of harmonic functions at infinity, we have

$$|u\|_{\partial B_R(0)} \leq \frac{C}{R^{n-2}}, \qquad \left|\frac{\partial u}{\partial \nu}\right\|_{\partial B_R(0)} \leq \frac{C}{R^{n-1}}.$$

Then

$$\int_T \sum_{k=1}^n \left(\frac{\partial u}{\partial x_k}\right)^2 dV \leq \int_{\partial B_R(0)} |u| \left|\frac{\partial u}{\partial \nu}\right| d\sigma$$

$$\leq \frac{C^2}{R^{2n-3}} |\partial B_R(0)| = \frac{\operatorname{Const} \cdot R^{n-1}}{R^{2n-3}} = \frac{\operatorname{Const}}{R^{n-2}}.$$

Letting $R \to \infty$, we obtain $T \to V_-$ and

$$\int_{V_-} \sum_{k=1}^n \left(\frac{\partial u}{\partial x_k}\right)^2 dV \leq 0, \quad n \geq 3.$$

Hence $\frac{\partial u}{\partial x_k} = 0$, $k = 1, \ldots, n$ and $u \equiv \operatorname{Const}$. Since for $n \geq 3$ a harmonic function decays at infinity, we obtain $\operatorname{Const} = 0$ and $u \equiv 0$.

Let $n = 2$. Then, due to Theorem 128,

$$\left|\frac{\partial u}{\partial \nu}\right| < \frac{\operatorname{Const}}{R^2}.$$

Thus,

$$\int_T \sum_{k=1}^{2} \left(\frac{\partial u}{\partial x_k} \right)^2 dV \leq \frac{\text{Const}}{R^2} 2\pi R = \frac{\text{Const}}{R} \to 0 \quad \text{when } R \to \infty.$$

Again this implies $u = \text{Const}$. However, differently from the case $n \geq 3$, now solutions that are only bounded at infinity are admissible. Hence, in general, u is not necessarily zero, and $u_1 = u_2 + \text{Const}$.   ∎

Similarly, to the uniqueness, the question on stability of solutions of the Dirichlet and Neumann boundary value problems can be studied. We restrict ourselves to the consideration of Problem D_+.

Theorem 134 *A solution of Problem D_+ is stable (i.e., small changes in boundary data produce small changes in the solution).*

Proof Let u_1 and u_2 be two solutions of Problem D_+ corresponding to the different boundary conditions:

$$\begin{cases} \Delta u_1 = 0 \quad \text{in } V_+, \\ u_1|_\Sigma = \varphi_1, \end{cases} \quad \text{and} \quad \begin{cases} \Delta u_2 = 0 \quad \text{in } V_+, \\ u_2|_\Sigma = \varphi_2. \end{cases}$$

Due to the maximum principle, we have

$$\max_{M \in \overline{V}_+} |u_1(M) - u_2(M)| \leq \max_{P \in \Sigma} |u_1(P) - u_2(P)|,$$

i.e.,

$$\max_{M \in \overline{V}_+} |u_1(M) - u_2(M)| \leq \max_{P \in \Sigma} |\varphi_1(P) - \varphi_2(P)|.$$

Hence, if $|\varphi_1(P) - \varphi_2(P)| < \varepsilon$ for all $P \in \Sigma$, then $|u_1(M) - u_2(M)| < \varepsilon$ for all $M \in \overline{V}_+$.   ∎

10.3   Green's Function of the Dirichlet Problem

Consider the interior Dirichlet problem D_+.

$$\begin{cases} \Delta u = 0 \quad \text{in } V_+, \\ u(P) = \varphi(P), \quad P \in \Sigma. \end{cases}$$

According to Property 2 of harmonic functions, at any point $M \in V_+$, we have the equality

$$u(M) = C_n \int_\Sigma \left(\frac{1}{r_{PM}^{n-2}} \frac{\partial u(P)}{\partial v} - \varphi(P) \frac{\partial}{\partial v} \left(\frac{1}{r_{PM}^{n-2}} \right) \right) d\sigma_P. \qquad (10.2)$$

This formula does not yet provide a solution of Problem D_+, because the values of $\frac{\partial u(P)}{\partial v}$, $P \in \Sigma$, are still unknown. Consider a new function

$$G(P, M) = \frac{C_n}{r_{PM}^{n-2}} + v(P),$$

where a harmonic function v is still to be found. We suppose v to be continuous up to the boundary. Let us substitute it into (10.2),

$$u(M) = \int_\Sigma \left((G(P, M) - v(P)) \frac{\partial u(P)}{\partial v} - \varphi(P) \frac{\partial}{\partial v} (G(P, M) - v(P)) \right) d\sigma_P$$

$$= \int_\Sigma G(P, M) \frac{\partial u(P)}{\partial v} d\sigma_P - \int_\Sigma \varphi(P) \frac{\partial G(P, M)}{\partial v} d\sigma_P$$

$$+ \int_\Sigma \left(u(P) \frac{\partial v(P)}{\partial v} - v(P) \frac{\partial u(P)}{\partial v} \right) d\sigma_P.$$

The last integral equals zero due to the second Green identity. Let us choose the arbitrary until now harmonic function v in the following form:

$$v(P) = -\frac{C_n}{r_{PM}^{n-2}} \quad \text{for all } P \in \Sigma. \qquad (10.3)$$

This leads to the equality

$$G(P, M) = 0 \quad \text{for all } P \in \Sigma.$$

Then

$$u(M) = -\int_\Sigma \varphi(P) \frac{\partial G(P, M)}{\partial v_P} d\sigma_P. \qquad (10.4)$$

The subindex P in the normal means that the normal is considered at the point P.

All the reasonings were applied to $n \geq 3$. When $n = 2$, the function $G(P, M)$ is sought in the form

$$G(P, M) = \frac{1}{2\pi} \ln \frac{1}{r_{PM}} + v(P).$$

Choosing the harmonic function v satisfying the boundary condition

$$v(P) = -\frac{1}{2\pi} \ln \frac{1}{r_{PM}} \quad \text{for all } P \in \Sigma \tag{10.5}$$

leads again to the formula (10.4).

Thus, the function $G(P, M)$ is defined by the properties.

1. The function $G(P, M)$ is harmonic in the domain V_+ everywhere, except for the point $M \in V_+$, where $G(P, M)$ has a singularity of the form

$$G(P, M) \sim_{P \to M} \begin{cases} \dfrac{C_n}{r_{PM}^{n-2}}, & n \geq 3, \\ \dfrac{1}{2\pi} \ln \dfrac{1}{r_{PM}}, & n = 2. \end{cases}$$

2. $G(P, M) = 0$ for all $P \in \Sigma$.

Definition 135 A function $G(P, M)$ satisfying these two properties is called **Green's function of the Dirichlet problem**.

If a Green's function is known for a domain V_+, then with the aid of (10.4) a solution of Problem D_+ can be found for any given boundary data φ. In turn, Green's function is known if there is constructed the solution $v(P)$, i.e., the Dirichlet problem with the special boundary conditions (10.3) for $n \geq 3$ or (10.5) for $n = 2$ is solved.

Thus, the concept of Green's function allows one to reduce solution of Problem D_+ with arbitrary boundary data φ to solution of Problem D_+ with a special boundary condition (10.3) or (10.5).

10.4   Properties of Green's Function

Property 1 Green's function is strictly positive inside the domain, and the following inequalities hold:

$$0 < G(P, M) < \frac{C_n}{r_{PM}^{n-2}}, \quad n \geq 3, \quad P, M \in V_+, \quad P \neq M, \tag{10.6}$$

$$0 < G(P, M) < \frac{1}{2\pi} \ln \frac{1}{r_{PM}}, \quad n = 2, \quad P, M \in V_+, \quad P \neq M. \tag{10.7}$$

Proof Let $n \geq 3$. We have

$$G(P, M) = \frac{C_n}{r_{PM}^{n-2}} + v(P, M).$$

For every M fixed, the function v as a function of P is harmonic in V_+ and satisfies (10.3) on the boundary, which shows that

$$v(P, M) < 0 \quad \text{for all } P \in \Sigma.$$

Hence, by the maximum principle, $v(P, M) < 0$ everywhere in V_+. This implies the inequality

$$G(P, M) < \frac{C_n}{r_{PM}^{n-2}} \quad \text{everywhere in } V_+. \tag{10.8}$$

It remains to show that $G(P, M) > 0$. Let $B_\varepsilon(M)$ be a ball of radius ε centered at M. Consider the domain T_ε enclosed between the surfaces Σ and $\partial B_\varepsilon(M)$: $T_\varepsilon = V_+ \backslash \overline{B}_\varepsilon(M)$. Green's function is harmonic in T_ε (with respect to P), and

$$G(P, M) = 0 \quad \text{for all } P \in \Sigma,$$

$$G(P, M) = \frac{C_n}{\varepsilon_{PM}^{n-2}} + v(P, M) \quad \text{for all } P \in \partial B_\varepsilon(M).$$

Since $v(P, M)$ is harmonic in the neighborhood of M, it is bounded. Hence, for a sufficiently small ε, the term $\frac{C_n}{\varepsilon_{PM}^{n-2}}$ becomes dominant, and $G(P, M) > 0$ for all $P \in \partial B_\varepsilon(M)$. Then application of the maximum and minimum principle to the function $G(P, M)$ in the domain T_ε leads to the inequality $G(P, M) > 0$ for all $P \in T_\varepsilon$. Since ε can be chosen arbitrarily small, we conclude that $G(P, M)$ is positive for all $P \in V_+$ except for the point $P = M$. This result together with (10.8) gives us (10.6). The proof of (10.7) is analogous. ∎

Property 2 The normal derivative of Green's function satisfies the inequality

$$\left. \frac{\partial G(P, M)}{\partial \nu_P} \right|_\Sigma \leq 0, \tag{10.9}$$

and moreover,

$$\int_\Sigma \frac{\partial G(P, M)}{\partial \nu_P} d\sigma = -1. \tag{10.10}$$

Proof Due to Property 1, Green's function decays when P tends to the boundary from inside of the domain. This observation leads to (10.9).

In order to prove (10.10), let us consider Problem D_+ with the constant boundary data $\varphi \equiv 1$. The harmonic function $u \equiv 1$ satisfies this Dirichlet problem, and due to Theorem 129, this solution is unique. Formula (10.4) written for $u \equiv 1$ has the form

$$1 \equiv -\int_{\Sigma} \frac{\partial G(P, M)}{\partial v_P} d\sigma.$$

∎

Property 3 Green's function is symmetric, i.e.,

$$G(P, M) = G(M, P). \tag{10.11}$$

Proof Let M_0' and M_0'' be two different points of the domain V_+. Consider two Green functions

$$G' = G(P, M_0') \quad \text{and} \quad G'' = G(P, M_0'').$$

Let us apply to them the second Green identity in a domain $T_\varepsilon = V_+ \setminus \left(\overline{B}_\varepsilon(M_0') \cup \overline{B}_\varepsilon(M_0'') \right)$, where $\varepsilon > 0$ is sufficiently small. The domain T_ε is depicted in Fig. 10.2. Both Green's functions G' and G'' are harmonic in T_ε. Thus, from

$$\int_{\Sigma + \partial B_\varepsilon(M_0') + \partial B_\varepsilon(M_0'')} \left(G' \frac{\partial G''}{\partial v} - G'' \frac{\partial G'}{\partial v} \right) d\sigma - \int_{T_\varepsilon} (G' \Delta G'' - G'' \Delta G') dV = 0,$$

we obtain

$$\int_{\partial B_\varepsilon(M_0')} \left(G' \frac{\partial G''}{\partial v} - G'' \frac{\partial G'}{\partial v} \right) d\sigma = \int_{\partial B_\varepsilon(M_0'')} \left(G'' \frac{\partial G'}{\partial v} - G' \frac{\partial G''}{\partial v} \right) d\sigma,$$

where besides the harmonicity of G' and G'' we took into account that they satisfy the homogeneous Dirichlet condition on Σ.

Let $\varepsilon \to 0$ applying the mean value theorem. We have

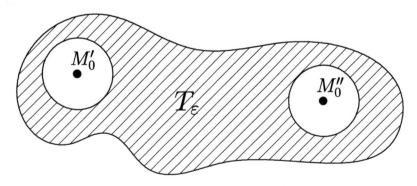

Fig. 10.2 Domain T_ε

$$I_1 = \int_{\partial B_\varepsilon(M_0')} G' \frac{\partial G''}{\partial \nu} d\sigma = G' \frac{\partial G''}{\partial \nu}\Big|_{P=\overline{P}} \cdot |\partial B_\varepsilon(M_0')|$$

$$= \left(v'(\overline{P}, M') + \frac{C_n}{\varepsilon^{n-2}} \right) \frac{\partial G''(\overline{P}, M_0')}{\partial \nu} \cdot \text{Const} \cdot \varepsilon^{n-1}, \quad \overline{P} \in \partial B_\varepsilon(M_0').$$

Since G'' is harmonic in the neighborhood of the point M_0', the function $\frac{\partial G''(\overline{P}, M_0')}{\partial \nu}$ is bounded there. Hence

$$|I_1| = \text{Const} \cdot \frac{\varepsilon^{n-1}}{\varepsilon^{n-2}} \to 0 \quad \text{when } \varepsilon \to 0.$$

Similarly,

$$|I_2| = \int_{\partial B_\varepsilon(M_0'')} G'' \frac{\partial G'}{\partial \nu} d\sigma \to 0 \quad \text{when } \varepsilon \to 0.$$

Thus,

$$\lim_{\varepsilon \to 0} \int_{\partial B_\varepsilon(M_0')} G'' \frac{\partial G'}{\partial \nu} d\sigma = \lim_{\varepsilon \to 0} \int_{\partial B_\varepsilon(M_0'')} G' \frac{\partial G''}{\partial \nu} d\sigma.$$

That is,

$$\lim_{\varepsilon \to 0} \int_{\partial B_\varepsilon(M_0')} G'' \frac{\partial}{\partial r} \left(v' + \frac{C_n}{r^{n-2}} \right) d\sigma = \lim_{\varepsilon \to 0} \int_{\partial B_\varepsilon(M_0'')} G' \frac{\partial}{\partial r} \left(v'' + \frac{C_n}{r^{n-2}} \right) d\sigma.$$

We differentiate and apply the mean value theorem

$$\lim_{\varepsilon \to 0} G''(\overline{P}', M_0'') \left(\frac{\partial v'}{\partial r} + \frac{(2-n) C_n}{\varepsilon^{n-1}} \right) \varepsilon^{n-1} |\partial B_\varepsilon(M_0')|$$

$$= \lim_{\varepsilon \to 0} G'(\overline{P}'', M_0') \left(\frac{\partial v''}{\partial r} + \frac{(2-n) C_n}{\varepsilon^{n-1}} \right) \varepsilon^{n-1} |\partial B_\varepsilon(M_0')|,$$

where $\overline{P}' \in \partial B_\varepsilon(M_0')$, $\overline{P}'' \in \partial B_\varepsilon(M_0'')$, and $\overline{P}' \to M_0'$, $\overline{P}'' \to M_0''$ when $\varepsilon \to 0$. We took also into account that the areas of the spheres $\partial B_\varepsilon(M_0')$ and $\partial B_\varepsilon(M_0'')$ coincide. From this equality at the limit, we obtain that $G''(M_0', M_0'') = G'(M_0'', M_0')$. ∎

10.5 Physical Meaning of Green's Function

Let $n = 3$. In this case Green's function has the form

$$G(P, M) = \frac{1}{4\pi r_{PM}} + v(P).$$

It has a simple physical meaning of a potential of the field created by a point source. According to Coulomb's law, at an arbitrary point P, the potential of a unitary point source (electric charge) located at a point M in free space equals $1/(4\pi r_{PM})$. Put this charge inside a hollow grounded conductor. Then, on the boundary Σ of such domain, i.e., on the conducting surface, there will be induced charges. Thus, the potential of the field generated by the point source inside the domain will be a sum of the potential $1/(4\pi r_{PM})$ and the potential $v(P)$ of the field generated by the charges induced on Σ. The total potential on the grounded boundary equals zero. Thus,

$$\frac{1}{4\pi r_{PM}} + v(P) = 0 \quad \text{for all } P \in \Sigma.$$

This is the homogeneous Dirichlet condition satisfied by Green's function $G(P, M)$, and hence Green's function can be interpreted as the potential of the field generated by a point source located inside a closed grounded conducting surface.

10.6   Method of Electrostatic Images and Solution of the Dirichlet Problem for a Ball

For constructing Green's function, the following method can be used. Let

$$G(P, M) = \frac{1}{4\pi r_{PM}} + v(P),$$

where a harmonic function v needs to be found, satisfying the condition $v(P) = -\frac{1}{4\pi r_{PM}}$, $P \in \Sigma$. The function v can be considered as the potential of a field generated by charges located outside the domain and such that $v|_{\Sigma} = -\frac{1}{4\pi r_{PM}}$. The exterior charges are called the **electrostatic images** of the unitary point source. Constructing Green's function is equivalent to calculating the potential of the electrostatic images. The method based on these considerations is called the **method of electrostatic images**. Let us see how it works when applied to Problem D_+ in a ball.

Let $B_R(0)$ be a ball of radius R centered at the origin and $M \in B_R(0)$ be a point. We denote its distance from the origin by ρ. Together with M, consider an exterior point $M^* = \text{In } V_R M$ located at a distance ρ^* from the origin, $\rho\rho^* = R^2$.

Let P be a point on $\partial B_R(0)$. Consider the segments $r = MP$ and $r^* = M^*P$. Let us verify that the distances r and r^* are related by the formula

$$\frac{1}{r^*}\bigg|_{\partial B_R(0)} = \frac{\rho}{R}\frac{1}{r}\bigg|_{\partial B_R(0)}. \tag{10.12}$$

Indeed, consider the triangles OPM and OPM^* with one of the vertices being the origin (denoted by O). These triangles are similar, since they have one common angle and

$$\frac{OM}{R} = \frac{R}{OM^*}.$$

From the similarity of the triangles, the equality follows

$$\frac{MP}{M^*P} = \frac{OM}{OP},$$

i.e.,

$$\frac{r}{r^*} = \frac{\rho}{R},$$

which is (10.12).

Now consider the function

$$v(P, M) = -\frac{1}{4\pi}\frac{R}{\rho}\frac{1}{r_{PM^*}}.$$

It is harmonic everywhere except for the point $P = M^*$. In particular, it is harmonic in $B_R(0)$. On the boundary of the ball, due to (10.12), we have

$$v(P, M) = -\frac{1}{4\pi r_{PM}} \quad \text{for all } P \in \partial B_R(0).$$

Hence

$$G(P, M) = \frac{1}{4\pi r_{PM}} + v(P, M) = \frac{1}{4\pi}\left(\frac{1}{r_{PM}} - \frac{R}{\rho}\frac{1}{r_{PM^*}}\right)$$

is Green's function of the Dirichlet problem

$$\begin{cases} \Delta u = 0 & \text{in } B_R(0), \\ u|_{\partial B_R(0)} = \varphi. \end{cases}$$

Thus, its solution has the form

$$u(M) = -\int_{\partial B_R(0)} \varphi(P)\frac{\partial G(P, M)}{\partial \nu_P}d\sigma$$

$$= -\frac{1}{4\pi} \int_{\partial B_R(0)} \varphi(P) \frac{\partial}{\partial \nu_P} \left(\frac{1}{r_{PM}} - \frac{R}{\rho} \frac{1}{r_{PM^*}} \right) d\sigma.$$

Let us perform the differentiation. To simplify the notations, we write $r = r_{PM}$ and $r^* = r_{PM^*}$. We have

$$\frac{\partial}{\partial \nu} \left(\frac{1}{r} \right) = -\frac{1}{r^2} \frac{\partial r}{\partial \nu} = -\frac{1}{r^2} \cos(\widehat{r, \nu}),$$

$$\frac{\partial}{\partial \nu} \left(\frac{1}{r^*} \right) = -\frac{1}{r^{*2}} \frac{\partial r^*}{\partial \nu} = -\frac{1}{r^{*2}} \cos\left(\widehat{r^*, \nu}\right),$$

where

$$\frac{\partial r}{\partial \nu} = \frac{\partial r}{\partial x} \cos(\widehat{\nu, x}) + \frac{\partial r}{\partial y} \cos(\widehat{\nu, y}) + \frac{\partial r}{\partial z} \cos(\widehat{\nu, z}) = \cos(\widehat{r, \nu}).$$

Thus,

$$\frac{\partial G}{\partial \nu} = \frac{1}{4\pi} \left(\frac{\cos(\widehat{r^*, \nu})}{r^{*2}} \frac{R}{\rho} - \frac{\cos(\widehat{r, \nu})}{r^2} \right). \tag{10.13}$$

Considering again the triangles OPM and OPM^*, we find that

$$\rho^2 = R^2 + r^2 - 2Rr \cos(\widehat{r, \nu})$$

and

$$\rho^{*2} = R^2 + r^{*2} - 2Rr^* \cos(\widehat{r^*, \nu}).$$

Let us substitute the expressions for $\cos(\widehat{r, \nu})$ and $\cos(\widehat{r^*, \nu})$ derived from here into (10.13):

$$\frac{\partial G}{\partial \nu} = \frac{1}{4\pi} \left(\frac{R}{\rho r^{*2}} \frac{R^2 + r^{*2} - \rho^{*2}}{2Rr^*} - \frac{R^2 + r^2 - \rho^2}{2Rrr^2} \right).$$

Substituting

$$r^* = \frac{R}{\rho} r,$$

we obtain

$$\frac{\partial G}{\partial v} = \frac{1}{4\pi}\left(\frac{R}{\rho}\frac{\rho^2}{R^2r^2}\frac{R^2 + \frac{R^2}{\rho^2}r^2 - \frac{R^4}{\rho^2}}{2R\frac{R}{\rho}r} - \frac{R^2 + r^2 - \rho^2}{2Rr^3}\right)$$

$$= \frac{1}{4\pi}\left(\frac{R^2\rho^2 + R^2r^2 - R^4}{2R^3r^3} - \frac{R^2 + r^2 - \rho^2}{2Rr^3}\right)$$

$$= \frac{1}{8\pi R^3r^3}\left(R^2\rho^2 + R^2r^2 - R^4 - R^4 - R^2r^2 + R^2\rho^2\right)$$

$$= \frac{1}{4\pi R^3r^3}\left(R^2\rho^2 - R^4\right) = \frac{1}{4\pi Rr^3}\left(\rho^2 - R^2\right)$$

$$= -\frac{1}{4\pi R}\frac{R^2 - \rho^2}{r^3}.$$

Thus,

$$u(M) = \frac{1}{4\pi R}\int_{\partial B_R(0)} \varphi(P)\frac{R^2 - \rho^2}{r^3}d\sigma_P$$

$$= \frac{1}{4\pi R}\int_{\partial B_R(0)} \varphi(P)\frac{R^2 - r_{OM}^2}{r_{PM}^3}d\sigma_P.$$

This integral is called the **Poisson integral for the ball**. Let us write it in spherical coordinates

$$\begin{cases} x = r\sin\theta\cos\phi, \\ y = r\sin\theta\sin\phi, \\ z = r\cos\theta. \end{cases}$$

Then $d\sigma = R^2\sin\theta d\theta d\phi$, and

$$u(M) = \frac{R}{4\pi}\int_0^{2\pi}\int_0^{\pi} \varphi(\theta, \phi)\frac{(R^2 - \rho^2)\sin\theta d\theta d\phi}{(R^2 - 2R\rho\cos\gamma + \rho^2)^{3/2}}, \tag{10.14}$$

where γ is the angle between the vectors $\overrightarrow{OP}$ and $\overrightarrow{OM}$. Since

$$\frac{\overrightarrow{OM}}{|OM|} = \{\sin\theta_M\cos\phi_M, \ \sin\theta_M\sin\phi_M, \ \cos\theta_M\},$$

$$\frac{\overrightarrow{OP}}{|OP|} = \{\sin\theta\cos\phi, \ \sin\theta\sin\phi, \ \cos\theta\},$$

we have

$$\cos \gamma = \sin \theta_M \sin \theta \left(\cos \phi_M \cos \phi + \sin \phi_M \sin \phi \right) + \cos \theta_M \cos \theta$$
$$= \sin \theta_M \sin \theta \cos (\phi - \phi_M) + \cos \theta_M \cos \theta.$$

The same formula (10.14) but with an opposite sign gives us the solution for the exterior Dirichlet problem in the domain exterior to the sphere.

10.7   Solution of the Dirichlet Problem in a Disk

In the case of a ball, Green's function was obtained in the form

$$G(P, M) = \frac{1}{4\pi} \left(\frac{1}{r_{PM}} - \frac{R}{\rho} \frac{1}{r_{PM^*}} \right).$$

Similar reasonings in the planar case lead to Green's function of the form

$$G(P, M) = \frac{1}{2\pi} \left(\ln \frac{1}{r_{PM}} - \ln \left(\frac{R}{\rho} \frac{1}{r_{PM^*}} \right) \right).$$

After transformations analogous to those for a ball, we obtain

$$\left. \frac{\partial G}{\partial \nu} \right|_{C_R} = -\frac{1}{2\pi R} \frac{R^2 - \rho^2}{r_{PM}^2},$$

and the solution of Problem D_+

$$\begin{cases} \Delta u = 0, \\ u|_{C_R} = f \end{cases}$$

has the form

$$u(M) = \frac{1}{2\pi R} \int_{C_R} f(P) \frac{R^2 - \rho^2}{r_{PM}^2} dl.$$

This is the **Poisson integral for a disk**. In the polar coordinates, it has the form

$$u(\rho_M, \theta_M) = \frac{1}{2\pi} \int_0^{2\pi} f(\theta) \frac{R^2 - \rho_M^2}{R^2 + \rho_M^2 - 2R\rho_M \cos(\theta - \theta_M)} d\theta.$$

The same formula but with an opposite sign represents the solution for the exterior problem.

10.8 Solution of the Dirichlet Problem for a Half-Plane

Let us solve the Dirichlet problem for the upper half-plane

$$\begin{cases} u_{xx} + u_{yy} = 0, & y > 0, \\ u(x, 0) = f(x). \end{cases}$$

Let $M_0 = (x_0, y_0)$ be a point of the upper half-plane. Following the method of electrostatic images, we introduce a symmetric point $M_0^* = (x_0, -y_0)$. Denote $r = MM_0$ and $r^* = MM_0^*$. Then the function

$$G(M, M_0) = \frac{1}{2\pi} \left(\ln \frac{1}{r} - \ln \frac{1}{r^*} \right)$$

is Green's function of the problem.

Obviously,

$$\left. \frac{\partial G}{\partial v} \right|_{\Sigma} = -\left. \frac{\partial G}{\partial y} \right|_{y=0} = \frac{1}{2\pi} \left. \left(\frac{1}{r} \frac{\partial r}{\partial y} - \frac{1}{r^*} \frac{\partial r^*}{\partial y} \right) \right|_{y=0}$$

$$= \frac{1}{2\pi} \left. \left(\frac{1}{r} \frac{y - y_0}{r} - \frac{1}{r^*} \frac{y + y_0}{r^*} \right) \right|_{y=0} = -\frac{y_0}{\pi} \frac{1}{(x - x_0)^2 + y_0^2}.$$

Hence the solution has the form

$$u(x_0, y_0) = \frac{1}{\pi} \int_{-\infty}^{\infty} \frac{y_0}{(x - x_0)^2 + y_0^2} f(x) dx.$$

This is the **Poisson integral for a half-plane**.

10.9 Solution of the Dirichlet Problem for a Half-Space

The Dirichlet problem for a half-space

$$\begin{cases} u_{xx} + u_{yy} + u_{zz} = 0, & z > 0, \\ u(x, y, 0) = f(x, y) \end{cases}$$

is solved similarly to the two-dimensional case.

We have

$$G(M, M_0) = \frac{1}{4\pi} \left(\frac{1}{r} - \frac{1}{r^*} \right), \quad r = r_{MM_0}, \quad r^* = r_{MM_0^*},$$

where

$$M_0 = (x_0, y_0, z_0) \,, \quad M_0^* = (x_0, y_0, -z_0) \,,$$

$$r = \sqrt{(x - x_0)^2 + (y - y_0)^2 + (z - z_0)^2}, \quad r^* = \sqrt{(x - x_0)^2 + (y - y_0)^2 + (z + z_0)^2},$$

$$\left. \frac{\partial G}{\partial \nu} \right|_{\Sigma} = - \left. \frac{\partial G}{\partial z} \right|_{z=0} = \frac{1}{4\pi} \left. \left(\frac{1}{r^2} \frac{z - z_0}{r} - \frac{1}{(r^*)^2} \frac{z + z_0}{r^*} \right) \right|_{z=0} = \frac{1}{2\pi} \frac{z_0}{r^3},$$

and thus,

$$u(x_0, y_0, z_0) = \frac{1}{2\pi} \int_{-\infty}^{\infty} \int_{-\infty}^{\infty} \frac{z_0}{\left((x - x_0)^2 + (y - y_0)^2 + z_0^2 \right)^{3/2}} f(x, y) dx dy.$$

This is the **Poisson integral for a half-space**.

Summary 136 *We derived formulas for the Dirichlet problems for a ball, disk, half-plane, and half-space, under the assumption that corresponding solutions exist. It remains to clarify the question on their existence for which corresponding existence theorem should be proved. For this purpose, we continue with studying the potential theory and the theory of integral equations. With the aid of the potential theory, we reduce the boundary value problems to integral equations and applying the theory of integral equations prove the existence theorems.*

Chapter 11
Potential Theory

We begin by considering some analogies from field theory.

(I) If a charge q is located at a point P of $\mathbb{R}^3$, the field potential at a point M equals $\frac{q}{4\pi r_{PM}}$. If charges are distributed over some surface Σ, the field potential equals

$$\frac{1}{4\pi} \int_\Sigma \frac{q(P)}{r_{PM}} d\sigma_P,$$

where $q(P)$ is the surface charge distribution density.

(II) Another example is given by the gravitational field potential. If a mass m is concentrated at a point P, its field potential at a point M equals $\gamma \frac{m}{r_{PM}}$, where γ is the gravitational constant. If the mass is distributed over a volume V, the field potential equals

$$\gamma \int_V \frac{q(P)}{r_{PM}} dV_P,$$

where $q(P)$ is the mass distribution density. This integral is called **Newton potential** or volume mass potential.

(III) Now suppose that the charges are distributed on both sides of a surface, positive ones from one side and negative ones from the other. They form a so-called **double layer**. In other words, the surface is filled by dipoles. As it is proved in electrostatics, the potential of such double layer equals

$$\int_\Sigma \mu(P) \frac{\partial}{\partial \nu} \left(\frac{1}{r_{PM}} \right) d\sigma_P,$$

where ν is the exterior normal vector and $\mu(P)$ is the distribution density of the dipoles.

© The Author(s), under exclusive license to Springer Nature Switzerland AG 2022
A. N. Karapetyants, V. V. Kravchenko, *Methods of Mathematical Physics*,
https://doi.org/10.1007/978-3-031-17845-0_11 243

Definition 137 Let Σ be an arbitrary oriented surface and V an arbitrary domain in an n-dimensional space, $n \geq 3$. Let $r = r_{PM}$. The integrals

$$u(M) = \int_V \frac{q(P)}{r^{n-2}} dV_P, \tag{11.1}$$

$$v(M) = \int_\Sigma \frac{q(P)}{r^{n-2}} d\sigma_P, \tag{11.2}$$

and

$$w(M) = \int_\Sigma q(P) \frac{\partial}{\partial v} \left(\frac{1}{r^{n-2}} \right) d\sigma_P \tag{11.3}$$

are called the **volume potential**, **single layer potential**, and **double layer potential**, respectively.

The function $q(P)$ is called the **density** of the corresponding potential. Sometimes the potentials (11.1)–(11.3) will be denoted as $u_q(M)$, $v_q(M)$, and $w_q(M)$ to emphasize that their corresponding density is q.

Obviously, the integral representation of a harmonic function in terms of the potentials and their densities can be written as follows:

$$u(M) = C_n \left(v_{\frac{\partial u}{\partial v}}(M) - w_u(M) \right).$$

In a two-dimensional case ($n = 2$) instead of (11.1)–(11.3), we study the integrals

$$u(M) = \int_S q(P) \ln \frac{1}{r} dS_P, \tag{11.4}$$

$$v(M) = \int_L q(P) \ln \frac{1}{r} dl_P, \tag{11.5}$$

and

$$w(M) = \int_L q(P) \frac{\partial}{\partial v} \ln \frac{1}{r} dl_P. \tag{11.6}$$

The integral (11.4) is called **logarithmic potential**, and the integrals (11.5) and (11.6) are known as **logarithmic potentials of single and double layer**, respectively.

Differentiating the fundamental solution of the Laplace equation in formulas (11.3) and (11.6), we obtain

$$w(M) = \begin{cases} (2-n)\int_\Sigma q(P)\frac{\cos(r,v)}{r^{n-1}}d\sigma_P, & n \geq 3, \\[2mm] -\int_L q(P)\frac{\cos(r,v)}{r}dl_P, & n = 2. \end{cases}$$

11.1 Theorems on the Volume Potential

Let V be a bounded domain in $\mathbb{R}^n$, $n \geq 3$, with a piecewise smooth boundary Σ.

Theorem 138 *If the density of the volume potential is continuous in a closed domain $\overline{V}$, the volume potential is continuous in the whole space together with its partial derivatives of first order, and differentiation can be performed under the integral sign*

$$\frac{\partial u}{\partial x_k} = \int_V q(P)\frac{\partial}{\partial x_k}\left(\frac{1}{r^{n-2}}\right)dV_P, \quad k = 1, 2, \ldots, n. \tag{11.7}$$

Proof Let us divide the proof into three parts: (I) $M \notin \overline{V}$, (II) $M \in V$, and (III) $M \in \Sigma$.

(I) In the first case the statement is obvious because when $M \notin \overline{V}$ the integral

$$u(M) = \int_V \frac{q(P)}{r_{PM}^{n-2}}dV_P$$

is not improper and can be differentiated under the integral sign any number of times.

In cases (II) and (III) for simplicity, let us restrict our consideration to $n = 3$. Consider an auxiliary function

$$g_\varepsilon(P, M) = \begin{cases} \frac{1}{r_{PM}}, & r_{PM} > \varepsilon, \\[2mm] \alpha + \beta r_{PM}^2, & r_{PM} \leq \varepsilon, \end{cases}$$

where the numbers α and β are chosen such that the function $g_\varepsilon(P, M)$ be continuously differentiable:

$$\begin{cases} \left.\frac{1}{r}\right|_{r=\varepsilon} = \alpha + \beta r^2\big|_{r=\varepsilon}, \\[3mm] \left.-\frac{1}{r^2}\right|_{r=\varepsilon} = 2\beta r|_{r=\varepsilon}. \end{cases}$$

From here, we obtain that

$$
\begin{cases}
\alpha + \beta \varepsilon^2 = \frac{1}{\varepsilon}, \\[2mm]
2\beta \varepsilon = -\frac{1}{\varepsilon^2}.
\end{cases}
$$

Thus,

$$
\alpha = \frac{3}{2\varepsilon} \quad \text{and} \quad \beta = -\frac{1}{2\varepsilon^3}.
$$

Consider a new integral

$$
u_\varepsilon(M) = \int_V q(P) g_\varepsilon(P, M) dV_P.
$$

Let us show that

$$
u_\varepsilon(M) \to u(M) \tag{11.8}
$$

uniformly when $\varepsilon \to 0$. Let us begin with case (II).

(II) We have

$$
\begin{aligned}
|u(M) - u_\varepsilon(M)| &= \left| \int_V q(P) \left(\frac{1}{r} - g_\varepsilon \right) dV \right| \\[2mm]
&= \left| \int_{B_\varepsilon(M)} q(P) \left(\frac{1}{r} - g_\varepsilon \right) dV \right| \\[2mm]
&\leq \max_{P \in \overline{V}} |q(P)| \int_{B_\varepsilon(M)} \left(\frac{1}{r} + |g_\varepsilon| \right) dV \\[2mm]
&= \text{Const} \int_0^\varepsilon \int_0^{2\pi} \int_0^\pi \left(\frac{1}{r} + |\alpha + \beta r^2| \right) r^2 \sin\theta \, d\theta \, d\phi \, dr \\[2mm]
&\leq \text{Const} \int_0^\varepsilon \int_0^{2\pi} \int_0^\pi \left(r + \frac{3r^2}{2\varepsilon} + \frac{r^4}{2\varepsilon^3} \right) d\theta \, d\phi \, dr \\[2mm]
&= \text{Const} \int_0^\varepsilon \left(r + \frac{3r^2}{2\varepsilon} + \frac{r^4}{2\varepsilon^3} \right) dr \leq \text{Const} \cdot \varepsilon^2,
\end{aligned}
$$

and hence (11.8) is true. Note that the convergence is uniform because Const here does not depend on M.

(III) The point M is located on the boundary Σ. Consider a ball $B_\varepsilon(M)$, and by D_ε denote its part belonging to V: $D_\varepsilon = B_\varepsilon(M) \cap V$ (Fig. 11.1). Then

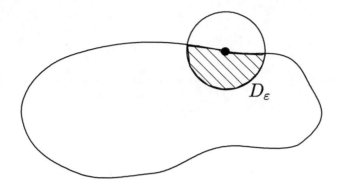

Fig. 11.1 Domain D_ε

$$|u(M) - u_\varepsilon(M)| \le \text{Const} \int_{D_\varepsilon} \left(\frac{1}{r} + |g_\varepsilon| \right) dV \le \text{Const} \int_{B_\varepsilon(M)} \left(\frac{1}{r} + |g_\varepsilon| \right) dV,$$

and repeating the reasonings from case (II), we obtain the result.

Thus, the potential $u(M)$ is a uniform limit of the function $u_\varepsilon(M)$. Obviously, all $u_\varepsilon(M)$ are continuous functions of M, and hence $u(M)$ is also continuous.

It remains to prove the continuous differentiability of $u(M)$ and the validity of (11.7). Denote

$$A(M) = \int_V q(P) \frac{\partial}{\partial x_k} \left(\frac{1}{r_{PM}} \right) dV.$$

We need to prove that $A(M) = \frac{\partial u}{\partial x_k}$. Obviously, the function $u_\varepsilon(M)$ is differentiable under the integral sign

$$\frac{\partial u_\varepsilon}{\partial x_k} = \int_V q(P) \frac{\partial g_\varepsilon}{\partial x_k} dV.$$

From calculus, we know that if

1. $u_\varepsilon \to u$ uniformly
2. $\frac{\partial u_\varepsilon}{\partial x_k} \to A$ uniformly

then $A = \frac{\partial u}{\partial x_k}$. Hence, it remains to prove that $A - \frac{\partial u}{\partial x_k} \to 0$ uniformly when $\varepsilon \to 0$.

To simplify the notation, we write $x = x_k$. We have then

$$\left| A - \frac{\partial u_\varepsilon}{\partial x} \right| = \left| \int_V q(P) \frac{\partial}{\partial x} \left(\frac{1}{r} - g_\varepsilon \right) dV \right| \le \text{Const} \int_{B_\varepsilon(M)} \left| \frac{\partial}{\partial x} \left(\frac{1}{r} - g_\varepsilon \right) \right| dV$$

$$\le \text{Const} \int_{B_\varepsilon(M)} \left| \frac{1}{r^2} \frac{\partial r}{\partial x} + \frac{1}{\varepsilon^3} r \frac{\partial r}{\partial x} \right| dV.$$

Since

$$\left| \frac{\partial r}{\partial x} \right| = |\cos(r, x)| \leq 1,$$

we obtain

$$\left| A - \frac{\partial u_\varepsilon}{\partial x} \right| \leq \text{Const} \int_{B_\varepsilon(M)} \left(\frac{1}{r^2} + \frac{r}{\varepsilon^3} \right) dV$$

$$= \text{Const} \int_0^\varepsilon \int_0^{2\pi} \int_0^\pi \left(\frac{1}{r^2} + \frac{r}{\varepsilon^3} \right) r^2 \sin \theta \, d\theta \, d\phi \, dr$$

$$\leq \text{Const} \int_0^\varepsilon \left(1 + \frac{r^3}{\varepsilon^3} \right) dr \leq \text{Const} \cdot \varepsilon \to 0$$

uniformly.   ∎

Theorem 139 *If the density of the volume potential $u(M)$ is continuous together with its first partial derivatives in a closed domain $\overline{V}$, then the potential is twice continuously differentiable inside and outside the domain, and the differentiation can be performed under the integral sign. Moreover,*

$$\Delta u(M) = \begin{cases} 0, & M \notin \overline{V}, \\ -\frac{q(M)}{C_n}, & M \in V. \end{cases}$$

Proof If $M \notin \overline{V}$, then as it was discussed above, the differentiation under the integral sign is admissible any number of times. Thus,

$$\Delta u(M) = \int_V q(P) \Delta \left(\frac{1}{r_{PM}^{n-2}} \right) dV_P = 0.$$

It remains to consider $M \in V$. Consider a ball $B_\varepsilon(M)$ of radius ε centered at M. We have

$$u(M) = \int_{B_\varepsilon(M)} \frac{q(P)}{r_{PM}^{n-2}} dV_P + \int_{V \setminus B_\varepsilon(M)} \frac{q(P)}{r_{PM}^{n-2}} dV_P = u_\varepsilon^1(M) + u_\varepsilon^2(M).$$

Since $M \notin \overline{V \setminus B_\varepsilon(M)}$, for the function $u_\varepsilon^2(M)$, we have again that $\Delta u_\varepsilon^2(M) = 0$. Hence it remains to study the function $u_\varepsilon^1(M)$. Due to Theorem 138, we have

$$\frac{\partial u_\varepsilon^1}{\partial x_k} = \int_{B_\varepsilon(M)} q(P) \frac{\partial}{\partial x_k} \left(\frac{1}{r_{PM}^{n-2}} \right) dV_P,$$

where $P = (\xi_1, \xi_2, \ldots, \xi_n)$, $M = (x_1, x_2, \ldots, x_n)$,

$$r = r_{PM} = \sqrt{(x_1 - \xi_1)^2 + (x_2 - \xi_2)^2 + \ldots + (x_n - \xi_n)^2},$$

and hence

$$\frac{\partial r}{\partial x_k} = -\frac{\partial r}{\partial \xi_k}. \tag{11.9}$$

Taking this equality into account, we obtain

$$\frac{\partial u_\varepsilon^1}{\partial x_k} = \int_{B_\varepsilon(M)} \frac{\partial q}{\partial \xi_k} \frac{1}{r^{n-2}} dV - \int_{B_\varepsilon(M)} \frac{\partial}{\partial \xi_k} \left(\frac{q}{r^{n-2}} \right) dV$$

$$= \int_{B_\varepsilon(M)} \frac{\partial q}{\partial \xi_k} \frac{1}{r^{n-2}} dV - \int_{\partial B_\varepsilon(M)} \frac{q}{r^{n-2}} \cos(\nu, \xi_k)\, d\sigma,$$

where the Gauss–Ostrogradski theorem was applied to the second integral. Here both terms are continuously differentiable with respect to x_k. The first one due to Theorem 138 as a volume potential with a continuous density $\frac{\partial q}{\partial \xi_k}$ and the second one because it is not already an improper integral, the integration is performed on a surface while the singularity is located inside the domain.

Differentiation under the integral sign and taking into account (11.9) give

$$\frac{\partial^2 u_\varepsilon^1}{\partial x_k^2} = \int_{B_\varepsilon(M)} \frac{\partial q}{\partial \xi_k} \frac{\partial}{\partial x_k} \left(\frac{1}{r_{PM}^{n-2}} \right) dV - \int_{\partial B_\varepsilon(M)} q \cos(\nu, \xi_k) \frac{\partial}{\partial x_k} \left(\frac{1}{r_{PM}^{n-2}} \right) d\sigma$$

$$= (2 - n) \int_{B_\varepsilon(M)} \frac{\partial q}{\partial \xi_k} \frac{1}{r_{PM}^{n-1}} \cos(r_{MP}, x_k)\, dV + (2 - n) \int_{\partial B_\varepsilon(M)} q \cos(\nu, \xi_k) \frac{1}{r_{PM}^{n-1}} \cos(r_{MP}, \xi_k)\, d\sigma.$$

Denote the first term as $\alpha_k(\varepsilon, M)$. Let us show that $\alpha_k(\varepsilon, M) \to 0$ when $\varepsilon \to 0$ uniformly with respect to M. Indeed,

$$|\alpha_k(\varepsilon, M)| \leq \text{Const} \int_{B_\varepsilon(M)} \frac{dV}{r^{n-1}}$$

or in terms of spherical coordinates ($dV = r^{n-1} dr d\sigma$),

$$|\alpha_k(\varepsilon, M)| \leq \text{Const} \int_0^\varepsilon \int_{\Sigma_1} \frac{1}{r^{n-1}} r^{n-1} dr d\sigma,$$

where Σ_1 is a unitary sphere. Hence

$$|\alpha_k(\varepsilon, M)| \leq \text{Const} \int_0^\varepsilon dr = \varepsilon \cdot \text{Const} \to 0$$

uniformly, when $\varepsilon \to 0$.

Thus,

$$\frac{\partial^2 u_\varepsilon^1}{\partial x_k^2} = \alpha_k(\varepsilon, M) + (2 - n) \int_{\partial B_\varepsilon(M)} \frac{q(P)}{r^{n-1}} \cos^2(r, \xi_k) \, d\sigma_P.$$

In the surface integral, we took into account that the directions of ν and r coincide.

Since the last equality is valid for all $k = 1, 2, \ldots, n$, summing up the corresponding derivatives, we obtain

$$\Delta u_\varepsilon^1 = \sum_{k=1}^n \alpha_k(\varepsilon, M) + (2 - n) \int_{\partial B_\varepsilon(M)} \frac{q(P)}{r^{n-1}} d\sigma_P$$

$$= \sum_{k=1}^n \alpha_k(\varepsilon, M) + (2 - n) q(\overline{P}) \frac{1}{\varepsilon^{n-1}} \varepsilon^{n-1} |\Sigma_1|,$$

where $\overline{P}$ is a point of the sphere $\partial B_\varepsilon(M)$ (and $\overline{P} \to M$ when $\varepsilon \to 0$).

Finally we have

$$\Delta u = \Delta u_\varepsilon^1 + \Delta u_\varepsilon^2 = \sum_{k=1}^n \alpha_k(\varepsilon, M) + (2 - n) q(\overline{P}) |\Sigma_1|.$$

Letting $\varepsilon \to 0$, we obtain $\Delta u = (2 - n) q(M) |\Sigma_1| = -q(M)/C_n$.   ∎

Corollary 140 *The Poisson equation*

$$\Delta u = f$$

with a continuously differentiable right hand side admits a particular solution

$$u(M) = u_{-C_n f}(M) = -C_n \int_V \frac{f(P)}{r_{PM}^{n-2}} dV_P.$$

Analogous theorems can be proved for the logarithmic potential.

Theorem 141 *If the density of a logarithmic potential is continuous in a closed domain $\overline{S}$, then the logarithmic potential*

$$u(M) = \int_S q(P) \ln \frac{1}{r} dS_P$$

is continuous on the whole plane together with its first order partial derivatives and admits differentiation under the integral sign.

Theorem 142 *If the density of a logarithmic potential is continuously differentiable in a closed domain $\overline{S}$, then the logarithmic potential is twice continuously differentiable inside and outside the domain S, and*

$$\Delta u(M) = \begin{cases} 0, & M \notin \overline{S}, \\ -2\pi q(M), & M \in S. \end{cases}$$

The proofs of these two theorems are analogous to those of Theorems 138 and 139.

Next we are going to study the single and double layer potentials for which the natural requirement on the corresponding surface is its belonging to the class of so-called Lyapunov surfaces. Thus, we begin with a study of this important notion.

11.2 Lyapunov Surfaces and Their Properties

Definition 143 A function $y = f(x)$ is said to satisfy the **Hölder condition** on a set Γ if there exists a constant C and a number $0 < \lambda \leq 1$ such that

$$|f(x_1) - f(x_2)| \leq C |x_1 - x_2|^\lambda \tag{11.10}$$

for any two points $x_1, x_2 \in \Gamma$.

Note that if $\lambda > 1$, and f satisfies (11.10), then $\frac{|f(x_1)-f(x_2)|}{|x_1-x_2|} \leq C |x_1 - x_2|^{\lambda-1}$, from where letting $x_1 \to x_2$, we would obtain $|f'(x_2)| = 0 \Rightarrow f'(x_2) = 0$ and $f \equiv$ Const.

The Hölder condition characterizes the smoothness of the function and is something in between the continuity of a function and its differentiability. Continuous functions are not necessarily Hölder ones, while continuously differentiable functions satisfy necessarily the Hölder condition with $\lambda = 1$.

Lyapunov surfaces are surfaces whose normal vectors satisfy the Hölder condition.

Definition 144 A surface Σ in $\mathbb{R}^n$ is called a Lyapunov surface if

1. A normal vector exists at each point of the surface.
2. Let P_1 and P_2 be two points of Σ, $r = r_{P_1 P_2}$, and γ an angle generated by the normal vectors at points P_1 and P_2. Then there exist a positive number A and a number $0 < \alpha \leq 1$, such that $\gamma \leq Ar^\alpha$ (Hölder's condition).

The angle γ is called the **contingence angle** (see Fig. 11.2).

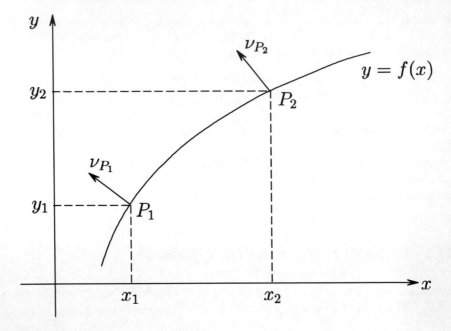

Fig. 11.2 The contingence angle γ between the normal vectors ν_{P_1} and ν_{P_2} is relatively small

Theorem 145 (Sufficient Condition of a Lyapunov Surface) *Let a surface Σ in $\mathbb{R}^n$ be defined by an explicit equation $x_n = f(x_1, x_2, \ldots, x_{n-1})$. If the partial derivatives $\frac{\partial f}{\partial x_k}$ ($k = 1, 2, \ldots, n - 1$) satisfy the Hölder condition*

$$\left| \frac{\partial f(P)}{\partial x_k} - \frac{\partial f(M)}{\partial x_k} \right| \leq C_k r_{PM}^{\lambda_k},$$

the surface Σ is a Lyapunov surface.

Proof In order to simplify calculations, let us consider the case $n = 2$ (of a Lyapunov curve). Let $y = f(x)$. Thus,

$$\left| f'(x_1) - f'(x_2) \right| \leq C \left| x_1 - x_2 \right|^{\lambda}.$$

We need to show that

$$\gamma = (\widehat{\nu_{P_1}, \nu_{P_2}}) \leq A r^{\alpha},$$

where

$$r = r_{P_1 P_2} = |P_1 - P_2| = \sqrt{(x_1 - x_2)^2 + (y_1 - y_2)^2}.$$

We have

$$v_{P_i} = \left\{ -\frac{y_i'}{\sqrt{1 + (y_i')^2}}, \frac{1}{\sqrt{1 + (y_i')^2}} \right\}, \quad i = 1, 2,$$

$$\cos \gamma = \cos(\widehat{v_{P_1}, v_{P_2}}) = \frac{y_1' y_2' + 1}{\sqrt{\left(1 + (y_1')^2\right)\left(1 + (y_2')^2\right)}},$$

$$\sin^2 \gamma = 1 - \cos^2 \gamma = 1 - \frac{(1 + y_1' y_2')^2}{\left(1 + (y_1')^2\right)\left(1 + (y_2')^2\right)} = \frac{(y_1' - y_2')^2}{\left(1 + (y_1')^2\right)\left(1 + (y_2')^2\right)} \le (y_1' - y_2')^2.$$

Hence

$$|\sin \gamma| \le |y_1' - y_2'| \le C |x_1 - x_2|^\lambda.$$

It is known that

$$\frac{2}{\pi} |\gamma| \le |\sin \gamma| \le |\gamma|. \tag{11.11}$$

With the aid of the left inequality, we obtain

$$|\gamma| \le \frac{\pi}{2} |\sin \gamma| \le \frac{\pi C}{2} |x_1 - x_2|^\lambda = A r^\alpha, \quad A = \frac{\pi C}{2}, \alpha = \lambda.$$

■

Lemma 146 (Main Property of Lyapunov Surfaces) *For any point M of a Lyapunov surface, there can be chosen a sphere of a small enough radius d centered at M, such that all straight lines parallel to the normal vector v_M intersect the surface inside the sphere at most once.*

Proof Let d be the radius of the sphere. We choose it small enough to fulfill the condition

$$A d^\alpha < 1.$$

Let us prove that this choice of d guarantees the validity of the statement. Suppose the opposite. Let a straight line parallel to the normal v_M intersect the surface inside the sphere at two points M_1 and M_2. At M_2, consider the tangent plane and the normal vector v_{M_2}. We note that the normal vectors v_{M_1} and v_{M_2} are directed to the different sides with respect to that tangent plane, and v_{M_2} is orthogonal to it. This means that necessarily $\left(\widehat{v_{M_1}, v_{M_2}}\right) \ge \frac{\pi}{2} > 1$. But, since the surface is of Lyapunov type, $\left(\widehat{v_{M_1}, v_{M_2}}\right) \le A r^\alpha \le A d^\alpha < 1$. The obtained contradiction proves the lemma.

■

Corollary 147 *If Σ_d are spheres of radius d such that $Ad^\alpha < 1$, then according to the just proved lemma, for any point of the surface, the sphere Σ_d centered at that point satisfies the requirement of Lemma 146.*

We will call all such spheres Σ_d with $d < A^{-\frac{1}{\alpha}}$ as **Lyapunov spheres**.

From this corollary and Lemma 146, we conclude that inside any Lyapunov sphere in local coordinates, the surface equation can be written in an explicit form.

Let M be a fixed point of a Lyapunov surface. In its neighborhood, let us introduce local coordinates $\xi_1, \xi_2, \ldots, \xi_n$ directing ξ_n along the normal v_M and locating the axes $\xi_1, \xi_2, \ldots, \xi_{n-1}$ in the tangent plane. Let P be an arbitrary point in the neighborhood of M, $P \in \Sigma$. Let

$$r = r_{PM} = \sqrt{\sum_{k=1}^{n} \xi_k^2}, \quad \rho = \sqrt{\sum_{k=1}^{n-1} \xi_k^2} \text{ is a projection of } r \text{ onto the tangent plane.}$$

Then $r^2 = \rho^2 + \xi_n^2$. Consider a Lyapunov sphere centered at M. The part of the surface located inside the sphere admits the explicit representation

$$\xi_n = f(\xi_1, \xi_2, \ldots, \xi_{n-1}).$$

We have

$$f(0, 0, \ldots, 0) = 0$$

and

$$\frac{\partial f(0, 0, \ldots, 0)}{\partial \xi_k} = 0, \quad k = 1, 2, \ldots, n - 1. \tag{11.12}$$

While the first equality is obvious, the second follows from the equality

$$\cos(v_P, \xi_k) = -\frac{\dfrac{\partial f(P)}{\partial \xi_k}}{\sqrt{1 + \left(\dfrac{\partial f}{\partial \xi_1}\right)^2 + \left(\dfrac{\partial f}{\partial \xi_2}\right)^2 + \ldots + \left(\dfrac{\partial f}{\partial \xi_{n-1}}\right)^2}}, \quad k = 1, 2, \ldots, n - 1.$$

In particular, when $P = M$, we have that $\cos(v_P, \xi_k) = \cos(v_M, \xi_k) = 0$ and hence

$$\frac{\partial f(M)}{\partial \xi_k} = 0.$$

Since in the local coordinates $M = (0, 0, \ldots, 0)$, we obtain (11.12).

Thus, the function f together with its first partial derivatives $\frac{\partial f}{\partial \xi_k} = 0$, $k = 1, 2, \ldots, n - 1$ vanishes at the origin in the local coordinates. Let us find out the order of smallness of these functions at the origin.

Lemma 148 (Estimates of f and $\frac{\partial f}{\partial \xi_k}$) *If Σ is a Lyapunov surface, $\gamma \le Ar^\alpha$ and $\xi_n = f(\xi_1, \xi_2, \dots, \xi_{n-1})$ is its equation in local coordinates, then inside the Lyapunov sphere the following inequalities hold:*

$$\left| \frac{\partial f(P)}{\partial \xi_k} \right| \le A\sqrt{3}r^\alpha, \quad k = 1, 2, \dots, n-1, \tag{11.13}$$

$$|f(P)| \le Cr^{\alpha+1} \text{ with } C = \frac{2^\alpha A\sqrt{3}}{1+\alpha}. \tag{11.14}$$

Proof Let γ be the angle between v_P and v_M: $\gamma = \left(\widehat{v_P, v_M} \right)$. Then

$$\cos(v_P, \xi_n) = \cos\gamma = \frac{1}{\sqrt{1 + \left(\frac{\partial f}{\partial \xi_1}\right)^2 + \left(\frac{\partial f}{\partial \xi_2}\right)^2 + \dots + \left(\frac{\partial f}{\partial \xi_{n-1}}\right)^2}}.$$

Let us make use of the inequality

$$\cos\gamma \ge 1 - \frac{\gamma^2}{2}. \tag{11.15}$$

Note that it follows from the right inequality in (11.11). Indeed,

$$1 - \cos\gamma = 2\sin^2\frac{\gamma}{2} \le 2\left(\frac{\gamma}{2}\right)^2 = \frac{\gamma^2}{2}.$$

Due to (11.15),

$$\cos\gamma \ge 1 - \frac{A^2 r^{2\alpha}}{2}$$

because $\gamma \le Ar^\alpha$. Note that inside the Lyapunov sphere $Ad^\alpha < 1$ and hence $0 < \gamma < \pi/2$.

We have

$$\frac{1}{\cos\gamma} = \frac{1}{\cos(v_P, v_M)} = \sqrt{1 + \left(\frac{\partial f}{\partial \xi_1}\right)^2 + \left(\frac{\partial f}{\partial \xi_2}\right)^2 + \dots + \left(\frac{\partial f}{\partial \xi_{n-1}}\right)^2}$$

$$\le \frac{1}{1 - \frac{A^2 r^{2\alpha}}{2}} \le 1 + A^2 r^{2\alpha}.$$

The last inequality is due to the following reasoning:

$$\left(1 - \frac{A^2 r^{2\alpha}}{2}\right)\left(1 + A^2 r^{2\alpha}\right) = 1 - \frac{A^2 r^{2\alpha}}{2} + A^2 r^{2\alpha} - \frac{A^4 r^{4\alpha}}{2}$$

$$= 1 + \frac{A^2 r^{2\alpha}}{2}\left(1 - A^2 r^{2\alpha}\right) > 1 + \frac{A^2 r^{2\alpha}}{2}\left(1 - A^2 d^{2\alpha}\right) > 1.$$

Thus,

$$1 + \left(\frac{\partial f}{\partial \xi_1}\right)^2 + \left(\frac{\partial f}{\partial \xi_2}\right)^2 + \ldots + \left(\frac{\partial f}{\partial \xi_{n-1}}\right)^2 \le \left(1 + A^2 r^{2\alpha}\right)^2,$$

i.e.,

$$\left(\frac{\partial f}{\partial \xi_1}\right)^2 + \left(\frac{\partial f}{\partial \xi_2}\right)^2 + \ldots + \left(\frac{\partial f}{\partial \xi_{n-1}}\right)^2 \le 2A^2 r^{2\alpha} + A^4 r^{4\alpha} \le 2A^2 r^{2\alpha} + A^2 r^{2\alpha} A^2 d^{2\alpha} < 3A^2 r^{2\alpha}.$$

From here, we obtain

$$\left|\frac{\partial f}{\partial \xi_k}\right| \le A\sqrt{3} r^{\alpha}, \quad k = 1, 2, \ldots, n-1.$$

Thus, the estimate (11.13) is proved.

Note that the direction ξ_k in (11.13) can be arbitrary. In particular, the estimate (11.13) remains valid for the directional derivative

$$\left|\frac{\partial f}{\partial \rho}\right| \le A\sqrt{3} r^{\alpha}.$$

Taking into account that $Ar^{\alpha} < 1$, we see that $\left|\frac{\partial f}{\partial \rho}\right| < \sqrt{3}$. Then

$$|f(P)| = |f(P) - f(M)| \le \max \left|\frac{\partial f}{\partial \rho}\right| \rho < \sqrt{3}\rho, \quad f(P) = \xi_n.$$

Hence

$$r = \sqrt{\rho^2 + \xi_n^2} < \sqrt{\rho^2 + 3\rho^2} = 2\rho,$$

and

$$\left|\frac{\partial f}{\partial \rho}\right| \le 2^{\alpha} A\sqrt{3} \rho^{\alpha}.$$

From here, we obtain

$$|f| \le \int_0^{\rho} \left|\frac{\partial f}{\partial t}\right| dt \le \frac{2^{\alpha} A\sqrt{3}}{1+\alpha} \rho^{1+\alpha} \le \frac{2^{\alpha} A\sqrt{3}}{1+\alpha} r^{1+\alpha}.$$

∎

Remark 149 When proving this lemma, we showed that inside the Lyapunov sphere the inequality holds $r < 2\rho$. Hence the following inequalities are valid:

$$\rho < r < 2\rho, \quad \frac{1}{2}r < \rho < r.$$

Then the estimate (11.13) can be written in the form

$$\left| \frac{\partial f(P)}{\partial \xi_k} \right| \leq 2^\alpha A\sqrt{3}\rho^\alpha, \quad k = 1, 2, \ldots, n - 1.$$

Lemma 150 (An Estimate for $\cos(r, v)$) *Let Σ be a Lyapunov surface and P and M two points of the surface located inside a Lyapunov sphere. Let $r = r_{PM}$ and $v = v_P$. Then*

$$|\cos(r, v)| \leq br^\alpha,$$

where $b = (n-1)A\sqrt{3} + \frac{2^\alpha A\sqrt{3}}{1+\alpha} = $ Const.

Proof We have

$$\cos(v, \xi_k) = -\frac{\frac{\partial f(P)}{\partial \xi_k}}{\sqrt{1 + \left(\frac{\partial f}{\partial \xi_1}\right)^2 + \left(\frac{\partial f}{\partial \xi_2}\right)^2 + \ldots + \left(\frac{\partial f}{\partial \xi_{n-1}}\right)^2}}, \quad k = 1, 2, \ldots, n - 1.$$

Hence

$$|\cos(v, \xi_k)| \leq \left| \frac{\partial f}{\partial \xi_k} \right|,$$

and due to Lemma 148,

$$|\cos(v, \xi_k)| \leq A\sqrt{3}r^\alpha, \quad k = 1, 2, \ldots, n - 1.$$

Then

$$|\cos(r, v)| = |\cos(r, \xi_1)\cos(v, \xi_1) + \ldots + \cos(r, \xi_n)\cos(v, \xi_n)|$$

$$\leq |\cos(v, \xi_1)| + \ldots + |\cos(v, \xi_{n-1})| + |\cos(r, \xi_n)|$$

$$\leq \underbrace{A\sqrt{3}r^\alpha + \ldots + A\sqrt{3}r^\alpha}_{n-1} + \frac{|\xi_n|}{r} \leq (n-1)A\sqrt{3}r^\alpha + \frac{2^\alpha A\sqrt{3}}{1+\alpha}r^\alpha = br^\alpha. \quad \blacksquare$$

Lemma 151 (An Estimate for $\cos(\nu_P, \nu_M) = \cos(\nu, \xi_n)$) *Let Σ be a Lyapunov surface and P and M two points of the surface located inside a Lyapunov sphere. Then the inequality holds*

$$|\cos(\nu_P, \nu_M)| \geq \frac{1}{2}.$$

Proof Using (11.15), we have

$$|\cos(\nu_P, \nu_M)| = |\cos \gamma| \geq 1 - \frac{\gamma^2}{2} \geq 1 - \frac{A^2 r^{2\alpha}}{2} \geq 1 - \frac{A^2 d^{2\alpha}}{2} \geq 1 - \frac{1}{2} = \frac{1}{2}.$$

∎

11.3　Solid Angle of Lyapunov Surface

In geometry, a solid angle (symbol: Ω) is a measure of how large the object appears to an observer looking from some particular point M. When Σ is a part of a sphere in a three-dimensional space centered at a point M, then $\Omega = |\Sigma|/R^2$, where $|\Sigma|$ is the area of Σ and R is the radius of the sphere.

Now let Σ be an arbitrary surface and $d\sigma$ a small part of the surface containing a point P. Consider a sphere centered at (observer's) point M and crossing the point P. Denote

$$\Omega_{d\sigma} = \frac{\cos(\nu, r) d\sigma_P}{r^2},$$

where $\nu = \nu_P$ and $r = r_{PM}$. For the whole surface, we set

$$\Omega_{\Sigma} = \int_{\Sigma} \frac{\cos(\nu, r) d\sigma_P}{r^2}.$$

More generally, the following concept is introduced.

Definition 152 A **solid angle** under which the surface Σ in $\mathbb{R}^n$ is seen from the point M is the integral

$$\Omega_{\Sigma} = \int_{\Sigma} \frac{\cos(\nu_P, r) d\sigma_P}{r^{n-1}}.$$

Thus, the solid angle is the double layer potential with the density $1/(2 - n)$:

$$\Omega_{\Sigma}(M) = \frac{1}{2 - n} \int_{\Sigma} \frac{\partial}{\partial \nu} \left(\frac{1}{r^{n-2}} \right) d\sigma_P.$$

Lemma 153 *If Σ is a Lyapunov surface, then for any point M the integral defining the solid angle converges absolutely and is bounded,*

$$\int_{\Sigma} \left| \frac{\cos(\nu_P, r)}{r^{n-1}} \right| d\sigma_P \leq K < \infty.$$

Proof We will prove that the integral converges absolutely for any point M skipping the proof of its boundedness.

(I) Let $M \notin \Sigma$. Then $C = \inf_{P \in \Sigma} r_{PM} \neq 0$. We have

$$\int_{\Sigma} \left| \frac{\cos(\nu_P, r)}{r^{n-1}} \right| d\sigma_P \leq \frac{1}{r^{n-1}} \int_{\Sigma} d\sigma_P = \text{Const}.$$

(II) Let $M \in \Sigma$. Construct a Lyapunov sphere $\partial B_\varepsilon(M)$ centered at M. Denote $\Sigma_\varepsilon = \Sigma \cap B_\varepsilon(M)$. Then

$$\int_{\Sigma} \frac{|\cos(\nu_P, r)|}{r^{n-1}} d\sigma_P = \int_{\Sigma_\varepsilon} \frac{|\cos(\nu_P, r)|}{r^{n-1}} d\sigma_P + \int_{\Sigma \setminus \Sigma_\varepsilon} \frac{|\cos(\nu_P, r)|}{r^{n-1}} d\sigma_P = I_1 + I_2.$$

The second integral exists as a proper integral, while to the first one we apply the lemma on the estimate of $|\cos(r, \nu)|$ (Lemma 151)

$$I_1 \leq b \int_{\Sigma_\varepsilon} \frac{r^\alpha}{r^{n-1}} d\sigma_P = b \int_{\Sigma_\varepsilon} \frac{d\sigma_P}{r^{n-1-\alpha}}$$

$$= b \int_{\text{Pr}_\tau \Sigma_\varepsilon} \frac{1}{r^{n-1-\alpha}} \frac{d\xi_1 d\xi_2 \ldots d\xi_{n-1}}{|\cos(\nu_P, \xi_n)|}$$

$$\leq 2b \int_{\text{Pr}_\tau \Sigma_\varepsilon} \frac{d\xi_1 d\xi_2 \ldots d\xi_{n-1}}{\rho^{n-1-\alpha}},$$

where τ is the tangent plane at the point M, and $\text{Pr}_\tau \Sigma_\varepsilon$ denotes the projection of Σ_ε onto the plane τ, $r \geq \rho = \text{Pr}_\tau r$. Then

$$I_1 \leq 2b \int_0^\varepsilon \frac{\rho^{(n-1)-1}}{\rho^{n-1-\alpha}} d\rho \int_{\Sigma_1} d\sigma,$$

and hence

$$I_1 \leq 2b \int_0^\varepsilon \rho^{\alpha-1} d\rho \cdot \text{Const} = \text{Const} \cdot \left. \frac{\rho^\alpha}{\alpha} \right|_0^\varepsilon = \frac{\text{Const}}{\alpha} \cdot \varepsilon^\alpha < \infty, \quad 0 < \alpha < 1.$$

∎

11.4   Surface Potentials on Lyapunov Surfaces

Theorem 154 *The single layer potential with a continuous density is continuous in the whole space and harmonic outside the surface.*

Proof First of all, let us verify the existence of $v(M) = \int_\Sigma \frac{q(P)}{r_{PM}^{n-2}} d\sigma_P$. If $M \notin \Sigma$, the integral is not improper, and

$$|v(M)| \leq \max_{P \in \Sigma} |q(P)| \int_\Sigma \frac{d\sigma_P}{r_{PM}^{n-2}} \leq \frac{1}{C^{n-2}} \max_{P \in \Sigma} |q(P)| \int_\Sigma d\sigma$$

with $C = \text{dist}(M, \Sigma)$. The harmonicity in this case follows from the possibility of differentiation under the integral sign any number of times:

$$\Delta v(M) = \int_\Sigma q(P) \Delta \left(\frac{1}{r_{PM}^{n-2}} \right) d\sigma = 0, \quad M \notin \Sigma.$$

If $M \in \Sigma$, we have

$$|v(M)| \leq \max_{P \in \Sigma} |q(P)| \left(\int_{\Sigma_\varepsilon} \frac{d\sigma_P}{r^{n-2}} + \int_{\Sigma \setminus \Sigma_\varepsilon} \frac{d\sigma_P}{r^{n-2}} \right),$$

where Σ_ε is defined in the proof of Lemma 153. The second integral is finite, and in the first one we make use of the inequality $r \geq \rho = \text{Pr}_\tau r$:

$$\int_{\Sigma_\varepsilon} \frac{d\sigma_P}{r^{n-2}} \leq \int_{\text{Pr}_\tau \Sigma_\varepsilon} \frac{1}{\rho^{n-2}} \frac{d\xi_1 d\xi_2 \ldots d\xi_{n-1}}{|\cos(\nu_P, \xi_n)|} < \infty.$$

Thus, $v(M)$ exists for all $M \in \Sigma$ as an improper integral.

Let us prove the continuity of $v(M)$. When $M \notin \Sigma$, this is obvious, and we have already considered this case. Let $M \in \Sigma$ and M_1 be any point of space close to M (and not necessarily belonging to Σ). Let us show that $v(M) - v(M_1) \to 0$ when $M_1 \to M$. Denote

$$v'(M) = \int_{\Sigma_\varepsilon} q(P) \frac{d\sigma_P}{r_{PM}^{n-2}}, \quad v''(M) = \int_{\Sigma \setminus \Sigma_\varepsilon} q(P) \frac{d\sigma_P}{r_{PM}^{n-2}}.$$

Hence $v(M) = v'(M) + v''(M)$. Obviously, $v''(M)$ is continuous. We have

$$|v(M) - v(M_1)| \leq |v'(M)| + |v'(M_1)| + |v''(M) - v''(M_1)|.$$

We can always choose a neighborhood of M such that for M_1 belonging to it the inequality holds $|v''(M) - v''(M_1)| < \epsilon/3$. Let us estimate $|v'(M)|$ and $|v'(M_1)|$. We have

$$\left|v'(M)\right| \le \int_{\Sigma_\varepsilon} q(P) \frac{d\sigma_P}{r_{PM}^{n-2}} \le \max_{P\in\Sigma} |q(P)| \int_{\mathrm{Pr}_\tau \Sigma_\varepsilon} \frac{1}{\rho^{n-2}} \frac{d\xi_1 d\xi_2 \dots d\xi_{n-1}}{|\cos(v_P, \xi_n)|}$$

$$\le 2 \max_{P\in\Sigma} |q(P)| \int_{\mathrm{Pr}_\tau \Sigma_\varepsilon} \frac{d\xi_1 d\xi_2 \dots d\xi_{n-1}}{\rho^{n-2}}$$

$$\le 2 \max_{P\in\Sigma} |q(P)| \left(\int_0^\varepsilon \frac{d\rho}{\rho^{n-2}} \int_{\Sigma_1} \rho^{n-2} d\sigma \right) = \mathrm{Const} \cdot \varepsilon.$$

Choosing the radius of the Lyapunov sphere small enough, we obtain $\left|v'(M)\right| < \epsilon/3$. Similarly (if necessary, choosing ε still smaller) $\left|v'(M_1)\right| < \epsilon/3$. Thus, $|v(M) - v(M_1)| < \epsilon$ for M_1 sufficiently close to M. Due to the arbitrariness of $\epsilon > 0$, we obtain the continuity of $v(M)$. $\blacksquare$

In contrast to the single layer potential, the double layer potential has a jump discontinuity on the surface Σ.

Definition 155 Let $M \in \Sigma$. The value of the double layer potential $w(M)$ at a point of the surface is called **direct value of the potential**.

We will show that generally the direct value does not coincide with the limit value, when a point approaches the surface from one or the other side of it, i.e., $\lim_{M_1 \to M} w(M_1) \ne w(M)$ where $M \in \Sigma$ and $M_1 \notin \Sigma$.

First we prove the following statement.

Theorem 156 *The double layer potential with a continuous density is harmonic outside the surface, and its direct value on the surface exists.*

Proof When $M \notin \Sigma$, the integral

$$w(M) = \int_\Sigma \frac{\partial}{\partial v_P} \left(\frac{1}{r_{PM}^{n-2}} \right) q(P) d\sigma_P$$

admits differentiation with respect to the coordinates of the point M any number of times, thus,

$$\Delta w(M) = \int_\Sigma \Delta_M \frac{\partial}{\partial v_P} \left(\frac{1}{r_{PM}^{n-2}} \right) q(P) d\sigma_P$$

$$= \int_\Sigma \frac{\partial}{\partial v_P} \left(\Delta_M \left(\frac{1}{r_{PM}^{n-2}} \right) \right) q(P) d\sigma_P = 0.$$

Now let $M \in \Sigma$. Then

$$|w(M)| \leq \max_{P \in \Sigma} |q(P)| (2-n) \int_{\Sigma} \frac{|\cos(\nu_P, r)|}{r_{PM}^{n-1}} d\sigma_P.$$

This integral exists due to Lemma 153 on the solid angle. ∎

Prior to prove the theorem on the jump discontinuity on the surface of the double layer potential, let us calculate a special integral, the double layer potential with a density $q \equiv 1$ in the case of a closed Lyapunov surface Σ,

$$w_0(M) = (2-n) \int_{\Sigma} \frac{\cos(\nu_P, r)}{r_{PM}^{n-1}} d\sigma_P = \int_{\Sigma} \frac{\partial}{\partial \nu_P} \left(\frac{1}{r_{PM}^{n-2}} \right) d\sigma_P.$$

This integral is called the **Gauss integral**.

11.5 Calculation of the Gauss Integral

(I) Let $M \in V_-$ (Fig. 11.3).Then the function r_{PM}^{2-n} is harmonic with respect to P everywhere inside V_+ and on Σ. Hence, due to the property of the normal derivative of a harmonic function, we have

$$w_0(M) = \int_{\Sigma} \frac{\partial}{\partial \nu_P} \left(\frac{1}{r_{PM}^{n-2}} \right) d\sigma_P = 0, \quad M \in V_-.$$

(II) Let $M \in V_+$ and $B_\varepsilon(M)$ be a ball of radius ε centered at M. Consider the domain $T_\varepsilon = V_+ \setminus \overline{B}_\varepsilon(M)$ (Fig. 11.4). Then

Fig. 11.3 $M \in V_-$

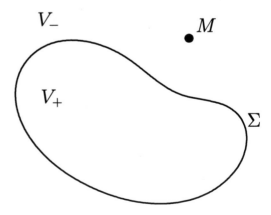

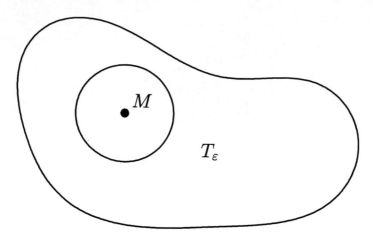

Fig. 11.4 $M \in V_+, T_\varepsilon = V_+ \setminus \overline{B}_\varepsilon(M)$

$$\int_{\Sigma + \partial B_\varepsilon(M)} \frac{\partial}{\partial \nu_P} \left(\frac{1}{r_{PM}^{n-2}} \right) d\sigma_P = 0.$$

Hence

$$w_0(M) = \int_\Sigma \frac{\partial}{\partial \nu_P} \left(\frac{1}{r_{PM}^{n-2}} \right) d\sigma_P = \left(\int_{\Sigma + \partial B_\varepsilon(M)} - \int_{\partial B_\varepsilon(M)} \right) \frac{\partial}{\partial \nu_P} \left(\frac{1}{r_{PM}^{n-2}} \right) d\sigma_P$$

$$= - \int_{\partial B_\varepsilon(M)} \frac{\partial}{\partial \nu_P} \left(\frac{1}{r_{PM}^{n-2}} \right) d\sigma_P = \int_{\partial B_\varepsilon(M)} \frac{\partial}{\partial r} \left(\frac{1}{r^{n-2}} \right) d\sigma_P$$

$$= (2 - n) \int_{\partial B_\varepsilon(M)} \frac{1}{r^{n-1}} d\sigma_P = \frac{(2-n)}{r^{n-1}} \int_{\partial B_\varepsilon(M)} d\sigma_P = (2 - n) |\Sigma_1|.$$

Note that the results obtained in the first and second cases (I) and (II) are valid also for an arbitrary piecewise smooth closed surface.

(III) Let $M \in \Sigma$ and $B_\varepsilon(M)$ be a ball of radius ε centered at M. Denote $\Sigma_\varepsilon' = \partial B_\varepsilon(M) \cap \overline{V}_+$ and $\Sigma_\varepsilon'' = \Sigma \cap \overline{B}_\varepsilon(M)$ (see Fig. 11.5). Then

$$\int_{(\Sigma \setminus \Sigma_\varepsilon'') + \Sigma_\varepsilon'} \frac{\partial}{\partial \nu_P} \left(\frac{1}{r_{PM}^{n-2}} \right) d\sigma_P = 0.$$

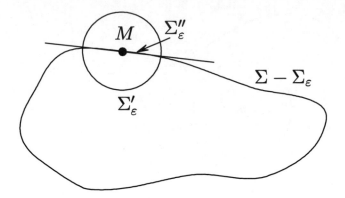

Fig. 11.5 $M \in \Sigma$

Hence

$$w_0(M) = \int_\Sigma \frac{\partial}{\partial v_P} \left(\frac{1}{r_{PM}^{n-2}} \right) d\sigma_P = \lim_{\varepsilon \to 0} \int_{\Sigma \setminus \Sigma_\varepsilon''} \frac{\partial}{\partial v_P} \left(\frac{1}{r_{PM}^{n-2}} \right) d\sigma_P$$

$$= -\lim_{\varepsilon \to 0} \int_{\Sigma_\varepsilon'} \frac{\partial}{\partial v_P} \left(\frac{1}{r_{PM}^{n-2}} \right) d\sigma_P = \lim_{\varepsilon \to 0} \int_{\Sigma_\varepsilon'} \frac{\partial}{\partial r} \left(\frac{1}{r^{n-2}} \right) d\sigma$$

$$= (2 - n) \lim_{\varepsilon \to 0} \frac{1}{\varepsilon^{n-1}} \int_{\Sigma_\varepsilon'} d\sigma.$$

Thus,

$$w_0(M) = (2 - n) \lim_{\varepsilon \to 0} \frac{|\Sigma_\varepsilon'|}{\varepsilon^{n-1}}.$$

The area $|\Sigma_\varepsilon'|$ differs from the area of a hemisphere resting on the tangent plane by a magnitude of an order higher than ε^{n-1}, that is,

$$|\Sigma_\varepsilon'| = \frac{1}{2} \frac{2\pi^{n/2}}{\Gamma(n/2)} \varepsilon^{n-1} + o\left(\varepsilon^{n-1} \right). \tag{11.16}$$

For simplicity, let us show this fact in the case $n = 3$. Introducing curvilinear coordinates ϕ and θ on Σ_ε', we obtain

$$|\Sigma_\varepsilon'| = \int_{\Sigma_\varepsilon'} d\sigma = \int_0^{2\pi} \int_0^{\theta(\phi)} \sin\theta \, \varepsilon^2 \, d\phi d\theta = \varepsilon^2 \int_0^{2\pi} \int_0^{\pi/2} \sin\theta \, d\phi d\theta - \varepsilon^2 \int_0^{2\pi} \int_{\theta(\phi)}^{\pi/2} \sin\theta \, d\phi d\theta$$

$$= 2\pi\varepsilon^2 - \varepsilon^2 \int_0^{2\pi} \int_{\theta(\phi)}^{\pi/2} \sin\theta \, d\phi d\theta = 2\pi\varepsilon^2 + \varepsilon^2 \int_0^{2\pi} \cos\theta\,(\phi) \, d\phi.$$

To prove (11.16), it remains to verify that letting $\varepsilon \to 0$, we obtain

$$\int_0^{2\pi} \cos\theta\,(\phi) \, d\phi \to 0.$$

We have $\cos\theta\,(\phi) = \zeta/r$. Due to the estimate $|\zeta| \leq Ar^{\alpha+1}$ inside the Lyapunov sphere,

$$|\cos\theta\,(\phi)| \leq \frac{Cr^{\alpha+1}}{r} = Cr^\alpha \leq C\varepsilon^\alpha.$$

Hence $\left| \int_0^{2\pi} \cos\theta\,(\phi) \, d\phi \right| \leq 2\pi C\varepsilon^\alpha \to 0$ when $\varepsilon \to 0$ that proves the closeness of the areas of Σ'_ε and hemisphere. Thus, we obtain

$$w_0(M) = -\frac{n-2}{2} |\Sigma_1|, \quad M \in \Sigma.$$

We summarize the obtained results in the formula

$$w_0(M) = \int_\Sigma \frac{\partial}{\partial v_P} \left(\frac{1}{r_{PM}^{n-2}} \right) d\sigma_P = \begin{cases} -(n-2)|\Sigma_1|, & M \in V_+, \\ 0, & M \in V_-, \\ -\frac{n-2}{2}|\Sigma_1|, & M \in \Sigma, \end{cases}$$

where $|\Sigma_1|$ is the area of a unitary sphere in an n-dimensional space.
  In particular, for $n = 3$, we have

$$w_0(M) = \int_\Sigma \frac{\partial}{\partial v} \left(\frac{1}{r} \right) d\sigma = \begin{cases} -4\pi, & M \in V_+, \\ 0, & M \in V_-, \\ -2\pi, & M \in \Sigma. \end{cases} \tag{11.17}$$

For $n = 2$, similarly to the previous one, we obtain

$$w_0(M) = \int_S \frac{\partial}{\partial v} \left(\ln\frac{1}{r} \right) dl = \begin{cases} -2\pi, & M \in S_+, \\ 0, & M \in S_-, \\ -\pi, & M \in S. \end{cases}$$

  Thus, the double layer potential with a constant density is already piecewise continuous and not a continuous function.

11.6   Jump Discontinuity of the Double Layer Potential

Now let $q(P)$ be an arbitrary continuous function defined on Σ. Denote

$$w_+(M) = \lim_{M_1 \to M} w(M_1), \quad M_1 \in V_+,$$

$$w_-(M) = \lim_{M_1 \to M} w(M_1), \quad M_1 \in V_-.$$

The values $w_+(M)$ and $w_-(M)$ are called the **limit values of the double layer potential at the point** $M \in \Sigma$ **from inside and from outside the domain** (Fig. 11.6).

Theorem 157 (On the Jump of the Double Layer Potential) *Let Σ be a closed Lyapunov surface. The limit values $w_\pm(M)$ and the direct value on Σ of a double layer potential with a continuous density $q(P)$, $P \in \Sigma$, are related by the equalities*

$$w_+(M) = w(M) - \frac{n-2}{2}|\Sigma_1|q(M), \quad M \in \Sigma, \tag{11.18}$$

and

$$w_-(M) = w(M) + \frac{n-2}{2}|\Sigma_1|q(M), \quad M \in \Sigma. \tag{11.19}$$

In particular, for $n = 3$,

$$w_\pm(M) = w(M) \mp 2\pi q(M), \quad M \in \Sigma, \tag{11.20}$$

and for $n = 2$,

$$w_\pm(M) = w(M) \mp \pi q(M), \quad M \in L.$$

Fig. 11.6 $M \in \Sigma$

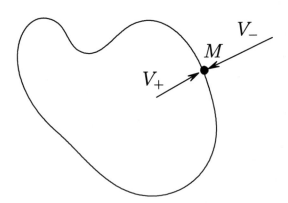

Proof Let $n \geq 3$ and M_1 be an arbitrary point. We have

$$w(M_1) = \int_\Sigma q(P) \frac{\partial}{\partial \nu_P} \left(\frac{1}{r_{PM_1}^{n-2}} \right) d\sigma_P$$

$$= \int_\Sigma (q(P) - q(M)) \frac{\partial}{\partial \nu_P} \left(\frac{1}{r_{PM_1}^{n-2}} \right) d\sigma_P + \int_\Sigma q(M) \frac{\partial}{\partial \nu_P} \left(\frac{1}{r_{PM_1}^{n-2}} \right) d\sigma_P$$

$$= A(M, M_1) + q(M) w_0(M_1),$$

where $w_0(M_1)$ is the Gauss integral. Hence

$$w(M_1) = \begin{cases} A(M, M_1) - (n-2) |\Sigma_1| q(M), & M_1 \in V_+, \\ A(M, M_1), & M_1 \in V_-. \end{cases}$$

Let us show that the function $A(M, M_1)$ is continuous when $M_1 \to M$ from inside or outside the surface. Let Σ_ε be a Lyapunov sphere of radius ε centered at M, and let Σ_ε' be the part of the surface Σ lying inside the sphere Σ_ε. Then

$$A(M, M_1) = \int_{\Sigma_\varepsilon'} + \int_{\Sigma \backslash \Sigma_\varepsilon'} = A_1(M, M_1) + A_2(M, M_1).$$

Obviously, the function $A_2(M, M_1)$ is continuous. Let us estimate $A_1(M, M_1)$,

$$|A_1(M, M_1)| \leq \sup_{P \in \Sigma_\varepsilon'} |q(P) - q(M)| \int_{\Sigma_\varepsilon'} \frac{(n-2) |\cos(\nu, r)|}{r_{PM_1}^{n-1}} d\sigma_P.$$

Due to Lemma 153 on a solid angle,

$$\int_{\Sigma_\varepsilon'} \frac{|\cos(\nu, r)|}{r_{PM_1}^{n-1}} d\sigma_P \leq K < \infty$$

for any point M_1. Choose ε such that $|q(P) - q(M)| < \frac{\varepsilon}{4K(n-2)}$ for all $P \in \Sigma_\varepsilon'$. Then $|A_1(M, M_1)| < \varepsilon/4$ for all M_1. Hence

$$|A(M, M_1) - A(M, M)| = |A_1(M, M_1) + A_2(M, M_1) - A_1(M, M) - A_2(M, M)|$$

$$\leq |A_1(M, M_1)| + |A_1(M, M)| + |A_2(M, M_1) - A_2(M, M)|$$

$$< \frac{\varepsilon}{4} + \frac{\varepsilon}{4} + |A_2(M, M_1) - A_2(M, M)|.$$

The last difference is less than $\varepsilon/2$ because $A_2(M, M_1)$ is continuous. Thus, for an arbitrary point M_1, we have

$$w(M_1) = A(M, M_1) + q(M)w_0(M_1), \tag{11.21}$$

where $w_0(M_1)$ is the Gauss integral and the function $A(M, M_1)$ is continuous when $M_1 \to M$.

Let $M_1 \in V_+$. Then $w_0(M_1) = -(n-2)|\Sigma_1|$, so that

$$w(M_1) = \int_\Sigma (q(P) - q(M)) \frac{\partial}{\partial \nu_P} \left(\frac{1}{r_{PM_1}^{n-2}} \right) d\sigma_P - (n-2)|\Sigma_1|q(M), \quad M_1 \in V_+.$$

Let $M_1 \to M$, $M_1 \in V_+$. Then, due to the continuity of the first term, we have

$$\lim_{V_+ \ni M_1 \to M} w(M_1) = w_+(M) = \int_\Sigma (q(P) - q(M)) \frac{\partial}{\partial \nu_P} \left(\frac{1}{r_{PM}^{n-2}} \right) d\sigma_P - (n-2)|\Sigma_1|q(M)$$

$$= \int_\Sigma q(P) \frac{\partial}{\partial \nu_P} \left(\frac{1}{r_{PM}^{n-2}} \right) d\sigma_P - q(M)w_0(M) - (n-2)|\Sigma_1|q(M)$$

$$= \int_\Sigma q(P) \frac{\partial}{\partial \nu_P} \left(\frac{1}{r_{PM}^{n-2}} \right) d\sigma_P + q(M) \frac{(n-2)}{2}|\Sigma_1| - (n-2)|\Sigma_1|q(M)$$

$$= w(M) - q(M) \frac{(n-2)}{2}|\Sigma_1|,$$

which proves (11.18).

Now choosing $M_1 \in V_-$ in (11.21) and letting $M_1 \to M$, similarly to the previous one, we obtain the formula (11.19)

$$w_-(M) = \int_\Sigma (q(P) - q(M)) \frac{\partial}{\partial \nu_P} \left(\frac{1}{r_{PM}^{n-2}} \right) d\sigma_P$$

$$= \int_\Sigma q(P) \frac{\partial}{\partial \nu_P} \left(\frac{1}{r_{PM}^{n-2}} \right) d\sigma_P - q(M)w_0(M)$$

$$= w(M) + q(M) \frac{(n-2)}{2}|\Sigma_1|.$$

■

Corollary 158 *When crossing the surface* Σ, *the double layer potential makes a jump equal to*

$$w_-(M) - w_+(M) = (n-2)\,|\Sigma_1|\,q(M), \quad n \geq 3.$$

In particular, $w_-(M) - w_+(M) = 4\pi q(M)$ *for* $n = 3$. *Similarly,* $w_-(M) - w_+(M) = 2\pi q(M)$ *for* $n = 2$.

We proceed to study the single layer potential.

11.7   Normal Derivative of the Single Layer Potential

In Theorem 154, we proved that the single layer potential

$$v(M) = \int_\Sigma \frac{q(P)}{r_{PM}^{n-2}}\,d\sigma_P$$

with a continuous density q is harmonic everywhere except for the surface Σ and remains continuous when crossing Σ. Here we are going to study the behavior of the normal derivative of the single layer potential on the surface Σ. We will show that it has a jump discontinuity similar to that of the double layer potential.

Let M be a point of the surface Σ and ν_M the exterior normal vector at M. Then the normal derivative of a function v at a point M_1 which lies on a straight line containing ν_M and does not necessarily belong to Σ is understood as a directional derivative of v in the direction ν_M.

Assume first that $M_1 \notin \Sigma$. Then the differentiation under the integral sign is allowed, and we have

$$\frac{\partial v(M_1)}{\partial \nu_M} = \int_\Sigma q(P)\frac{\partial}{\partial \nu_M}\left(\frac{1}{r_{PM_1}^{n-2}}\right)d\sigma_P, \quad M_1 \notin \Sigma.$$

We have seen earlier that

$$\frac{\partial}{\partial \nu_P}\left(\frac{1}{r_{PM_1}^{n-2}}\right) = -\frac{n-2}{r_{PM_1}^{n-1}}\cos\left(\nu_P, r_{PM_1}\right),$$

where the direction of r_{PM_1} in the argument of the cosine is from M_1 to P. Hence

$$\frac{\partial}{\partial \nu_M}\left(\frac{1}{r_{PM_1}^{n-2}}\right) = -\frac{n-2}{r_{PM_1}^{n-1}}\cos\left(\nu_M, r_{PM_1}\right) = \frac{n-2}{r_{PM_1}^{n-1}}\cos\left(\nu_M, r_{M_1P}\right).$$

Definition 159 The integral

$$
\frac{\partial v(M)}{\partial v_M} = \int_\Sigma q(P) \frac{\partial}{\partial v_M} \left(\frac{1}{r_{PM}^{n-2}} \right) d\sigma_P = (n-2) \int_\Sigma q(P) \frac{\cos(v_M, r_{MP})}{r_{PM}^{n-1}} d\sigma_P
$$

(11.22)

when $M \in \Sigma$ is called **direct value of the normal derivative of the single layer potential**.

Denote

$$
\left(\frac{\partial v}{\partial v_M} \right)_- (M) = \lim_{V_- \ni M_1 \to M} \frac{\partial v(M_1)}{\partial v_M},
$$

$$
\left(\frac{\partial v}{\partial v_M} \right)_+ (M) = \lim_{V_+ \ni M_1 \to M} \frac{\partial v(M_1)}{\partial v_M},
$$

where $M \in \Sigma$ and $M_1 \to M$ along the normal v_M. These are limit values of the normal derivative.

Due to (11.22), the normal derivative $\frac{\partial v}{\partial v}$ of a single layer potential does not differ that much from the double layer potential w, and it is natural to expect that $\frac{\partial v}{\partial v}$ has a jump discontinuity when crossing the surface Σ. Prior to proving that, let us verify the existence of the direct value of $\frac{\partial v}{\partial v}$ as an improper integral.

Let as before Σ_ε be a Lyapunov sphere of radius ε centered at M, and let Σ'_ε be the part of the surface Σ lying inside the sphere Σ_ε (Fig. 11.7). We have

$$
\frac{\partial v(M)}{\partial v_M} = \int_{\Sigma'_\varepsilon} + \int_{\Sigma \setminus \Sigma'_\varepsilon}.
$$

Fig. 11.7 Σ'_ε is the part of the surface Σ lying inside the Lyapunov sphere Σ_ε

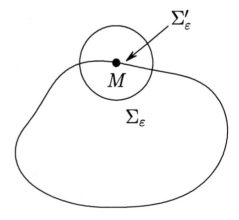

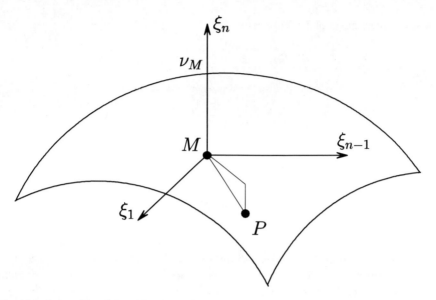

Fig. 11.8 Points M and P and local coordinates at M

The second integral

$$\int_{\Sigma \setminus \Sigma'_\varepsilon} = (n-2) \int_{\Sigma \setminus \Sigma'_\varepsilon} q(P) \frac{\cos(\nu_M, r_{MP})}{r_{PM}^{n-1}} d\sigma_P$$

obviously exists. Let us prove the existence of the first one (see Fig. 11.8)

$$\left| \int_{\Sigma'_\varepsilon} \right| = (n-2) \left| \int_{\Sigma'_\varepsilon} q(P) \frac{\cos(\nu_M, r_{MP})}{r_{PM}^{n-1}} d\sigma_P \right|$$

$$\leq (n-2) \int_{\Sigma'_\varepsilon} |q(P)| \frac{|\xi_n|}{r \cdot r^{n-1}} d\sigma_P$$

$$\leq (n-2) \int_{\Sigma'_\varepsilon} |q(P)| \frac{|f|}{r^n} d\sigma_P.$$

Applying Lemma 148, we obtain

$$\left| \int_{\Sigma'_\varepsilon} \right| \leq (n-2) \int_{\Sigma'_\varepsilon} |q(P)| \frac{A\sqrt{3} r^{\alpha+1}}{r^n} d\sigma_P$$

$$\leq \text{Const} \cdot \int_{\Sigma'_\varepsilon} \frac{d\sigma_P}{r^{n-\alpha-1}} = \text{Const} \cdot \int_{\text{Pr}_\tau \Sigma'_\varepsilon} \frac{1}{r^{n-\alpha-1}} \frac{d\xi_1 d\xi_2 \ldots d\xi_{n-1}}{|\cos(\nu_P, \nu_M)|},$$

where τ is the tangent plane. Furthermore,

$$\left| \int_{\Sigma_\varepsilon'} \right| \le 2 \, \text{Const} \cdot \int_{\text{Pr}_\tau \Sigma_\varepsilon'} \frac{d\xi_1 d\xi_2 \dots d\xi_{n-1}}{r^{n-\alpha-1}}$$

$$\le 2 \, \text{Const} \cdot \int_{\text{Pr}_\tau \Sigma_\varepsilon'} \frac{d\xi_1 d\xi_2 \dots d\xi_{n-1}}{\rho^{n-\alpha-1}}$$

$$\le \text{Const} \cdot \int_0^\varepsilon \frac{\rho^{(n-1)-1} d\rho}{\rho^{n-1-\alpha}} \int_{\Sigma_1} d\sigma \le \text{Const} \cdot \int_0^\varepsilon \rho^{-1+\alpha} d\rho < \text{Const} \cdot \varepsilon^\alpha < \infty.$$

Theorem 160 (On the Jump Discontinuity of the Normal Derivative of a Single Layer Potential) *Let Σ be a closed Lyapunov surface and q a continuous function defined on Σ. The limit values $\left(\frac{\partial v}{\partial v} \right)_\pm$ of the normal derivative of the single layer potential with the density q are related with the direct value of the normal derivative by the formulas*

$$\left(\frac{\partial v}{\partial v} \right)_+ (M) = \frac{\partial v(M)}{\partial v} + \frac{n-2}{2} |\Sigma_1| q(M), \quad n \ge 3 \tag{11.23}$$

and

$$\left(\frac{\partial v}{\partial v} \right)_- (M) = \frac{\partial v(M)}{\partial v} - \frac{n-2}{2} |\Sigma_1| q(M), \quad n \ge 3. \tag{11.24}$$

In particular, when $n = 3$,

$$\left(\frac{\partial v}{\partial v} \right)_\pm (M) = \frac{\partial v(M)}{\partial v} \pm 2\pi q(M),$$

while in the case $n = 2$,

$$\left(\frac{\partial v}{\partial v} \right)_\pm (M) = \frac{\partial v(M)}{\partial v} \pm \pi q(M).$$

Proof Let $M_1 \notin \Sigma$ and lie (outside or inside) on the straight line containing v_M. We have

$$\frac{\partial v(M_1)}{\partial v_M} = (n-2) \int_\Sigma q(P) \frac{\cos \left(v_M, r_{M_1 P} \right)}{r_{M_1 P}^{n-1}} d\sigma_P.$$

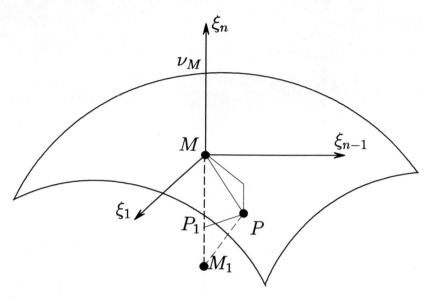

Fig. 11.9 Local coordinates

Consider a double layer potential with the same density $q(P)$,

$$w(M_1) = -(n-2) \int_\Sigma q(P) \frac{\cos\left(\nu_P, r_{PM_1}\right)}{r_{PM_1}^{n-1}} d\sigma_P$$

and the sum

$$F(M_1) = \frac{\partial v(M_1)}{\partial \nu_M} + w(M_1) = (n-2) \int_\Sigma q(P) \frac{\cos\psi - \cos\varphi}{r^{n-1}} d\sigma_P,$$

where $r = r_{PM_1}$, $\cos\psi = \cos\left(\nu_M, r_{M_1 P}\right)$, and $\cos\varphi = \cos\left(\nu_P, r_{M_1 P}\right)$. Let us show that the function $F(M_1)$ is continuous when $M_1 \to M$. For this purpose, we first transform the difference $\cos\psi - \cos\varphi$ into a more convenient form.

Let $\xi_1, \xi_2, \ldots, \xi_n$ denote the local coordinate system in the neighborhood of the point M, $M = M(0, 0, \ldots, 0)$, $M_1 = M_1(0, 0, \ldots, x_n)$, $P = P(\xi_1, \xi_2, \ldots, \xi_n)$ (Fig. 11.9). We have

$$\cos\psi = \cos\left(\nu_M, r_{M_1 P}\right) = \frac{r_{M_1 P_1}}{r_{M_1 P}},$$

where $P_1 = P_1(0, 0, \ldots, \xi_n)$. Hence

$$\cos\psi = \frac{\xi_n - x_n}{r}.$$

On the other hand,

$$\cos \varphi = \cos \left(\nu_P, r_{M_1 P} \right) = \cos(r, \xi_1) \cos(\nu_P, \xi_1) + \ldots + \cos(r, \xi_n) \cos(\nu_P, \xi_n)$$

$$= \frac{\xi_1}{r} \cos(\nu_P, \xi_1) + \ldots + \frac{\xi_{n-1}}{r} \cos(\nu_P, \xi_{n-1}) + \frac{\xi_n - x_n}{r} \cos(\nu_P, \nu_M),$$

and thus,

$$\cos \psi - \cos \varphi = - \left(\frac{\xi_1}{r} \cos(\nu_P, \xi_1) + \ldots + \frac{\xi_{n-1}}{r} \cos(\nu_P, \xi_{n-1}) \right) + \frac{\xi_n - x_n}{r} \left(1 - \cos(\nu_P, \nu_M) \right).$$

Let again Σ_ε be a Lyapunov sphere of radius ε centered at M and Σ'_ε the part of the surface Σ lying inside the sphere Σ_ε. Then

$$F(M_1) = (n - 2) \left(\int_{\Sigma'_\varepsilon} q(P) \frac{\cos \psi - \cos \varphi}{r^{n-1}} d\sigma_P + \int_{\Sigma \setminus \Sigma'_\varepsilon} q(P) \frac{\cos \psi - \cos \varphi}{r^{n-1}} d\sigma_P \right)$$

$$= F_1(M_1) + F_2(M_1).$$

The function $F_2(M_1)$ is obviously continuous. Let us estimate $F_1(M_1)$:

$$|F_1(M_1)| = (n - 2) \left| \int_{\Sigma'_\varepsilon} q(P) \frac{\cos \psi - \cos \varphi}{r^{n-1}} d\sigma_P \right|$$

$$\leq \text{Const} \int_{\Sigma'_\varepsilon} \left(|\cos(\nu_P, \xi_1)| + \ldots + |\cos(\nu_P, \xi_{n-1})| + (1 - \cos(\nu_P, \nu_M)) \right) \frac{d\sigma_P}{r_{PM_1}^{n-1}}.$$

Since

$$|\cos(\nu_P, \xi_k)| = \frac{\left| \frac{\partial f(P)}{\partial \xi_k} \right|}{\sqrt{1 + \left(\frac{\partial f}{\partial \xi_1} \right)^2 + \left(\frac{\partial f}{\partial \xi_2} \right)^2 + \ldots + \left(\frac{\partial f}{\partial \xi_{n-1}} \right)^2}}, \quad k = 1, 2, \ldots, n - 1,$$

we have $|\cos(\nu_P, \xi_k)| \leq \left| \frac{\partial f(P)}{\partial \xi_k} \right| \leq \text{Const} \cdot r^\alpha$, $k = 1, 2, \ldots, n - 1$. Furthermore,

$$1 - \cos(\nu_P, \nu_M) = 1 - \cos \gamma = 2 \sin^2 \frac{\gamma}{2} \leq \frac{\gamma^2}{2} \leq A \rho^{2\alpha} \leq A r^\alpha$$

and hence

$$|F_1(M_1)| \leq \text{Const} \int_{\Sigma_\varepsilon'} \frac{d\sigma_P}{r_{PM_1}^{n-1-\alpha}}.$$

This integral converges, so for $\varepsilon < \varepsilon_0$ small enough, we obtain $|F_1(M_1)| < \eta/4$, where $\eta > 0$ can be made arbitrarily small. Similarly, $|F_1(M)| < \eta/4$. Hence

$$|F(M_1) - F(M)| < |F_1(M_1)| + |F_1(M)| + |F_2(M_1) - F_2(M)| < \frac{\eta}{2} + |F_2(M_1) - F_2(M)|.$$

From here, due to the continuity of $F_2(M_1)$, we obtain that

$$|F(M_1) - F(M)| \to 0 \quad \text{when } M_1 \to M.$$

Thus,

$$\frac{\partial v(M_1)}{\partial \nu_M} = F(M_1) - w(M_1),$$

where $w(M_1)$ is the double layer potential and $F(M_1)$ is a continuous function.

Now let $M_1 \in V_+$ and $M_1 \to M$ along ν_M from inside. We have

$$\lim_{V_+ \ni M_1 \to M} \frac{\partial v(M_1)}{\partial \nu_M} = F(M) - \lim_{V_+ \ni M_1 \to M} w(M_1)$$

$$= \frac{\partial v(M)}{\partial \nu_M} - w(M) - w_+(M)$$

$$= \frac{\partial v(M)}{\partial \nu_M} + \frac{n-2}{2} |\Sigma_1| q(M),$$

which proves (11.23). Similarly, letting $V_- \ni M_1 \to M$ along ν_M, we obtain (11.24). $\blacksquare$

Corollary 161 *For the limit values of the normal derivative of a single layer potential with a continuous density q defined on a closed Lyapunov surface Σ, the following equalities are valid*

$$\left(\frac{\partial v}{\partial \nu}\right)_+ (M) - \left(\frac{\partial v}{\partial \nu}\right)_- (M) = (n-2) |\Sigma_1| q(M), \quad n \geq 3$$

and

$$\left(\frac{\partial v}{\partial \nu}\right)_+ (M) - \left(\frac{\partial v}{\partial \nu}\right)_- (M) = 2\pi q(M), \quad n = 2.$$

11.8   Reduction of Boundary Value Problems for the Laplace Equation to Integral Equations

Let us recall the boundary value problem statements.

1. **Interior Dirichlet problem** (Problem D_+)

     Given a continuous function f defined on Σ. Find a harmonic function u in V_+, such that

$$u(M) = f(M) \quad \text{for all } M \in \Sigma.$$

2. **Interior Neumann problem** (Problem N_+)

     Given a continuous function f defined on Σ. Find a harmonic function u in V_+, such that

$$\frac{\partial u(M)}{\partial v} = f(M) \quad \text{for all } M \in \Sigma.$$

3. **Exterior Dirichlet problem** (Problem D_-)

     Given a continuous function f defined on Σ. Find a harmonic function u in V_-, such that

$$u(M) = f(M) \quad \text{for all } M \in \Sigma,$$

and at infinity u satisfies the decay condition (see Definition 120).

4. **Exterior Neumann problem** (Problem N_-)

     Given a continuous function f defined on Σ. Find a harmonic function u in V_-, such that

$$\frac{\partial u(M)}{\partial v} = f(M) \quad \text{for all } M \in \Sigma,$$

and at infinity u satisfies the decay condition (see Definition 120).

Let us look for the solutions of problems $D_\pm$ in the form of a double layer potential and the solutions of problems $N_\pm$ in the form of a single layer potential. Then the condition $\Delta u = 0$ and the conditions at infinity are fulfilled automatically, and it remains to fulfill the corresponding boundary conditions:
Problem D_+

$$w_+(M) = f(M), \quad M \in \Sigma,$$

Problem D_-

$$w_-(M) = f(M), \quad M \in \Sigma,$$

Problem N_+

$$\left(\frac{\partial v}{\partial v}\right)_+ (M) = f(M), \quad M \in \Sigma,$$

Problem N_-

$$\left(\frac{\partial v}{\partial v}\right)_- (M) = f(M), \quad M \in \Sigma.$$

Due to Theorems 157 and 160, we obtain that these equalities can be written in the form:

Problem D_+

$$w(M) - \frac{1}{2C_n} q(M) = f(M), \quad M \in \Sigma,$$

Problem D_-

$$w(M) + \frac{1}{2C_n} q(M) = f(M), \quad M \in \Sigma,$$

Problem N_+

$$\frac{\partial v(M)}{\partial v} + \frac{1}{2C_n} q(M) = f(M), \quad M \in \Sigma,$$

Problem N_-

$$\frac{\partial v(M)}{\partial v} - \frac{1}{2C_n} q(M) = f(M), \quad M \in \Sigma,$$

where

$$C_n = \frac{1}{(n-2)|\Sigma_1|}.$$

More explicitly,

Problem D_+

$$q(M) + \frac{2}{|\Sigma_1|} \int_\Sigma \frac{\cos(r, v_P)}{r^{n-1}} q(P) d\sigma_P = -2C_n f(M), \quad M \in \Sigma,$$

Problem D_-

$$q(M) - \frac{2}{|\Sigma_1|} \int_\Sigma \frac{\cos(r, \nu_P)}{r^{n-1}} q(P) d\sigma_P = 2C_n f(M), \quad M \in \Sigma,$$

Problem N_+

$$q(M) + \frac{2}{|\Sigma_1|} \int_\Sigma \frac{\cos(r, \nu_M)}{r^{n-1}} q(P) d\sigma_P = 2C_n f(M), \quad M \in \Sigma,$$

Problem N_-

$$q(M) - \frac{2}{|\Sigma_1|} \int_\Sigma \frac{\cos(r, \nu_M)}{r^{n-1}} q(P) d\sigma_P = -2C_n f(M), \quad M \in \Sigma.$$

Thus, solution of Problems $D_\pm$ and $N_\pm$ is reduced to solution of integral equations with an unknown function $q(M)$. The obtained integral equations are known as integral equations of potential theory. In particular, for $n = 3$, the equations take the form
Problem D_+

$$q(M) + \frac{1}{2\pi} \int_\Sigma \frac{\cos(r, \nu_P)}{r^2} q(P) d\sigma_P = -\frac{1}{2\pi} f(M), \quad M \in \Sigma,$$

Problem D_-

$$q(M) - \frac{1}{2\pi} \int_\Sigma \frac{\cos(r, \nu_P)}{r^2} q(P) d\sigma_P = \frac{1}{2\pi} f(M), \quad M \in \Sigma,$$

Problem N_+

$$q(M) + \frac{1}{2\pi} \int_\Sigma \frac{\cos(r, \nu_M)}{r^2} q(P) d\sigma_P = \frac{1}{2\pi} f(M), \quad M \in \Sigma,$$

Problem N_-

$$q(M) - \frac{1}{2\pi} \int_\Sigma \frac{\cos(r, \nu_M)}{r^2} q(P) d\sigma_P = -\frac{1}{2\pi} f(M), \quad M \in \Sigma.$$

Finally, in the case $n = 2$, the integral equations take the form
Problem D_+

$$q(M) + \frac{1}{\pi} \int_L \frac{\cos(r, \nu_P)}{r} q(P) dl_P = -\frac{1}{\pi} f(M), \quad M \in \Sigma,$$

Problem D_-

$$q(M) - \frac{1}{\pi} \int_L \frac{\cos(r, \nu_P)}{r} q(P) dl_P = \frac{1}{\pi} f(M), \quad M \in \Sigma,$$

Problem N_+

$$q(M) + \frac{1}{\pi} \int_L \frac{\cos(r, \nu_M)}{r} q(P) dl_P = \frac{1}{\pi} f(M), \quad M \in \Sigma,$$

Problem N_-

$$q(M) - \frac{1}{\pi} \int_L \frac{\cos(r, \nu_M)}{r} q(P) dl_P = -\frac{1}{\pi} f(M), \quad M \in \Sigma.$$

Chapter 12
Elements of Theory of Integral Equations

Let Ω be a domain in $\mathbb{R}^n$. Equalities of the form

$$\int_\Omega K(P, M)\varphi(P)d\sigma_P = f(M), \quad M \in \Omega,$$

and

$$\varphi(M) - \int_\Omega K(P, M)\varphi(P)d\sigma_P = f(M), \quad M \in \Omega, \tag{12.1}$$

are called **linear integral equations** of the first and second kind, respectively.

We will study the equations of the second kind, and for the sake of simplicity, suppose that $\Omega = [a, b]$ is a segment. In this case equation, (12.1) takes the form

$$\varphi(x) - \lambda \int_a^b K(x, t)\varphi(t)dt = f(x), \quad a < x < b \quad (-\infty \le a < b \le \infty), \tag{12.2}$$

where λ is a numerical parameter, which is introduced deliberately for the convenience of the presentation.

Definition 162 The function $K(x, t)$ is called the **kernel of the integral equation**, and the number λ is called the **parameter of the integral equation**. The square $Q = \{a < x, t < b\}$ is called **the main square of the integral equation**.

It can be shown that the study of the equation

$$\varphi(x) - \lambda \int_a^b K(x, t)\varphi(t)dt = f(x), \quad c < x < d,$$

where $\varphi(x)$ is defined on $(a, b) \cup (c, d)$ does not make much sense unless $(a, b) \cap (c, d) \ne \varnothing$. Indeed, when $(a, b) \cap (c, d) = \varnothing$, this equation has a solution for any

© The Author(s), under exclusive license to Springer Nature Switzerland AG 2022
A. N. Karapetyants, V. V. Kravchenko, *Methods of Mathematical Physics*,
https://doi.org/10.1007/978-3-031-17845-0_12

function $\varphi(x)$ defined on (a, b), while if $(a, b) \cap (c, d) = (r, s)$, solution of such equation is equivalent to solution of (12.2) on the interval (r, s). We leave this as an exercise for the interested reader.

Definition 163 The integral equation (12.2) is called **Fredholm equation** (of the second kind) if its kernel is square integrable in the main square:

$$\int_a^b \int_a^b |K(x, t)|^2 \, dx dt < \infty. \tag{12.3}$$

Remark 164 In what follows, the condition (12.3) is always assumed to be fulfilled.

12.1 Space $L_2 (a, b)$ and Its Properties

Denote by $L_2 (a, b) = L_2$ the space of (Lebesgue) measurable functions on (a, b) which are (Lebesgue) square integrable on (a, b):

$$L_2 = L_2 (a, b) = \left\{ \varphi(x) : \quad \int_a^b |\varphi(x)|^2 \, dx < \infty \right\}.$$

The norm in L_2 is defined by

$$\|\varphi\| = \|\varphi\|_{L_2} = \left(\int_a^b |\varphi(x)|^2 \, dx \right)^{\frac{1}{2}}. \tag{12.4}$$

The convergence of a sequence of functions φ_n to a function φ with respect to this norm is called sometimes the mean square convergence and will be denoted as

$$\varphi = \lim_{n \to \infty} \varphi_n \quad \Longleftrightarrow \quad \lim_{n \to \infty} \|\varphi - \varphi_n\| = 0.$$

Let $\{\varphi_n\}$ be a Cauchy sequence in the sense of L_2. A question arises: is there a limit function $\varphi \in L_2$? Or, in other words, is the L_2-space **complete** with respect to the norm (12.4)? It occurs that if the integration in (12.4) is understood in the Riemann sense, the answer is negative. However, being understood in the Lebesgue sense, the limit function is necessarily square integrable on (a, b). This, for example, is one of the advantages of Lebesgue integration, which justifies our choice in favor of the Lebesgue integral.

We will not dwell upon the definition of the Lebesgue integration referring the interested reader to excellent books exposing this subject (e.g., [42]). Instead, we will remember that the integration is understood in the Lebesgue sense, and from the condition that for every $\varepsilon > 0$ there exists a natural number N such that for all natural numbers $m, n > N$ the inequality holds $\|\varphi_n - \varphi_m\| < \varepsilon$ ($\{\varphi_n\}$ is a Cauchy sequence), it follows that $\varphi_n \to \varphi \in L_2$ with respect to the norm (12.4).

The introduced space L_2 enjoys the following properties:

1. If $f, \varphi \in L_2$, then their product is integrable:

$$\int_a^b |f(x)\varphi(x)| \, dx < \infty.$$

Proof Obviously, $(|f| - |\varphi|)^2 \geq 0$, from which we obtain $|f(x)\varphi(x)| \leq \frac{1}{2}(|f|^2 + |\varphi|^2)$ that gives us the result. ∎

2. In the space L_2, an inner (or scalar) product (f, φ) can be defined as follows:

$$(f, \varphi) = \int_a^b f(x)\overline{\varphi(x)}dx.$$

Proof One should verify (a) $(f, \varphi) = \overline{(\varphi, f)}$, (b) $(f, f) \geq 0$ and $(f, f) = 0$ if only $f = 0$, (c) $(\alpha f + \beta\varphi, \psi) = \alpha(f, \psi) + \beta(\varphi, \psi)$ (which is obvious) as well as $(f, \alpha\varphi) = \overline{\alpha}(f, \varphi)$. ∎

A complete normed space, in which a scalar product exists, is called Hilbert space. It is proved in functional analysis that any Hilbert space is isomorphic to L_2.

3. **Cauchy–Bunyakovsky–Schwarz inequality**. For all $f, \varphi \in L_2$, the inequality holds

$$|(f, \varphi)| \leq \|f\| \, \|\varphi\| \tag{12.5}$$

or more explicitly

$$\left| \int_a^b f(x)\varphi(x)dx \right| \leq \left(\int_a^b |f(x)|^2 \, dx \right)^{\frac{1}{2}} \left(\int_a^b |\varphi(x)|^2 \, dx \right)^{\frac{1}{2}}.$$

Here, in the integral, we omit the conjugation over $\varphi(x)$ due to the invariance $|\varphi(z)| = |\overline{\varphi}(z)|$. The above inequality reflects the fact that the product of Lebesgue square integrable functions is a function Lebesgue integrable. This inequality is a special case of the so-called Hölder inequality.

Proof We assume that $\varphi \neq 0$ because otherwise there is nothing to prove. Denote

$$\psi = f - \frac{(f, \varphi)}{\|\varphi\|^2}\varphi.$$

We have

$$(\psi, \varphi) = \left(f - \frac{(f, \varphi)}{\|\varphi\|^2} \varphi, \varphi \right) = (f, \varphi) - \frac{(f, \varphi)}{\|\varphi\|^2} (\varphi, \varphi)$$

$$= (f, \varphi) - \frac{(f, \varphi)}{\|\varphi\|^2} \|\varphi\|^2 = 0.$$

Therefore ψ is orthogonal to φ, and $f = \psi + \lambda\varphi$, where $\lambda = \frac{(f,\varphi)}{\|\varphi\|^2}$. Consider

$$\|f\|^2 = (\psi + \lambda\varphi, \psi + \lambda\varphi) = \|\psi\|^2 + |\lambda|^2 \|\varphi\|^2$$

$$= \|\psi\|^2 + \left| \frac{(f, \varphi)}{\|\varphi\|^2} \right|^2 \|\varphi\|^2 = \|\psi\|^2 + \frac{|(f, \varphi)|^2}{\|\varphi\|^2}$$

$$\geq \frac{|(f, \varphi)|^2}{\|\varphi\|^2}$$

from where (12.5) follows.   ∎

12.2 Fredholm Operator and Its Iterated Kernels

Definition 165 The integral operator

$$K\varphi(x) = \int_a^b K(x, t)\varphi(t)dt$$

is called the **Fredholm operator** if its kernel $K(x, t)$ is square summable in the main square

$$\int_a^b \int_a^b |K(x, t)|^2 \, dxdt < \infty.$$

The operator K is obviously linear: $K(\alpha_1\varphi_1 + \alpha_2\varphi_2) = \alpha_1 K\varphi_1 + \alpha_2 K\varphi_2$. Denote

$$B = \left(\int_a^b \int_a^b |K(x, t)|^2 \, dxdt \right)^{\frac{1}{2}} < \infty.$$

Lemma 166 *Any Fredholm operator is bounded in L_2, moreover, its operator norm does not exceed the constant B, i.e.,*

$$\| K\varphi \| \leq B \, \| \varphi \| \, ,$$

for any $\varphi \in L_2$.

Proof We need to show that

$$\left(\int_a^b |K\varphi(x)|^2 \, dx \right)^{\frac{1}{2}} \leq B \left(\int_a^b |\varphi(x)|^2 \, dx \right)^{\frac{1}{2}} ,$$

for any $\varphi \in L_2$. Let us estimate the expression with the use of Cauchy–Bunyakovsky–Schwarz inequality

$$|K\varphi(x)|^2 = \left| \int_a^b K(x,t)\varphi(t)dt \right|^2 \leq \left(\int_a^b |K(x,t)\varphi(t)| \, dt \right)^2$$

$$\leq \int_a^b |K(x,t)|^2 \, dt \cdot \int_a^b |\varphi(t)|^2 \, dt.$$

Hence

$$\int_a^b |K\varphi(x)|^2 \, dx \leq \| \varphi \|^2 \int_a^b \int_a^b |K(x,t)|^2 \, dtdx = B^2 \, \| \varphi \|^2 ,$$

which finishes the proof.   ∎

Consider a composition (product) of two Fredholm operators $LK\varphi = L(K\varphi)$.

Lemma 167 *If both K and L are Fredholm operators, then their composition $LK\varphi$ is a Fredholm operator as well, and $\| LK\varphi \| \leq BB' \, \| \varphi \|$, for any $\varphi \in L_2$, where $B' = \left(\int_a^b \int_a^b |L(x,t)|^2 \, dxdt \right)^{\frac{1}{2}}$.*

Proof First of all, notice that the estimate $\| LK\varphi \| \leq BB' \, \| \varphi \|$ is obvious, because both K and L are Fredholm operators, and hence

$$\| LK\varphi \| \leq B' \, \| K\varphi \| \leq BB' \, \| \varphi \| ,$$

for any $\varphi \in L_2$. Let us calculate the composition $LK\varphi$:

$$(LK\varphi)(x) = \int_a^b L(x,t)(K\varphi)(t)dt = \int_a^b L(x,t) \left(\int_a^b K(t,s)\varphi(s)ds \right) dt$$

$$= \int_a^b \varphi(s) \int_a^b L(x,t)K(t,s)\,dt\,ds = \int_a^b M(x,s)\varphi(s)\,ds,$$

where

$$M(x,s) = \int_a^b L(x,t)K(t,s)\,dt, \tag{12.6}$$

and it remains to show that $M(x,s)$ is square integrable. If we show this, then we will arrive at the desired result, including justifying the earlier change in the order of integration by Fubini's theorem. Notice that

$$M(x,s) = \left(L(x,\cdot), \overline{K}(\cdot,s)\right),$$

and due to the Cauchy–Bunyakovsky–Schwarz inequality (12.5), we obtain

$$|M(x,s)|^2 \leq \|L(x,\cdot)\|^2\,\|K(\cdot,s)\|^2 = \int_a^b |L(x,t)|^2\,dt \int_a^b |K(t,s)|^2\,dt.$$

Then

$$B_M^2 = \int_a^b \int_a^b |M(x,s)|^2\,dx\,ds \leq \int_a^b \int_a^b \left(\int_a^b |L(x,t)|^2\,dt \int_a^b |K(t,s)|^2\,dt\right) dx\,ds$$

$$= \int_a^b \int_a^b |L(x,t)|^2\,dx\,dt \int_a^b \int_a^b |K(t,s)|^2\,dt\,ds = BB', \tag{12.7}$$

which finishes the proof.   ∎

Remark 168 In general, the operators L and K do not commute, i.e., $LK\varphi \neq KL\varphi$. Indeed,

$$LK\varphi(x) = \int_a^b M(x,s)\varphi(s)\,ds,$$

while

$$KL\varphi(x) = \int_a^b N(x,s)\varphi(s)\,ds,$$

where

$$M(x,s) = \int_a^b L(x,t)K(t,s)\,dt \neq N(x,s) = \int_a^b K(x,t)L(t,s)\,dt$$

for arbitrary Fredholm kernels $K(x,t)$ and $L(x,t)$.

Consider the following operator composition:

$$K^n \varphi(x) = \underbrace{KK \ldots K}_{n} \varphi(x).$$

Definition 169   The operator K^n is called the n-th power of the operator K.

Due to Lemma 167, we have that the operator K^n admits the form

$$K^n \varphi(x) = \int_a^b K_n(x, t) \varphi(t) dt,$$

where $K_n(x, t)$ is square integrable.

Definition 170   The kernel $K_n(x, t)$ is called the n-th iterated kernel of the Fredholm operator K.

Obviously, $K_1(x, t) = K(x, t)$, and due to operational equality $K^n = K K^{n-1}$ combined with (12.6), we have

$$K_n(x, t) = \int_a^b K(x, s) K_{n-1}(s, t) ds. \tag{12.8}$$

On the other hand, $K^n = K^{n-1} K$. Hence

$$K_n(x, t) = \int_a^b K_{n-1}(x, s) K(s, t) ds,$$

and therefore

$$\int_a^b K(x, s) K_{n-1}(s, t) ds = \int_a^b K_{n-1}(x, s) K(s, t) ds.$$

Denote

$$B_n = \left(\int_a^b \int_a^b |K_n(x, t)|^2 \, dx dt \right)^{\frac{1}{2}}.$$

Due to (12.7), we have

$$B_n \le B_{n-1} B \le B_{n-2} B^2 \le B^n.$$

Thus, by Lemma 166, we obtain

$$\| K^n \varphi \| \le B_n \| \varphi \| \le B^n \| \varphi \| ,$$

for any $\varphi \in L_2$. In other words, the operator norm of the operator K^n, as an operator acting in L_2, does not exceed B^n.

12.3   Method of Successive Approximations

Let us apply the method of successive approximations to the Fredholm integral equation of the second kind

$$\varphi(x) - \lambda \int_a^b K(x, t)\varphi(t)dt = f(x), \quad a < x < b, \tag{12.9}$$

where and below we assume the function f to be square integrable on (a, b). We have

$$\varphi = f + \lambda K\varphi.$$

Hence

$$\varphi = f + \lambda K (f + \lambda K\varphi) = f + \lambda Kf + \lambda^2 K^2\varphi$$

$$= f + \lambda Kf + \lambda^2 K^2 (f + \lambda K\varphi)$$

$$= f + \lambda Kf + \lambda^2 K^2 f + \lambda^3 K^3\varphi,$$

and after n such steps, we obtain

$$\varphi = f + \lambda Kf + \lambda^2 K^2 f + \ldots + \lambda^n K^n f + \lambda^{n+1} K^{n+1}\varphi.$$

Denote

$$\varphi_n = f + \lambda Kf + \lambda^2 K^2 f + \ldots + \lambda^n K^n f.$$

Obviously, φ_n is a partial sum of the (formal) series

$$\sum_{j=0}^{\infty} \lambda^j K^j f,$$

which is called the **Neumann series**.

Theorem 171 *If the kernel $K(x, t)$ is square integrable and $|\lambda| < B^{-1}$, then the Neumann series converges in the sense of the L_2-norm, and the sum of the series*

$$\varphi = \sum_{j=0}^{\infty} \lambda^j K^j f$$

is a solution of Eq. (12.9). This solution is unique in the space L_2.

Remark 172 The condition $|\lambda| < B^{-1}$ is only sufficient for the convergence of the Neumann series.

Proof Let us estimate the value of $\left\| \sum_{j=N+1}^{N+p} \lambda^j K^j f \right\|$ with the aid of the triangle inequality

$$\left\| \sum_{j=N+1}^{N+p} \lambda^j K^j f \right\| \le \sum_{j=N+1}^{N+p} |\lambda|^j \left\| K^j f \right\| \le \sum_{j=N+1}^{N+p} |\lambda|^j \|f\| B^j = \|f\| \sum_{j=N+1}^{N+p} (|\lambda| B)^j$$

$$= \|f\| \frac{(|\lambda| B)^{N+1} \left(1 - (|\lambda| B)^p\right)}{1 - |\lambda| B} \le \|f\| \frac{(|\lambda| B)^{N+1}}{1 - |\lambda| B} \to 0, \quad N \to \infty,$$

for all p. Thus, the Neumann series converges in L_2. Denote its sum by $\varphi(x)$. Let us show that $\varphi(x)$ satisfies equation (12.9). We have

$$\varphi_n = f + \lambda K \left(f + \lambda K f + \ldots + \lambda^{n-1} K^{n-1} f \right) = f + \lambda K \varphi_{n-1}.$$

Consider the L_2-limit of the equality $\varphi_n = f + \lambda K \varphi_{n-1}$ when $n \to \infty$. We have

$$\lim_{n \to \infty} \varphi_n = f + \lambda \lim_{n \to \infty} K \varphi_{n-1},$$

and thus,

$$\varphi = f + \lambda \lim_{n \to \infty} K \varphi_{n-1}.$$

It remains to show that $\lim_{n \to \infty} K \varphi_{n-1} = K \varphi$. Indeed,

$$\| K \varphi_{n-1} - K \varphi \| = \| K (\varphi_{n-1} - \varphi) \| \le B \| \varphi_{n-1} - \varphi \| \to 0.$$

Thus, $\varphi = f + \lambda K \varphi$, and hence φ is a solution of (12.9).

Let us prove its uniqueness in L_2. Suppose there exist two different solutions of (12.9): $\varphi_1(x)$ and $\varphi_2(x)$. Denote $\omega = \varphi_1 - \varphi_2$. Then $\omega - \lambda K \omega = 0$, i.e., $\omega = \lambda K \omega$. From here, we obtain

$$\|\omega\| = \|\lambda K \omega\| \le |\lambda| B \|\omega\|,$$

which means that $(1 - |\lambda| B) \|\omega\| \leq 0$. Since $1 - |\lambda| B > 0$, we obtain that $\|\omega\| = 0$ and hence $\omega = 0$, i.e., $\varphi_1 = \varphi_2$. ∎

Remark 173 Let the interval (a, b) be finite and the kernel $K(x, t)$ piecewise continuous. Then it is bounded, i.e., $|K(x, t)| < M$, $a \leq x, t \leq b$. Since any piecewise continuous function on a finite domain is square integrable, we can consider the number

$$B^2 = \int_a^b \int_a^b |K(x, t)|^2 \, dt dx.$$

It is easy to see that in this case $B \leq M (b - a)$, and as a corollary of Theorem 171, we obtain the following statement.

Theorem 174 *Let the interval (a, b) be finite and the kernel $K(x, t)$ piecewise continuous. Then its Neumann series converges absolutely and uniformly for $|\lambda| < \frac{1}{M(b-a)}$, where M is a constant such that $|K(x, t)| < M$, $a \leq x, t \leq b$.*

Proof We have

$$|K_2(x, t)| \leq \int_a^b |K(x, s)K(s, t)| \, ds \leq M^2 (b - a),$$

$$|K_3(x, t)| \leq M^3 (b - a)^2,$$

$$\dots$$

$$|K_n(x, t)| \leq M^n (b - a)^{n-1}.$$

Then

$$\left| \lambda^n K^n f \right| = |\lambda|^n \left| \int_a^b K_n(x, t) f(t) dt \right| \leq |\lambda|^n \left(\int_a^b |K_n(x, t)|^2 \, dt \right)^{\frac{1}{2}} \left(\int_a^b |f(t)|^2 \, dt \right)^{\frac{1}{2}},$$

where we used the Cauchy–Bunyakovsky–Schwarz inequality. Using the above estimate for $|K_n(x, t)|$, we obtain

$$\left| \lambda^n K^n f \right| \leq |\lambda|^n M^n (b - a)^{n-1/2} \|f\| = (|\lambda| M (b - a))^n \frac{\|f\|}{\sqrt{b - a}}.$$

Since $|\lambda| M (b - a) < 1$, the numerical series $\sum_{n=0}^{\infty} (|\lambda| M (b - a))^n$ converges, and hence, due to the Weierstrass criterion, the Neumann series converges absolutely and also uniformly.   ∎

12.4   Notion of the Resolvent of the Integral Equation

In the previous section we found that when $|\lambda| < 1/B$ the unique solution of Eq. (12.9) has the form

$$\varphi(x) = f(x) + \lambda K f(x) + \lambda^2 K^2 f(x) + \ldots$$

$$= f(x) + \sum_{n=1}^{\infty} \lambda^n \int_a^b K_n(x, t) f(t) dt.$$

Let us write this equality as follows:

$$\varphi(x) = f(x) + \lambda \int_a^b \sum_{n=1}^{\infty} \lambda^{n-1} K_n(x, t) f(t) dt.$$

Here we changed the order of summation and integration which would be possible if the series

$$\sum_{n=1}^{\infty} \lambda^{n-1} K_n(x, t)$$

converges with respect to the L_2-norm in the main square. Let us verify this fact. We have

$$\left\| \sum_{n=N+1}^{N+p} \lambda^{n-1} K_n(x, t) \right\| = \left(\int_a^b \int_a^b \left| \sum_{n=N+1}^{N+p} \lambda^{n-1} K_n(x, t) \right|^2 dx dt \right)^{\frac{1}{2}}$$

$$\leqslant \sum_{n=N+1}^{N+p} |\lambda|^{n-1} \left(\int_a^b \int_a^b |K_n(x, t)|^2 dx dt \right)^{\frac{1}{2}}$$

$$= \sum_{n=N+1}^{N+p} |\lambda|^{n-1} B_n$$

$$\leq \sum_{n=N+1}^{N+p} |\lambda|^{n-1} B^n \leq \frac{|\lambda|^N B^{N+1}}{1 - |\lambda| B} \to 0 \text{ when } N \to \infty \text{ for all } p,$$

which proves the statement.

Denote

$$R(x, t, \lambda) = \sum_{n=1}^{\infty} \lambda^{n-1} K_n(x, t). \tag{12.10}$$

This function is called the **resolvent** of the integral equation (12.9). Thus, when $|\lambda| B < 1$, the solution of (12.9) has the form

$$\varphi(x) = f(x) + \lambda \int_a^b R(x, t, \lambda) f(t) dt, \quad a < x < b, \tag{12.11}$$

where $R(x, t, \lambda)$ is the resolvent, for whose construction the iterated kernels $K_n(x, t)$ can be calculated from (12.8).

Example 175 Solve the equation

$$\varphi(x) - \lambda \int_0^1 e^{x+t} \varphi(t) dt = 1. \tag{12.12}$$

We have

$$K_1(x, t) = K(x, t) = e^{x+t},$$

$$K_2(x, t) = \int_0^1 K(x, s) K(s, t) ds$$

$$= \int_0^1 e^{x+s} e^{s+t} ds = e^{x+t} \int_0^1 e^{2s} ds = e^{x+t} \frac{e^2 - 1}{2},$$

$$K_3(x, t) = \int_0^1 K(x, s) K_2(s, t) ds$$

$$= e^{x+t} \frac{e^2 - 1}{2} \int_0^1 e^{2s} ds = e^{x+t} \left(\frac{e^2 - 1}{2} \right)^2.$$

Obviously,

$$K_n(x, t) = \left(\frac{e^2 - 1}{2} \right)^{n-1} e^{x+t}$$

and hence

$$R(x, t, \lambda) = \sum_{n=1}^{\infty} \lambda^{n-1} K_n(x, t) = e^{x+t} \sum_{n=1}^{\infty} \lambda^{n-1} \left(\frac{e^2 - 1}{2} \right)^{n-1}$$

$$= e^{x+t} \sum_{n=0}^{\infty} \left(\lambda \frac{e^2 - 1}{2} \right)^n = e^{x+t} \frac{1}{1 - \frac{\lambda}{2} \left(e^2 - 1 \right)}$$

under the condition $\frac{|\lambda|}{2} \left(e^2 - 1 \right) < 1$. Then

$$\varphi(x) = f(x) + \lambda \int_0^1 R(x, t, \lambda) f(t) dt$$

$$= 1 + \lambda \int_0^1 R(x, t, \lambda) dt$$

$$= 1 + \lambda \int_0^1 \frac{e^{x+t}}{1 - \frac{\lambda}{2} \left(e^2 - 1 \right)} dt$$

$$= 1 + \frac{\lambda}{1 - \frac{\lambda}{2} \left(e^2 - 1 \right)} e^x \int_0^1 e^t dt$$

$$= 1 + \frac{\lambda (e - 1)}{1 - \frac{\lambda}{2} \left(e^2 - 1 \right)} e^x, \quad |\lambda| < \frac{2}{e^2 - 1}.$$

Remark 176 Notice that it is easy to verify that the obtained formula gives us a solution of (12.12) for all λ except for $\lambda = \frac{2}{e^2-1}$, though the method of successive approximations is not applicable for large values of $|\lambda|$.

12.5   Volterra Integral Equation

Definition 177   Volterra's integral equation is an integral equation of the form

$$\varphi(x) - \lambda \int_a^x K(x, t) \varphi(t) dt = f(x), \quad a < x < b. \tag{12.13}$$

As above we suppose that $K(x, t)$ is a square integrable function. Obviously, the Volterra equation is a special case of the Fredholm equation. Indeed, we can write (12.13) in the form of a Fredholm equation:

$$\varphi(x) - \lambda \int_a^b K(x, t) \varphi(t) dt = f(x), \quad a < x < b,$$

where

Fig. 12.1 $K(x, t)$ extended
by zero

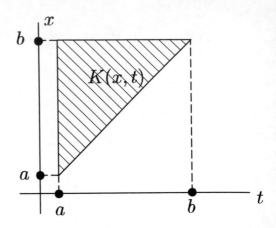

$$K(x,t) = \begin{cases} K(x,t), & t < x, \\ 0, & t > x \end{cases}$$

(see Fig. 12.1).

Let us deduce a recurrent formula for the iterated kernels. Denote, as before, $K_1(x, t) = K(x, t)$. We have

$$K_2(x,t) = \int_a^b K(x,s)K_1(s,t)ds = \int_a^b K(x,s)K(s,t)ds.$$

Let first $t > x$. Then

$$K_2(x,t) = \int_a^t K(x,s)K(s,t)ds + \int_t^b K(x,s)K(s,t)ds,$$

where the first integral equals zero because $K(s,t) = 0$ for $t > s$, and the second integral equals zero because $K(x,s) = 0$ ($x < t < s$). Thus,

$$K_2(x,t) = 0 \quad \text{for } t > x.$$

Now let $x > t$. Then

$$K_2(x,t) = \int_a^t K(x,s)K(s,t)ds + \int_t^x K(x,s)K(s,t)ds + \int_x^b K(x,s)K(s,t)ds.$$

Here the first integral equals zero because $K(s,t) = 0$ for $t > s$, and the third integral equals zero because $K(x,s) = 0$ for $x < s$. Thus,

$$K_2(x,t) = \begin{cases} \int_t^x K(x,s)K(s,t)ds, & t < x \\ 0, & t > x. \end{cases}$$

It is easy to obtain by induction that

$$K_n(x, t) = \begin{cases} \int_t^x K(x, s) K_{n-1}(s, t) ds, & t < x \\ 0, & t > x. \end{cases}$$

To simplify the subsequent reasonings, assume additionally that $K(x, t)$ is bounded: $|K(x, t)| \leq M$. Let us prove the estimate

$$|K_n(x, t)| \leq \frac{M^n}{(n-1)!} (x - t)^{n-1}, \quad t < x, \tag{12.14}$$

by induction. Obviously, for $n = 1$, it is valid. Suppose that it is valid for the number $n - 1$,

$$|K_{n-1}(x, t)| \leq \frac{M^{n-1}}{(n-2)!} (x - t)^{n-2}, \quad t < x.$$

Then

$$|K_n(x, t)| = \left| \int_t^x K(x, s) K_{n-1}(s, t) ds \right| \leq \int_t^x M \frac{M^{n-1}}{(n-2)!} (s - t)^{n-2} ds$$

$$= \frac{M^n}{(n-2)!} \left. \frac{(s - t)^{n-1}}{n - 1} \right|_t^x = \frac{M^n}{(n-1)!} (x - t)^{n-1}.$$

This proves the estimate (12.14).

With the aid of (12.14), let us prove that the Neumann series for the Volterra integral equation converges absolutely and uniformly for all values of the parameter λ.

Theorem 178 *Let the interval (a, b) be finite and the right hand side of the Volterra equation (12.13) absolutely integrable ($\int_a^b |f(x)| dx < \infty$). Let the kernel $K(x, t)$ be bounded: $|K(x, t)| \leq M$. Then the Volterra equation (12.13) possesses a unique absolutely integrable solution for all λ. This solution is the sum of the Neumann series, i.e., the solution can be obtained by the method of successive approximations for all λ.*

Proof We have

$$\left| \lambda^n K^n f \right| = |\lambda|^n \left| \int_a^x K_n(x, t) f(t) dt \right| \leq |\lambda|^n \int_a^x \frac{(x - t)^{n-1}}{(n-1)!} M^n |f(t)| dt$$

$$\leq \frac{|\lambda|^n M^n}{(n-1)!} (b - a)^{n-1} \int_a^b |f(t)| dt$$

$$= \left(|\lambda| \, M \int_a^b |f(t)| \, dt \right) \frac{(M \, |\lambda|)^{n-1}}{(n-1)!} \, (b-a)^{n-1} \, .$$

Thus, the Neumann series is majorized by the convergent number series

$$\text{Const} \cdot \sum_{n=1}^{\infty} \frac{(M \, |\lambda|)^{n-1}}{(n-1)!} \, (b-a)^{n-1} = \text{Const} \cdot e^{M|\lambda|(b-a)}$$

and hence converges absolutely and uniformly. It remains to show that an integrable solution is unique.

Suppose there exist two different solutions φ_1 and φ_2. Denote $\omega = \varphi_1 - \varphi_2$. Then

$$\omega = \lambda K \omega.$$

Hence

$$\omega = \lambda K \omega = \lambda^2 K^2 \omega = \ldots = \lambda^n K^n \omega.$$

Then

$$|\omega(x)| = |\lambda|^n \left| \left(K^n \omega \right) (x) \right| \le |\lambda|^n \int_a^x |K_n(x, t)| \, |\omega(t)| \, dt$$

$$\le |\lambda|^n \frac{M^n}{(n-1)!} \, (b-a)^{n-1} \int_a^b |\omega(t)| \, dt \to 0, \quad n \to \infty.$$

Since the left hand side is independent of n, we conclude that $|\omega(t)| \equiv 0$, and hence $\varphi_1 \equiv \varphi_2$. $\blacksquare$

Remark 179 When (a, b) is finite, the absolute integrability follows from the square integrability:

$$\int_a^b |f(x)| \, dx \le \left(\int_a^b |f(x)|^2 \, dx \right)^{\frac{1}{2}} \left(\int_a^b 1 \, dx \right)^{\frac{1}{2}} < \infty.$$

Remark 180 In the proof of the uniqueness, we used the supposition of absolute integrability. Without it, there can be more than one solution.

12.6   Integral Equations with Degenerate Kernels

Definition 181 The kernel $K(x, t)$ is called **degenerate** if it admits a representation in the form

$$K(x, t) = \sum_{k=1}^{n} a_k(x) b_k(t). \qquad (12.15)$$

Obviously, the kernels (1) $b_0(t)x^n + b_1(t)x^{n-1} + \ldots + b_n(t)$, (2) $a_0(x)t^n + a_1(x)t^{n-1} + \ldots + a_n(t)$, (3) $e^{x-t} = e^x e^{-t}$, and (4) $\sin(x+t) = \sin x \cos t + \cos x \sin t$ are degenerate, while the kernels $\ln(x+t)$, $\tan(x+t)$, and $e^{|x-t|}$ are not degenerate.

Let us show that an integral equation with a degenerate kernel can be fully studied. We have

$$\varphi(x) - \lambda \int_a^b \sum_{k=1}^{n} a_k(x) b_k(t) \varphi(t) dt = f(x), \quad a < x < b, \qquad (12.16)$$

or

$$\varphi(x) - \lambda \sum_{k=1}^{n} a_k(x) \int_a^b b_k(t) \varphi(t) dt = f(x), \quad a < x < b.$$

Let us multiply the equality by $b_j(x)$ $(j = 1, 2, \ldots, n)$ and integrate from a to b,

$$\int_a^b b_j(x) \varphi(x) dx = \int_a^b b_j(x) f(x) dx + \lambda \sum_{k=1}^{n} \left(\int_a^b b_j(x) a_k(x) dx \right) \left(\int_a^b b_k(t) \varphi(t) dt \right).$$

$$(12.17)$$

Denote

$$c_j = \int_a^b b_j(x) \varphi(x) dx \quad \text{and} \quad f_j = \int_a^b b_j(x) f(x) dx.$$

Obviously, the solution will be found if the constants c_j are found, because in such case

$$\varphi(x) = f(x) + \lambda \sum_{k=1}^{n} c_k a_k(x).$$

From (12.17) for the constants c_j, we obtain the system

$$c_j - \lambda \sum_{k=1}^{n} \alpha_{jk} c_k = f_j, \quad j = 1, 2, \ldots, n,$$

where

$$\alpha_{jk} = \int_a^b b_j(x)a_k(x)dx.$$

The system can be written as follows:

$$\begin{cases} (1 - \lambda\alpha_{11})\,c_1 - \lambda\alpha_{12}c_2 - \ldots - \lambda\alpha_{1n}c_n = f_1, \\ -\lambda\alpha_{21}c_1 + (1 - \lambda\alpha_{22})\,c_2 - \ldots - \lambda\alpha_{2n}c_n = f_2, \\ \qquad\qquad \ldots \\ -\lambda\alpha_{n1}c_1 + -\lambda\alpha_{n2}c_2 - \ldots + (1 - \lambda\alpha_{nn})\,c_n = f_n. \end{cases} \tag{12.18}$$

Remark 182 We assume all $a_k(x)$ be linearly independent and all $b_k(t)$ be linearly independent, otherwise the number of terms in the sum (12.15) can be reduced.

Remark 183 The system and the integral equation are equivalent in the sense that they are solvable or unsolvable simultaneously.

Proof The fact that system and the integral equation can be unsolvable only simultaneously is obvious. It is obvious as well that if the equation possesses a solution, then the system is also solvable. Let us show the opposite. Suppose that the system is solvable. We show that the function

$$\varphi(x) = f(x) + \lambda \sum_{k=1}^n c_k a_k(x)$$

is a solution of the equation. Substituting, we have

$$f(x) + \lambda \sum_{k=1}^n c_k a_k(x) - \lambda \sum_{k=1}^n a_k(x) \int_a^b \left(f(t) + \lambda \sum_{j=1}^n c_j a_j(t) \right) b_k(t)dt = f(x),$$

which is the same as

$$\lambda \sum_{k=1}^n a_k(x) \left(c_k - f_k - \lambda \sum_{j=1}^n c_j \alpha_{kj} \right) = 0,$$

from which we obtain the result.   ∎

Consider the determinant of the system (12.18)

$$D(\lambda) = \begin{vmatrix} 1 - \lambda\alpha_{11} & -\lambda\alpha_{12} & \ldots & -\lambda\alpha_{1n} \\ -\lambda\alpha_{21} & 1 - \lambda\alpha_{22} & \ldots & -\lambda\alpha_{2n} \\ & \ldots & & \\ -\lambda\alpha_{n1} & -\lambda\alpha_{n2} & \ldots & 1 - \lambda\alpha_{nn} \end{vmatrix},$$

which is a polynomial of degree not greater than n. Moreover, as $D(0) = 1$ we have that $D(\lambda)$ is not identically zero for all λ and hence possesses at most n zeros.

Conclusion 184 *If λ does not coincide with any of the roots of the polynomial $D(\lambda)$, the system (12.18) (as well as Eq. (12.16)) possesses a unique solution for any right hand side. In this case the homogeneous equation does not have non-trivial solutions.*

If λ is a root of $D(\lambda) = 0$, then, in general, the integral equation (12.16) is unsolvable. Let r be the rank of the matrix $D(\lambda)$. Then the homogeneous equation has $(n - r)$ linearly independent solutions.

Example 185 Solve the equation

$$\varphi(x) - \int_0^1 (1 + 6xt)\, \varphi(t)dt = 12x. \tag{12.19}$$

We have

$$\varphi(x) - 1 \cdot \underbrace{\int_0^1 1 \cdot \varphi(t)dt}_{c_1} - 6x \underbrace{\int_0^1 t \cdot \varphi(t)dt}_{c_2} = 12x.$$

Hence the equation takes the form

$$\varphi(x) = 12x + c_1 + 6c_2 x.$$

Multiplying first by 1 and next by x and integrating, we obtain

$$\begin{cases} \int_0^1 \varphi(x)dx - c_1 \int_0^1 dx - 6c_2 \int_0^1 xdx = 12 \int_0^1 xdx, \\ \int_0^1 x\varphi(x)dx - c_1 \int_0^1 xdx - 6c_2 \int_0^1 x^2dx = 12 \int_0^1 x^2dx, \end{cases}$$

which leads to the system

$$\begin{cases} c_1 - c_1 - 3c_2 = 6, \\ c_2 - \frac{c_1}{2} - 2c_2 = 4, \end{cases} \iff \begin{cases} c_2 = -2, \\ c_1 = -4. \end{cases}$$

Thus, the unique solution of (12.19) has the form

$$\varphi(x) = -4.$$

12.7   Integral Equations in a General Case

Let us consider the Fredholm equation in a general case, when $|\lambda|$ is not necessarily small, and $K(x, t)$ is not degenerate

$$\varphi(x) - \lambda \int_a^b K(x, t)\varphi(t)dt = f(x), \quad a < x < b. \tag{12.20}$$

Lemma 186 *Let* $\omega_k(x)$, $k = 1, 2, \ldots$, *be a complete orthonormal system of functions defined on the interval* (a, b). *Then the system of functions of two independent variables* $\Omega_{jk}(x, t) = \omega_j(x)\omega_k(t)$, $j, k = 1, 2, \ldots$, *is complete and orthonormal in the main square* $a < x, t < b$.

Proof We are given that $\omega_j(x)$, $j = 1, 2, \ldots$, is complete, and

$$(\omega_{j_1}, \omega_{j_2}) = \begin{cases} 0, & j_1 \neq j_2 \\ 1, & j_1 = j_2 \end{cases} = \delta_{j_1 j_2}.$$

We need to prove that Ω_{jk}, $j, k = 1, 2, \ldots$, enjoy these properties in the main square.

Let us check the orthonormality

$$\int_a^b \int_a^b \Omega_{jk}(x, t)\Omega_{pq}(x, t)dxdt = \int_a^b \int_a^b \omega_j(x)\omega_k(t)\omega_p(x)\omega_q(t)dxdt$$

$$= \int_a^b \omega_j(x)\omega_p(x)dx \cdot \int_a^b \omega_k(t)\omega_q(t)dt = \delta_{jp}\delta_{kq}.$$

Thus, Ω_{jk}, $j, k = 1, 2, \ldots$, represent an orthonormal system. Similarly the completeness can be shown: the equalities $\int_a^b \int_a^b \Omega_{jk}(x, t)\varphi(x, t)dxdt$ for all $j, k = 1, 2, \ldots$ imply that $\varphi(x, t) = 0$. ∎

From the mathematical analysis, the following theorem is known.

Theorem 187 *If* $\{\omega_k(x)\}_{k=1}^\infty$ *is a complete orthonormal system of functions in* L_2, *then any square integrable function* $\varphi(x)$ *admits a series expansion (Fourier series) in terms of* $\{\omega_k(x)\}_{k=1}^\infty$. *The series converges with respect to the* L_2*-norm and has the form*

$$\varphi(x) = \sum_{k=1}^\infty A_k\omega_k(x) = \sum_{k=1}^\infty (\varphi, \omega_k)\,\omega_k(x).$$

*Moreover, the following **Parseval identity** is valid:*

$$\|\varphi\|^2 = \int_a^b |\varphi(x)|^2 \, dx = \sum_{k=1}^{\infty} A_k^2.$$

This theorem is valid in the case of several independent variables as well.

Corollary 188 *Any square integrable kernel $K(x,t)$ admits a series expansion of the form*

$$K(x,t) = \sum_{k,j=1}^{\infty} A_{kj} \omega_k(x) \omega_j(t),$$

where $\{\omega_k(x)\}_{k=1}^{\infty}$ is a complete orthonormal system of functions in L_2. The series converges with respect to the L_2-norm, and the Parseval identity is valid

$$B^2 = \int_a^b \int_a^b |K(x,t)|^2 \, dt dx = \sum_{k,j=1}^{\infty} A_{kj}^2.$$

Returning to the Fredholm equation (12.20), we have that its kernel can be written in the form

$$K(x,t) = \sum_{k,j=1}^{n} A_{kj} \omega_k(x) \omega_j(t) + \sum_{k,j=n+1}^{\infty} A_{kj} \omega_k(x) \omega_j(t) = K_1(x,t) + K_2(x,t),$$

where the kernel

$$K_1(x,t) = \sum_{k,j=1}^{n} A_{kj} \omega_k(x) \omega_j(t)$$

is degenerate, and for $K_2(x,t)$, we have

$$B_2^2 = \int_a^b \int_a^b |K_2(x,t)|^2 \, dt dx = \sum_{k,j=n+1}^{\infty} A_{kj}^2.$$

The series $\sum_{k,j=1}^{\infty} A_{kj}^2$ converges, and hence its remainder $\sum_{k,j=n+1}^{\infty} A_{kj}^2$ can be made less than any predetermined number by choosing n large enough

$$\sum_{k,j=n+1}^{\infty} A_{kj}^2 \leq \frac{1}{4R^2}.$$

Then, for $|\lambda| \leq R$, we have $B_2 |\lambda| \leq \frac{1}{2R} R = \frac{1}{2} < 1$. Thus, any square integrable kernel can be represented as a sum of a degenerate kernel $K_1(x, t)$ and a kernel $K_2(x, t)$, for which $B_2 |\lambda| < 1$ for all $|\lambda| \leq R$.

Let us write the integral equation in the form

$$\varphi(x) - \lambda \int_a^b K_1(x, t)\varphi(t)dt - \lambda \int_a^b K_2(x, t)\varphi(t)dt = f(x), \quad a < x < b$$

or

$$\varphi - \lambda K_2\varphi = f + \lambda K_1\varphi.$$

Denote $g = f + \lambda K_1\varphi$, thus,

$$\varphi - \lambda K_2\varphi = g. \tag{12.21}$$

Since $B_2 |\lambda| < 1$, we obtain

$$\varphi(x) = g(x) + \lambda \int_a^b R_2(x, t, \lambda)g(t)dt, \quad |\lambda| \leq R,$$

where $R_2(x, t, \lambda)$ is the resolvent of Eq. (12.21). Then

$$\varphi(x) = f(x) + \lambda K_1\varphi(x) + \lambda \int_a^b R_2(x, t, \lambda)f(t)dt + \lambda^2 \int_a^b R_2(x, t, \lambda)K_1\varphi(t)dt.$$

Denote

$$f_1(x) = f(x) + \lambda \int_a^b R_2(x, t, \lambda)f(t)dt.$$

We have

$$\varphi(x) - \lambda \int_a^b K_1(x, t)\varphi(t)dt - \lambda^2 \int_a^b R_2(x, t, \lambda) \int_a^b K_1(t, s)\varphi(s)dsdt = f_1(x)$$

or

$$\varphi(x) - \lambda \int_a^b \left\{ K_1(x, t) + \lambda \int_a^b R_2(x, s, \lambda)K_1(s, t)ds \right\} \varphi(t)dt = f_1(x).$$

Denoting

$$K_3(x, t) = K_1(x, t) + \lambda \int_a^b R_2(x, s, \lambda)K_1(s, t)ds, \tag{12.22}$$

we obtain the equation

$$\varphi(x) - \lambda \int_a^b K_3(x, t)\varphi(t)dt = f_1(x). \tag{12.23}$$

Let us show that $K_3(x, t)$ is a degenerate kernel. Since $K_1(x, t)$ is degenerate, it is sufficient to prove that the second term in (12.22) is degenerate. We have

$$\int_a^b R_2(x, s, \lambda)K_1(s, t)ds = \int_a^b R_2(x, s, \lambda) \sum_{k=1}^n a_k(s)b_k(t)ds$$

$$= \sum_{k=1}^n b_k(t) \underbrace{\int_a^b R_2(x, s, \lambda)a_k(s)ds}_{c_k(x,\lambda)}$$

$$= \sum_{k=1}^n b_k(t)c_k(x, \lambda),$$

which shows that $K_3(x, t)$ is degenerate indeed.

Conclusion 189 *The Fredholm equation (12.20) reduces to subsequent solution of two integral equations: Eq. (12.21) with a small parameter and Eq. (12.23) with a degenerate kernel.*

As we know, Eq. (12.23) with a degenerate kernel reduces to a linear algebraic system

$$c_j - \lambda \sum_{k=1}^n \alpha_{jk}c_k = f_j, \quad j = 1, 2, \ldots, n. \tag{12.24}$$

In general, its determinant

$$D_R(\lambda) = \begin{vmatrix} 1 - \lambda\alpha_{11} & -\lambda\alpha_{12} & \ldots & -\lambda\alpha_{1n} \\ -\lambda\alpha_{21} & 1 - \lambda\alpha_{22} & \ldots & -\lambda\alpha_{2n} \\ & \ldots & & \\ -\lambda\alpha_{n1} & -\lambda\alpha_{n2} & \ldots & 1 - \lambda\alpha_{nn} \end{vmatrix} \tag{12.25}$$

depends on the choice of R, and the coefficients α_{jk} depend on λ. Indeed,

$$K_3(x, t) = \sum_{k=1}^n (a_k(x) + \lambda c_k(x, \lambda)) b_k(t).$$

Hence

$$\alpha_{jk} = \int_a^b b_j(t)\,(a_k(t) + \lambda c_k(t, \lambda))\,dt,$$

where

$$c_k(t, \lambda) = \int_a^b R_2(t, s, \lambda)a_k(s)ds.$$

Since the resolvent is a power series with respect to λ, for all $|\lambda| \le R$, the functions $c_k(t, \lambda)$ are analytic in λ, and this is true for α_{jk} as well.

Thus, $D_R(\lambda)$ is an analytic function with respect to λ in the disk $|\lambda| \le R$ and hence has there at most a finite number of zeros. Obviously, if $D_R(\lambda) \ne 0$ in the disk $|\lambda| \le R$, the system (12.24) has a unique solution for all f_k, and hence the integral equation is uniquely solvable for any right hand side.

12.8   Regular and Characteristic Values

Let

$$\varphi - \lambda K\varphi = f \tag{12.26}$$

be a Fredholm equation and

$$\varphi - \lambda K\varphi = 0 \tag{12.27}$$

its corresponding homogeneous equation.

Definition 190 A value of λ for which the homogeneous equation (12.27) has a nontrivial solution is called a **characteristic value** or an **eigenvalue**. The nontrivial solutions corresponding (belonging) to a corresponding characteristic value are called **eigenfunctions** of the integral equation.

Definition 191 A value of λ for which the homogeneous equation (12.27) has a trivial solution only $\varphi \equiv 0$ is called **regular.**

Theorem 192   *The following statements are valid:*

(a) *The characteristic values located in the disk $|\lambda| \leq R$ coincide with the roots of the determinant $D_R(\lambda)$.*
(b) *In any disk $|\lambda| \leq R$, there can be at most a finite number of characteristic values.*
(c) *To a given characteristic value, there corresponds a finite number of linearly independent eigenfunctions ($n - r_R$, where r_R is the rank of the matrix (12.25)).*
(d) *If λ is regular, then Eq. (12.26) is uniquely solvable for any right hand side f.*
(e) *If λ is characteristic, then, in general, Eq. (12.26) is unsolvable.*

Proof   Consider the disk $|\lambda| \leq R$. For such values of λ, Eqs. (12.26) and (12.27) are equivalent to the equations

$$\varphi(x) - \lambda \int_a^b K_3(x,t)\varphi(t)dt = f_1(x), \quad a < x < b \tag{12.28}$$

and

$$\varphi(x) - \lambda \int_a^b K_3(x,t)\varphi(t)dt = 0, \quad a < x < b, \tag{12.29}$$

respectively. Here we used the notations $K_3(x,t)$ and $f_1(x)$ as given earlier.

Hence the characteristic values for Eq. (12.29) coincide with those for Eq. (12.27) and vice versa. Equation (12.29) reduces to a linear algebraic system with the determinant $D_R(\lambda)$. Hence the characteristic values of (12.29) coincide with the roots of the function $D_R(\lambda)$. The statement (a) is proved.

Since $D_R(\lambda)$ can have at most a finite number of zeros in the disk $|\lambda| \leq R$ (in this disk $D_R(\lambda)$ is analytic, and $D_R(\lambda)$ is not identically zero because $D_R(0) = 1$), the number of eigenvalues in the disk $|\lambda| \leq R$ is finite as well. The statement (b) is proved.

To prove the statement (c), notice that the number of linearly independent eigenfunctions belonging to a certain eigenvalue equals the number of linearly independent solutions of the algebraic system, but this last number is finite and equals $n - r_R$, where r_R is the rank of the matrix (12.25).

The statement (d) is obvious, because when $D_R(\lambda) \neq 0$ the algebraic system and hence the integral equation is uniquely solvable.

The statement (e) follows from the equivalence of the algebraic system and the integral equation. They are solvable or not simultaneously. So, for a characteristic value of λ, the system is unsolvable ($D_R(\lambda) = 0$) and this remains true for the Fredholm integral equation as well. ∎

12.9   Adjoint Integral Equation

First, consider the linear algebraic system

$$\sum_{k=1}^{n} a_{jk}x_k = f_j, \quad j = 1, 2, \ldots, n.$$

The system

$$\sum_{k=1}^{n} \overline{a}_{kj}y_k = g_j, \quad j = 1, 2, \ldots, n,$$

is called its **adjoint**. In particular, the systems

$$c_j - \lambda \sum_{k=1}^{n} \alpha_{jk}c_k = f_j, \quad j = 1, 2, \ldots, n$$

and

$$d_j - \overline{\lambda} \sum_{k=1}^{n} \overline{\alpha}_{kj}d_k = g_j, \quad j = 1, 2, \ldots, n$$

are adjoint. The adjoint integral equations are defined in a similar way.

Definition 193 Let

$$\varphi(x) - \lambda \int_a^b K(x, t)\varphi(t)dt = f(x), \quad a < x < b, \tag{12.30}$$

be a Fredholm equation. The equation

$$\psi(x) - \overline{\lambda} \int_a^b \overline{K}(t, x)\psi(t)dt = g(x), \quad a < x < b, \tag{12.31}$$

is called the **adjoint** of (12.30). The kernel $K^*(x, t) = \overline{K}(t, x)$ is called the **adjoint kernel**, and the operator

$$K^*\psi(x) = \int_a^b K^*(x, t)\psi(t)dt$$

the adjoint operator of the operator K.

Note that if the kernel $K(x, t)$ is real valued, then $K^*(x, t) = K(t, x)$. Obviously,

$$(K^*)^* \varphi = K\varphi.$$

12.10 Properties of the Adjoint Operator

Property 1 If K is a Fredholm operator, then K^* is also Fredholm , and $B^* = B$.

Proof We have

$$(B^*)^2 = \int_a^b \int_a^b |K^*(x, t)|^2 \, dxdt = \int_a^b \int_a^b |\overline{K}(t, x)|^2 \, dxdt = \int_a^b \int_a^b |K(t, x)|^2 \, dxdt = B^2.$$

∎

Property 2 (Main Property) Let K be a Fredholm operator. For any functions φ, $\psi \in L_2$, the equality holds

$$(K\varphi, \psi) = (\varphi, K^*\psi).$$

Proof We have

$$(K\varphi, \psi) = \int_a^b \overline{\psi}(x) K\varphi(x) dx = \int_a^b \int_a^b K(x, s)\varphi(s)\overline{\psi}(x) dxds$$

$$= \int_a^b \int_a^b K(s, x)\varphi(x)\overline{\psi}(s) dxds = \int_a^b \varphi(x) \int_a^b K(s, x)\overline{\psi}(s) dsdx$$

$$= \int_a^b \varphi(x) \overline{\int_a^b \overline{K}(s, x)\psi(s) dsdx} = \int_a^b \varphi(x)\overline{K^*\psi(x)} dx = (\varphi, K^*\psi).$$

The change of the order of integration is due to Fubini's theorem. ∎

Remark 194 If K is an arbitrary operator (not necessarily Fredholm), then the equality

$$(K\varphi, \psi) = (\varphi, K^*\psi)$$

serves as a definition of its adjoint operator.

Property 3 (The Inverse Property) If two Fredholm operators K and L are related by the equality

$$(K\varphi, \psi) = (\varphi, L\psi)$$

for all $\varphi, \psi \in L_2$, then $L = K^*$.

Proof We have $(K\varphi, \psi) = (\varphi, L\psi)$ for all $\varphi, \psi \in L_2$, and at the same time, due to Property 2, $(K\varphi, \psi) = (\varphi, K^*\psi)$. Resting the second equality from the first, we obtain $0 = (\varphi, L\psi - K^*\psi)$. Take $\varphi = L\psi - K^*\psi$. Then $(L - K^*)\psi = 0$ for all $\psi \in L_2$, which implies $K^* = L$. ∎

Property 4 Let K be a Fredholm operator. Then $(\lambda K)^* = \overline{\lambda} K^*$.

Proof By Property 2, we have

$$(\lambda K\varphi, \psi) = (K\varphi, \overline{\lambda}\psi) = (\varphi, K^*(\overline{\lambda}\psi)) = (\varphi, \overline{\lambda} K^*\psi).$$

∎

Property 5 For any two Fredholm operators K_1 and K_2 the equality holds

$$(K_1 + K_2)^* = K_1^* + K_2^*.$$

Proof We have

$$(K_i\varphi, \psi) = (\varphi, K_i^*\psi), \quad i = 1, 2.$$

Then

$$
\begin{aligned}
(\varphi, (K_1 + K_2)^* \psi) &= ((K_1 + K_2)\varphi, \psi) = (K_1\varphi + K_2\varphi, \psi) \\
&= (K_1\varphi, \psi) + (K_2\varphi, \psi) = (\varphi, K_1^*\psi) + (\varphi, K_2^*\psi) \\
&= (\varphi, K_1^*\psi + K_2^*\psi) = (\varphi, (K_1^* + K_2^*)\psi).
\end{aligned}
$$

Using Property 3, we obtain the result. ∎

Property 6 Let K and L be two Fredholm operators. Then

$$(KL)^* = L^* K^* \quad \text{and} \quad (K^*)^n = (K^n)^*.$$

Proof We have

$$(KL\varphi, \psi) = (L\varphi, K^*\psi) = (\varphi, L^* K^*\psi).$$

Hence by Property 3, $(KL)^* = L^* K^*$. ∎

Lemma 195 *Let the equations* $\varphi - \lambda K \varphi = f$ *and* $\psi - \overline{\lambda} K^* \psi = g$ *be two adjoint Fredholm equations. Then in the disk* $|\lambda| \leq R$, *they reduce to the adjoint linear algebraic systems*

$$c_j - \lambda \sum_{k=1}^{n} \alpha_{jk} c_k = f_j, \quad j = 1, 2, \dots, n$$

and

$$d_j - \overline{\lambda} \sum_{k=1}^{n} \alpha_{jk}^* d_k = g_j, \quad j = 1, 2, \dots, n$$

with $\alpha_{jk}^* = \overline{\alpha}_{kj}$.

Proof For the proof, it is sufficient to repeat the calculations performed above for the operator K, taking into account that $K^*(x, t) = K(t, x)$, $R_2^*(x, t, \overline{\lambda}) = \overline{R}_2(t, x, \lambda), \dots$. ∎

12.11 Fredholm Theorems

Theorem 196 (First Fredholm Theorem) *A Fredholm equation can have at most a countable set of eigenvalues which can accumulate only at infinity.*

Proof Let us consider the infinite set of concentric disks $|\lambda| \leq R$ with $R = 1, 2, \dots$. In every such disk the number of eigenvalues is finite. Hence in every ring $n < |\lambda| \leq n + 1$, $n = 1, 2, \dots$ their number is also finite. The disk $|\lambda| \leq 1$ together with the rings $n < |\lambda| \leq n + 1$ cover the whole plane. The set whose elements are the disk $|\lambda| \leq 1$ and the rings $n < |\lambda| \leq n + 1$, $n = 1, 2, \dots$, is countable, and hence the set of the eigenvalues is at most countable. The accumulation point cannot be located at a finite distance because otherwise in some of the rings the number of the eigenvalues would be infinite. ∎

Theorem 197 (Second Fredholm Theorem) *If* λ *is a regular value, then the homogeneous Fredholm equation together with its adjoint does not admit nontrivial solutions, and the nonhomogeneous equations are uniquely solvable for any right hand side.*

Proof For the Fredholm equation, the theorem follows from the statement (d) of Theorem 192. For the adjoint equation, it becomes clear if one takes into account that the adjoint equation reduces to the adjoint system whose determinant is not zero (it is complex conjugate of the determinant of the system corresponding to the original Fredholm equation). Hence the adjoint equation is also uniquely solvable. ∎

Theorem 198 (Third Fredholm Theorem) *If λ is a characteristic value of the Fredholm equation, then both the given Fredholm equation and its adjoint are solvable or unsolvable simultaneously, and the corresponding homogenous equations admit an equal number of eigenfunctions.*

Proof The equations reduce to the adjoint systems for which it is known from algebra that they are solvable or unsolvable simultaneously, and hence it is true for the adjoint Fredholm equations. The number of solutions of the homogeneous equations equals the number of solutions of the homogeneous systems $n - r$ and $n - r^*$, where r and r^* are the ranks of the corresponding matrices, but they are equal because the matrices are adjoint. Hence $n - r = n - r^*$.   ∎

Theorem 199 (Fourth Fredholm Theorem) *For a given value of λ, the Fredholm equation*

$$\varphi - \lambda K \varphi = f$$

is solvable if and only if its right hand side f is orthogonal to all eigenfunctions of the adjoint equation, i.e.,

$$(f, \psi_l) = \int_a^b f(x)\overline{\psi_l(x)}dx = 0,$$

for all solutions $\psi = \psi_l$ of the equation

$$\psi - \overline{\lambda} K^* \psi = 0.$$

Proof For the necessity, we assume that $\varphi - \lambda K \varphi = f$ is solvable. Then

$$(f, \psi_l) = (\varphi - \lambda K \varphi, \psi_l) = \left(\varphi, (I - \lambda K)^* \psi_l\right) = \left(\varphi, \left(I - \overline{\lambda} K^*\right) \psi_l\right) = 0.$$

As for the sufficiency, for the sake of simplicity, we will give the proof in the case of degenerate kernels. Then the solution of $\varphi - \lambda K \varphi = f$ reduces to the algebraic system

$$c_j - \lambda \sum_{k=1}^n \alpha_{jk} c_k = f_j = \int_a^b f(x) b_j(x) dx. \tag{12.32}$$

Let $\left\{ \gamma_j^{(l)} \right\}_{j=1}^n$ be a complete set of linearly independent solutions of the adjoint system. It is known that (12.32) is solvable if and only if

$$\sum_{j=1}^n f_j \overline{\gamma}_j^{(l)} = 0 \quad \text{for all } l = 1, 2, \ldots, n - r.$$

Hence

$$\sum_{j=1}^{n} f_j \overline{\gamma}_j^{(l)} = \sum_{j=1}^{n} \overline{\gamma}_j^{(l)} \int_a^b f(x) b_j(x) dx = \int_a^b f(x) \left(\sum_{j=1}^{n} \overline{\gamma}_j^{(l)} b_j(x) \right) dx$$

$$= \int_a^b f(x) \overline{\psi_l(x)} dx = 0,$$

where $\psi_l(x) = \sum_{j=1}^{n} \gamma_j^{(l)} \overline{b_j(x)}$. It is easy to see that $\psi_l(x)$ are solutions of the adjoint equation

$$\psi - \overline{\lambda} K^* \psi = 0 \quad \Longleftrightarrow \quad \psi(x) - \overline{\lambda} \sum_{k=1}^{n} \overline{b_k(x)} \int_a^b \overline{a_k(t)} \psi(t) dt = 0.$$

Here it is sufficient to put

$$\gamma_k = \int_a^b \overline{a_k(t)} \psi(t) dt$$

after multiplying by $\overline{a_j(x)}$, $j = 1, 2, \ldots$, and integrating to verify that the obtained system is the adjoint one with the solutions $\left\{ \gamma_j^{(l)} \right\}_{j=1}^{n}$. ∎

Remark 200 The Fredholm theorems are often unified in the following **Fredholm alternative**: for every non-zero fixed complex number λ, either the homogeneous Fredholm equation (12.27) has a non-trivial solution, or the nonhomogeneous Fredholm equation (12.26) has a solution for all f.

12.12 Several Independent Variables

Let V be a domain in $\mathbb{R}^n$ and P and Q two points (vectors) in V. Similarly, to the previous one, the equation can be studied

$$\varphi(P) - \lambda \int_V K(P, Q) \varphi(Q) dV_Q = f(P). \tag{12.33}$$

Here $\varphi(P)$ is an unknown function and $f(P)$ is a known function, both from $L_2(V)$, where as usual for a Lebesgue measurable function φ

$$\varphi(P) \in L_2(V) \quad \Longleftrightarrow \quad \int_V |\varphi(P)|^2 dV_P < \infty.$$

Under the condition of the square integrability of the kernel

$$B^2 = \int_V \int_V |K(P, Q)|^2 \, dV_P dV_Q,$$

equation (12.33) is called a Fredholm equation. In a natural way, all the results obtained for the case $n = 1$ can be extended onto Eq. (12.33).

Similarly, the results for the Fredholm equations in the case $n = 1$ can be extended onto the case of surface integral equations

$$\varphi(P) - \lambda \int_S K(P, Q)\varphi(Q)dV_Q = f(P), \quad P \in S. \tag{12.34}$$

It is supposed that

$$\varphi(P) \in L_2(S) \quad \Longleftrightarrow \quad \int_S |\varphi(P)|^2 \, d\sigma_P < \infty,$$

and

$$\int_S \int_S |K(P, Q)|^2 \, d\sigma_P d\sigma_Q < \infty.$$

Again, in the definition, all functions are supposed to be Lebesgue measurable on the corresponding sets of integration.

Equation (12.34) reduces to (12.33) by considering the parameters determining the location of the points P and Q on the surface as the variables in (12.33).

12.13   Equations with a Weak Singularity

Let the kernel $K(P, Q)$ have the form

$$K(P, Q) = \frac{A(P, Q)}{r_{PQ}^\alpha}, \quad 0 \leq \alpha < n,$$

where $A(P, Q)$ is a bounded function

$$|A(P, Q)| \leq C.$$

Consider the integral equation

$$\varphi(P) - \lambda \int_V K(P, Q)\varphi(Q)dV_Q = f(P), \quad P \in V, \tag{12.35}$$

in a bounded domain $V \subset \mathbb{R}^n$.

Equation (12.35) is called the **integral equation with a weak singularity** and $K(P, Q)$ its **weakly singular kernel**.

Example 201

$$\varphi(x) - \lambda \int_0^a \frac{C(x, t)}{|x - t|^\alpha} dt = f(x), \quad 0 < x < a,$$

where $0 < \alpha < 1$ and $C(x, t)$ is bounded, is an integral equation with a weak singularity.

Notice that for $\alpha < n/2$ the integral operator with a weak singularity is a Fredholm operator. In this case the kernel satisfies the condition of the square integrability. Indeed, since V is bounded, we have that

$$\int_V |K(P, Q)|^2 \, dV_Q \leq C^2 \int_V \frac{dV_Q}{r_{PQ}^{2\alpha}}.$$

Introducing spherical coordinates centered at P, we obtain $dV_Q = r^{n-1} dr ds$. Denote by h the diameter of the domain V: $h = \sup_{P_1, P_2 \in V} r_{P_1 P_2}$. Then

$$\int_V \frac{dV_Q}{r_{PQ}^{2\alpha}} \leq \int_{r < h} \frac{dV_Q}{r^{2\alpha}} = \int_{\Sigma_1} ds \int_0^h r^{n-1} \frac{dr}{r^{2\alpha}} = |\Sigma_1| \frac{h^{n-2\alpha}}{n - 2\alpha}, \qquad (12.36)$$

where $|\Sigma_1|$ is the area of a unit sphere in $\mathbb{R}^n$. This proves that

$$\int_V |K(P, Q)|^2 \, dV_Q < \infty$$

and hence

$$B^2 = \int_V \int_V |K(P, Q)|^2 \, dV_P dV_Q < \infty.$$

Our aim is to prove that the integral operator with a weak singularity

$$\int_V K(P, Q)\varphi(Q) dV_Q$$

remains a Fredholm operator also in the case $\frac{n}{2} \leq \alpha < n$ in the sense that the results proved earlier for square integrable kernels remain valid in this more general situation.

Theorem 202 *Let $K(P, Q)$ be a weakly singular kernel and $\varphi \in L_2(V)$. The integral*

$$\psi(P) = \int_V K(P, Q)\varphi(Q) dV_Q$$

exists and is square integrable in V.

Proof Note that it follows from the previous proof that $\int_V \frac{dV_Q}{r_{PQ}^\alpha}$ is bounded for any $P \in \overline{V}$. Furthermore, due to Fubini's theorem, from the existence of the double integral

$$\int_V \int_V \frac{|\varphi(Q)|^2}{r_{PQ}^\alpha} dV_P dV_Q = \int_V |\varphi(Q)|^2 \int_V \frac{dV_P}{r_{PQ}^\alpha} dV_Q \le \|\varphi\|^2 \frac{|\Sigma_1| h^{n-\alpha}}{n - \alpha},$$

the existence of the integral

$$\int_V \frac{|\varphi(Q)|^2}{r_{PQ}^\alpha} dV_Q$$

follows. Moreover, this function of P is summable (integrable). Then the function $|K(P, Q)\varphi(Q)|$ is also summable. Indeed,

$$|K(P, Q)\varphi(Q)| \le \frac{C |\varphi(Q)|}{r_{PQ}^{\alpha/2}} \cdot \frac{1}{r_{PQ}^{\alpha/2}} \le \frac{C}{2} \left(\frac{|\varphi(Q)|^2}{r_{PQ}^\alpha} + \frac{1}{r_{PQ}^\alpha} \right).$$

Here the elementary inequality $a + \frac{1}{a} \ge 2$ for all $a > 0$ was used.

Both terms on the right hand side are summable for $P \in V$ a.e. Hence ψ is defined a.e. in V. Furthermore, by Cauchy–Bunyakovsky–Schwarz inequality,

$$|\psi(P)|^2 \le C^2 \left(\int_V \frac{|\varphi(Q)|}{r_{PQ}^{\alpha/2}} \frac{1}{r_{PQ}^{\alpha/2}} dV_Q \right)^2$$

$$\le C^2 \left(\int_V \frac{|\varphi(Q)|^2}{r_{PQ}^\alpha} dV_Q \right) \left(\int_V \frac{1}{r_{PQ}^\alpha} dV_Q \right)$$

$$\le C^2 |\Sigma_1| \frac{h^{n-\alpha}}{n - \alpha} \int_V \frac{|\varphi(Q)|^2}{r_{PQ}^\alpha} dV_Q.$$

Hence,

$$\|\psi\|^2 = \int_V |\psi(P)|^2 dV_P \le C^2 |\Sigma_1| \frac{h^{n-\alpha}}{n - \alpha} \int_V \int_V \frac{|\varphi(Q)|^2}{r_{PQ}^\alpha} dV_P dV_Q$$

$$\le C^2 \left(|\Sigma_1| \frac{h^{n-\alpha}}{n - \alpha} \right)^2 \|\varphi\|^2.$$

This finishes the proof.   ∎

Theorem 203 *Let K and M be integral operators with weakly singular kernels satisfying*

$$|K(P, Q)| \le \frac{C_1}{r^\alpha} \quad and \quad |L(P, Q)| \le \frac{C_2}{r^\beta},$$

where C_1 and C_2 are positive constants, and $0 \le \alpha, \beta < n$. Then the kernel $M(P, Q)$ of the product of the operators $KL = M$, given by

$$M(P, Q) = \int_V K(P, R)L(R, Q)dV_R,$$

admits the bound

$$|M(P, Q)| \le \text{Const} \begin{cases} c, & \alpha + \beta < n, \\ c\,|\ln r| + c_1, & \alpha + \beta = n, \\ \frac{c}{r^{\alpha+\beta-n}}, & \alpha + \beta > n, \end{cases}$$

where c and c_1 stand for some positive constants. That is, in all cases, the kernel $M(P, Q)$ is again a kernel with a weak singularity and even less singular than the initial kernels (in the worst case, $\alpha + \beta - n < \min(\alpha, \beta)$).

Proof Let h denote the diameter of V. Denote $r_0 = r_{PR}$ and $r_1 = r_{QR}$. Then

$$|M(P, Q)| \le C_1 C_2 \int_V \frac{dV_R}{r_0^\alpha r_1^\beta} \le C_1 C_2 \int_{r_0 < h} \frac{dV_R}{r_0^\alpha r_1^\beta}. \tag{12.37}$$

Let us locate the origin at the point P and choose the positive direction of the axis x_1 in such a way that the point Q belongs to the positive semi-axis. Then we have $P(0, 0, \ldots, 0)$, $Q(r, 0, \ldots, 0)$, $R(x_1, x_2, \ldots, x_n)$, and

$$r_0^2 = \sum_{k=1}^n x_k^2, \quad r_1^2 = (x_1 - r)^2 + \sum_{k=2}^n x_k^2.$$

Let us make the change of the variables $x_k = r y_k$ in (12.37), denoting $\rho^2 = \sum_{k=1}^n y_k^2$:

$$|M(P, Q)| \le \frac{C_1 C_2}{r^{\alpha+\beta-n}} \int_{\rho < \frac{h}{r}} \frac{dy_1 dy_2 \ldots dy_n}{\rho^\alpha \left(\rho^2 - 2y_1 + 1\right)^{\beta/2}}.$$

Let us estimate this integral. We have $dy_1 dy_2 \ldots dy_n = \rho^{n-1} d\rho ds$, and

$$\rho^2 - 2y_1 + 1 \ge (\rho - 1)^2.$$

Moreover, when $\rho > 2$, we obtain

$$(\rho - 1)^2 > \frac{\rho^2}{4}.$$

This inequality follows from the observation

$$(\rho - 1)^2 - \frac{\rho^2}{4} = \frac{1}{4}(3\rho - 2)(\rho - 2) > 0 \quad \text{when } \rho > 2.$$

Hence

$$|M(P, Q)| \le \frac{C_1 C_2}{r^{\alpha+\beta-n}} \left\{ \int_{\rho \le 2} \frac{\rho^{n-1-\alpha} d\rho ds}{(\rho^2 - 2y_1 + 1)^{\beta/2}} + 2^\beta \int_{2 < \rho < \frac{h}{r}} \rho^{n-1-\alpha-\beta} d\rho ds \right\}.$$

Notice that the first integral on the right hand side is some constant which we denote by c. Calculating the second integral, we obtain for $\alpha + \beta < n$

$$|M(P, Q)| \le C_1 C_2 |\Sigma_1| \left\{ cr^{n-\alpha-\beta} + \frac{2^\beta h^{n-\alpha-\beta}}{n-\alpha-\beta} \right\} \le C_1 C_2 |\Sigma_1| h^{n-\alpha-\beta} \left(c + \frac{2^\beta}{n-\alpha-\beta} \right).$$

When $\alpha + \beta = n$, we obtain

$$|M(P, Q)| \le C_1 C_2 |\Sigma_1| \left(c + 2^\beta \ln \frac{h}{2r} \right).$$

Finally, when $\alpha + \beta > n$, we obtain

$$|M(P, Q)| \le C_1 C_2 \frac{|\Sigma_1|}{r^{\alpha+\beta-n}} \left(c + 2^\beta \int_2^{\frac{h}{r}} \frac{d\rho}{\rho^{\alpha+\beta+1-n}} \right)$$

$$< C_1 C_2 \frac{|\Sigma_1|}{r^{\alpha+\beta-n}} \left(c + 2^\beta \int_2^\infty \frac{d\rho}{\rho^{\alpha+\beta+1-n}} \right)$$

$$= C_1 C_2 \frac{|\Sigma_1|}{r^{\alpha+\beta-n}} \left(c + \frac{2^{n-\alpha}}{\alpha+\beta-n} \right).$$

∎

The following statement is crucial for extending the Fredholm theory onto integral equations with weak singularities.

Corollary 204 *For a weakly singular kernel, all its associated iterated kernels starting from a certain number are bounded.*

Proof Obviously, l-th iterated kernel admits the bound

$$|K_l(P, Q)| \leq \begin{cases} C_l, & l\alpha - (l-1)n < 0, \\ \dfrac{C_l}{r^{l\alpha-(l-1)n}}, & l\alpha - (l-1)n > 0, \end{cases}$$

where C_l is a constant. Hence, for $l > \frac{n}{n-\alpha}$, the kernel $K_l(P, Q)$ together with all the subsequent iterated kernels is bounded. ∎

Let us write equation (12.35) in the operator form

$$(I - \lambda K)\varphi = f. \tag{12.38}$$

Let l be an integer number, such that $l > \frac{n}{n-\alpha}$. Denote $\varepsilon = e^{\frac{2\pi i}{l}}$, and apply the operator

$$(I - \varepsilon\lambda K)(I - \varepsilon^2\lambda K)\ldots(I - \varepsilon^{l-1}\lambda K) = I + \lambda K + \lambda^2 K^2 + \ldots + \lambda^{l-1}K^{l-1}$$

to equality (12.38). We obtain

$$(I - \lambda^l K^l)\varphi = f + \lambda Kf + \lambda^2 K^2 f + \ldots + \lambda^{l-1}K^{l-1}f.$$

Denote the right hand side here by F. Then

$$\varphi(P) - \lambda^l \int_V K_l(P, Q)\varphi(Q)dV_Q = F(P), \quad P \in V. \tag{12.39}$$

Due to the boundedness of the kernel $K_l(P, Q)$, this equation is Fredholm. Clearly, any solution of (12.38) is a solution of (12.39). The opposite is not true in general.

Now we are in a position to prove the Fredholm theorems for Eq. (12.38).

Theorem 205 *Equation (12.38) with a weakly singular kernel admits at most a countable set of eigenvalues with the only possible accumulation point at infinity.*

Proof We need to prove that in any finite part of the λ-plane there exist at most a finite number of the eigenvalues.

Let $\lambda = \lambda_0$ be an eigenvalue of (12.38). Then there exists a nontrivial solution of the equation $(I - \lambda_0 K)\varphi = 0$, which is also necessarily a solution of the iterated equations (12.39) with $F \equiv 0$:

$$(I - \lambda_0^l K^l)\varphi = 0. \tag{12.40}$$

Consider a disk $|\lambda| \leq R$. By Theorem 196, the kernel $K_l(P, Q)$ has at most a finite number of eigenvalues in the disk of radius R^l, but then Eq. (12.38) cannot have in the disk $|\lambda| \leq R$ more than a finite number of eigenvalues. ∎

Let us show that the integer l can be chosen in such a way that besides the boundedness of $K_l(P, Q)$ one will have the equivalence of the Eqs. (12.38) and (12.39).

First of all, let us show that l can be chosen such that the inequality holds $l > \frac{n}{n-\alpha}$, and for a given λ, none of the numbers $\varepsilon\lambda, \varepsilon^2\lambda,\ldots,\varepsilon^{l-1}\lambda$ is an eigenvalue, where $\varepsilon = e^{\frac{2\pi i}{l}}$. Suppose the opposite. Denote by $p_1, p_2,\ldots$ the prime numbers greater than $\frac{n}{n-\alpha}$, and let $\varepsilon_j = e^{\frac{2\pi i}{p_j}}$. According to our supposition for any j, there exists such exponent k_j, $1 \le k_j < p_j$, that for a given λ the number $\varepsilon_j^{k_j}\lambda$ is an eigenvalue of the kernel $K(P, Q)$. Since p_j is a prime number, all the numbers $\varepsilon_j^{k_j} = e^{\frac{2\pi i k_j}{p_j}}$ are different. But then on the circle centered at the origin and of radius $|\lambda|$, there exist infinitely many eigenvalues $\varepsilon_j^{k_j}\lambda$ of the kernel $K(P, Q)$ that contradicts the First Fredholm theorem. Hence, for any λ, there exists a prime number $p_j > \frac{n}{n-\alpha}$ such that the corresponding numbers $\varepsilon_j\lambda, \varepsilon_j^2\lambda,\ldots,\varepsilon_j^{p_j-1}\lambda$ are all regular. Let $l = p_j$.

Equation (12.39) can be written in the form

$$(I-\varepsilon\lambda K)(I-\varepsilon^2\lambda K)\ldots(I-\varepsilon^{l-1}\lambda K)(I-\lambda K)\varphi = (I-\varepsilon\lambda K)(I-\varepsilon^2\lambda K)\ldots(I-\varepsilon^{l-1}\lambda K)f$$

or equivalently

$$\prod_{j=1}^{l-1}(I - \varepsilon^j\lambda K)\,[(I - \lambda K)\varphi - f] = 0.$$

Denote $\omega = (I - \lambda K)\varphi - f$. Then

$$\prod_{j=1}^{l-1}(I - \varepsilon^j\lambda K)\omega = 0.$$

For

$$\omega_1 = \prod_{j=2}^{l-1}(I - \varepsilon^j\lambda K)\omega,$$

we have $(I - \varepsilon\lambda K)\omega_1 = 0$. Since $\varepsilon\lambda$ is a regular value for the kernel $K(P, Q)$, then $\omega_1 \equiv 0$, i.e.,

$$\prod_{j=2}^{l-1}(I - \varepsilon^j\lambda K)\omega = 0.$$

Denoting

$$\omega_2 = \prod_{j=3}^{l-1}(I - \varepsilon^j \lambda K)\omega$$

and taking into account that $\varepsilon^2\lambda$ is a regular value for the kernel $K(P, Q)$, from the equation $(I - \varepsilon^2\lambda K)\omega_2 = 0$, we obtain that $\omega_2 \equiv 0$. Repeating this procedure, we arrive at the equality $\omega \equiv 0$, i.e., $(I - \lambda K)\varphi = f$.

Thus, in this case, any solution of (12.39) satisfies (12.38) as well, that is, these equations are equivalent.

Let us show that the Third Fredholm theorem remains valid for equations with a weak singularity. Notice that the adjoint homogeneous equations for (12.38) and (12.39) have the form

$$\left(I - \bar{\lambda} K^*\right)\psi = 0 \tag{12.41}$$

and

$$\left(I - \bar{\lambda}^l \left(K^*\right)^l\right)\omega = 0, \tag{12.42}$$

respectively. Let λ_0 be an eigenvalue of (12.38). Then, since the equivalent equation (12.40) is Fredholm, the number of the associated linearly independent eigenfunctions is finite. Denote this number by r. Hence Eq. (12.42) also possesses r linearly independent solutions. Denote by $\tilde{r}$ the number of linearly independent solutions of the homogeneous equation $(I - \bar{\lambda}_0 K^*)\psi = 0$. Since every its solution satisfies also $\left(I - \bar{\lambda}_0^l (K^*)^l\right)\omega = 0$, we have $\tilde{r} \leq r$. If now we consider Eq. (12.41) as the original one and equation $(I - \lambda K)\varphi = 0$ as its adjoint, we obtain $r \leq \tilde{r}$. Thus, $r = \tilde{r}$. Hence the adjoint equations $(I - \lambda_0 K)\varphi = 0$ and $\left(I - \bar{\lambda}_0 K^*\right)\psi = 0$ have the same number of solutions.

Let us show the validity of the Fourth Fredholm theorem for the weakly singular integral equations of the second kind.

For the proof of necessity, we suppose that (12.38) is solvable, and φ is its solution, while ψ is a solution of the homogeneous adjoint equation (12.41). Let us show that $(f, \psi) = 0$. Indeed,

$$(f, \psi) = (\varphi - \lambda K\varphi, \psi) = (\varphi, \psi) - \lambda (K\varphi, \psi) = (\varphi, \psi) - \lambda \left(\varphi, K^*\psi\right)$$
$$= (\varphi, \psi) - \left(\varphi, \bar{\lambda} K^*\psi\right) = \left(\varphi, \psi - \bar{\lambda} K^*\psi\right) = 0.$$

For the proof of sufficiency, let l be chosen such that Eqs. (12.38) and (12.39) are equivalent. For the solvability of (12.39), according to the Fourth Fredholm theorem, it is sufficient (and necessary) the fulfillment of the condition

$$(F, \omega) = 0,$$

where ω is any solution of (12.42), and

$$F = f + \lambda K f + \lambda^2 K^2 f + \ldots + \lambda^{l-1} K^{l-1} f = \prod_{j=1}^{l-1}(I - \varepsilon^j \lambda K) f.$$

Equation (12.42) can be written in the form

$$0 = \left(I - \overline{\lambda}^l \left(K^*\right)^l\right)\omega = \prod_{j=0}^{l-1}(I - \overline{\varepsilon}^j \overline{\lambda} K^*)\omega = \left(I - \overline{\lambda} K^*\right)\prod_{j=1}^{l-1}(I - \overline{\varepsilon}^j \overline{\lambda} K^*)\omega.$$

Denoting $\psi = \prod\limits_{j=1}^{l-1}(I - \overline{\varepsilon}^j \overline{\lambda} K^*)\omega$, we obtain that $\left(I - \overline{\lambda} K^*\right)\psi = 0$, i.e., ψ is a solution of (12.41). Then we obtain the sufficient condition for the solvability of Eq. (12.38):

$$(F, \omega) = \left(\prod_{j=1}^{l-1}(I - \varepsilon^j \lambda K) f, \omega\right) = \left(f, \prod_{j=1}^{l-1}(I - \overline{\varepsilon}^j \overline{\lambda} K^*)\omega\right) = (f, \psi) = 0.$$

Finally, the Second Fredholm theorem is a corollary of the equivalence of the Eqs. (12.38) with (12.39) and (12.41) with (12.42).

Thus, the linear integral equation with a weak singularity (12.38) enjoys all the properties of the Fredholm integral equation of the second kind.

12.14 Continuity of Solutions of Integral Equations

12.14.1 Continuous in the Whole Kernels

In applications it is often important to know under which conditions solutions of integral equations are not only square integrable but also continuous functions. Let us consider the integral equation

$$\varphi(x) - \lambda \int_a^b K(x, s)\varphi(s)ds = f(x), \tag{12.43}$$

on a finite interval (a, b) with $f \in L_2(a, b)$, and suppose that apart from the square integrability on the main square, the kernel satisfies the condition

$$\int_a^b |K(x, s)|^2 ds \leq A. \tag{12.44}$$

That is, there exists such a constant A that for all $x \in (a, b)$ the inequality (12.44) holds.

Theorem 206 *If f is bounded on $[a, b]$ and Lebesgue measurable on $[a, b]$ and (12.43) is solvable, then any of its solution is bounded.*

Proof Write (12.43) in the form

$$\varphi(x) = f(x) + \lambda \int_a^b K(x, s)\varphi(s)ds.$$

Since f is bounded on $[a, b]$, it also belongs to $L_2(a, b)$. Thus, for the proof of the boundedness of φ, we need to verify the boundedness of the integral, which follows from the Cauchy–Bunyakovsky–Schwarz inequality

$$\left| \int_a^b K(x, s)\varphi(s)ds \right| \le \left(\int_a^b |K(x, s)|^2 \, ds \right)^{\frac{1}{2}} \left(\int_a^b |\varphi(s)|^2 \, ds \right)^{\frac{1}{2}} \le \sqrt{A} \, \|\varphi\|.$$

∎

Corollary 207 *If the square integrable in the main square kernel $K(x, s)$ satisfies additionally the condition (12.44), then its eigenfunctions are bounded (in this case $f \equiv 0$ is bounded).*

Definition 208 We say that the kernel $K(x, s)$ is continuous in the whole, if for any $\varepsilon > 0$ there can be proposed such $\delta > 0$ that $|x_1 - x_2| < \delta$ implies

$$\int_a^b |K(x_1, s) - K(x_2, s)| \, ds < \varepsilon.$$

Example 209

(a) If the kernel is continuous with respect to both variables x and s in the main square, then it is continuous in the whole.
(b) If the kernel satisfies (12.44) and is discontinuous in the main square only along a finite number r of lines represented by the equations $s = g_k(x)$, $k = 1, 2, \ldots, r$, with continuous functions g_k, then such kernel is continuous in the whole. The proof can be found in [71, p. 103].
(c) It is clear that in the case (b) one can still add a finite number of isolated points of discontinuity.

Theorem 210 *If f is continuous on $[a, b]$, and the kernel is continuous in the whole, then any solution of Eq. (12.43) is continuous on $[a, b]$.*

Proof It is sufficient to prove the continuity of the integral

$$u(x) = \int_a^b K(x, s)\varphi(s)ds.$$

Since f is continuous on $[a, b]$, it is bounded. Due to Theorem 206, the function φ is bounded as well, i.e., $|\varphi(x)| \le M$, for some constant $M > 0$ and for $a \le x \le b$. Given $\varepsilon > 0$, let us choose $\delta > 0$ such that $|x_1 - x_2| < \delta$ implies

$$\int_a^b |K(x_1, s) - K(x_2, s)|\, ds < \frac{\varepsilon}{M}.$$

Then

$$|u(x_1) - u(x_2)| = \left| \int_a^b (K(x_1, s) - K(x_2, s))\, \varphi(s)ds \right|$$

$$\le M \int_a^b |K(x_1, s) - K(x_2, s)|\, ds < \varepsilon.$$

∎

Corollary 211 *For a continuous in the whole kernel, the corresponding eigenfunctions are continuous.*

12.14.2   Weakly Singular Equations

Consider the equation with a weak singularity

$$\varphi(P) - \lambda \int_V K(P, Q)\varphi(Q)dV_Q = f(P), \quad P \in V, \tag{12.45}$$

in a bounded domain V of $\mathbb{R}^n$. The integral kernel in Eq. (12.45) has the form

$$K(P, Q) = \frac{A(P, Q)}{r_{PQ}^\alpha}, \quad 0 \le \alpha < n,$$

where the numerator $A(P, Q)$ is a bounded function

$$|A(P, Q)| \le C, \quad P, Q \in \overline{V}.$$

Similar results are valid in the case of weakly singular kernels considered in the previous section.

Lemma 212 *If the function f in (12.45) is bounded in $\overline{V}$, then the function*

$$F = f + \lambda K f + \lambda^2 K^2 f + \ldots + \lambda^{l-1} K^{l-1} f$$

is bounded in $\overline{V}$ *as well for all* $l \in \mathbb{N}$.

Proof Obviously, it suffices to verify the boundedness of the function Kf, and then the same is easily verified for $K^2 f, \ldots, K^{l-1} f$. We have

$$|Kf| = \left| \int_V \frac{A(P, Q)}{r_{PQ}^\alpha} f(Q) dV_Q \right| \le CC_1 \int_V \frac{dV_Q}{r_{PQ}^\alpha},$$

where $|f(Q)| \le C_1$ for all $Q \in \overline{V}$, and the last integral is bounded, see (12.36). ∎

Lemma 213 *If the function* f *in (12.45) is bounded in* $\overline{V}$, *and the function* $A(P, Q)$ *is continuous in* $\overline{V}$, *then the function* Kf *is continuous in* $\overline{V}$.

Proof Denote $g = Kf$. We have

$$g(P_1) - g(P_2) = \int_V (K(P_1, Q) - K(P_2, Q)) f(Q) dV_Q.$$

Consider a ball $B_\eta(P_1)$ centered at P_1 and of a small enough radius η. Let the distance between the points P_1 and P_2 be smaller than η. Then

$$|g(P_1) - g(P_2)| =$$

$$\left| \int_{V \setminus \overline{B}_\eta(P_1)} (K(P_1, Q) - K(P_2, Q)) f(Q) dV_Q + \int_{B_\eta(P_1)} (K(P_1, Q) - K(P_2, Q)) f(Q) dV_Q \right|$$

$$\le C_1 \int_{V \setminus \overline{B}_\eta(P_1)} |K(P_1, Q) - K(P_2, Q)| dV_Q + CC_1 \int_{\overline{B}_\eta(P_1)} \frac{dV_Q}{r_{P_1 Q}^\alpha} + CC_1 \int_{\overline{B}_\eta(P_1)} \frac{dV_Q}{r_{P_2 Q}^\alpha}.$$
$$(12.46)$$

Let us estimate each of the integrals. We have

$$\int_{\overline{B}_\eta(P_1)} \frac{dV_Q}{r_{P_1 Q}^\alpha} = \int_{r_{P_1 Q} \le \eta} \frac{dV_Q}{r_{P_1 Q}^\alpha} = |\Sigma_1| \int_0^\eta r^{n-1-\alpha} dr = \frac{|\Sigma_1| \eta^{n-\alpha}}{n - \alpha}.$$

Furthermore,

$$r_{P_2 Q} = |P_2 Q| \le |P_2 P_1| + |P_1 Q| \le \eta + r_{P_1 Q}.$$

Hence

$$\int_{\overline{B}_\eta(P_1)} \frac{dV_Q}{r^\alpha_{P_2 Q}} = \int_{r_{P_1 Q} \leq \eta} \frac{dV_Q}{r^\alpha_{P_2 Q}} \leq \int_{r_{P_2 Q} \leq 2\eta} \frac{dV_Q}{r^\alpha_{P_2 Q}} = \frac{|\Sigma_1| (2\eta)^{n-\alpha}}{n - \alpha}.$$

Thus, it is clear that the second and the third terms in (12.46) can be made arbitrarily small by the choice of a small enough η. Their sum can be made less than $\varepsilon/2$ for any $\varepsilon > 0$.

Consider the first integral in (12.46). In the domain $V \backslash \overline{B}_\eta(P_1)$, we have $r_{P_1 Q} \geq \eta > 0$. Due to the continuity of the function $A(P, Q)$, we can choose such $\delta > 0$, that for $|P_1 P_2| < \delta$ the inequality holds

$$|K(P_1, Q) - K(P_2, Q)| < \frac{\varepsilon}{2C_1 |V|},$$

where $|V|$ is the volume of $\overline{V}$. Then the first term in (12.46) is less than $\varepsilon/2$. Hence $|g(P_1) - g(P_2)| < \varepsilon$.   ∎

Theorem 214 *If the function f in (12.45) is bounded in $\overline{V}$, then any solution of (12.45) is bounded.*

Proof Choose l sufficiently large, so that the kernel $K_l(P, Q)$ be bounded. The solution of (12.45) satisfies also the Fredholm equation

$$\varphi(P) - \lambda^l \int_V K_l(P, Q)\varphi(Q)dV_Q = F(P), \quad P \in V.$$

Due to Lemma 212, the function F is bounded. Hence, by Theorem 206, any solution of the equation is bounded.   ∎

Theorem 215 *If the functions f and $A(P, Q)$ in (12.45) are continuous in $\overline{V}$, then any solution of (12.45) is continuous in $\overline{V}$.*

Proof Since f is continuous in the closed bounded domain $\overline{V}$, it is necessarily bounded there. Hence, due to the previous theorem, the solution φ of (12.45) is also bounded in $\overline{V}$. By Lemma 213, the function $K\varphi$ is continuous in $\overline{V}$. Hence the function $\varphi = f + \lambda K\varphi$ is continuous in $\overline{V}$.   ∎

Corollary 216 *Let the kernel $K(P, Q)$ be weakly singular:*

$$K(P, Q) = \frac{A(P, Q)}{r^\alpha_{PQ}}, \quad 0 \leq \alpha < n.$$

If the function $A(P, Q)$ is continuous in $\overline{V}$, then the corresponding eigenfunctions are continuous in $\overline{V}$.

12.15 Symmetric Integral Equations

Definition 217 The kernel $K(x, s)$ is called **symmetric** if

$$K(x, s) = K^*(x, s) = \overline{K}(s, x).$$

Example 218 The kernels $K(x, s) = 1$, $K(x, s) = \ln|x - s|$, $K(x, s) = x^2 + s^2$ are symmetric. The kernel $K(x, s) = i(x - s)$ is symmetric as well:

$$K^*(x, s) = \overline{i(s - x)} = -i(s - x) = i(x - s) = K(x, s).$$

A Fredholm operator with a symmetric kernel is called a **symmetric operator**, and an integral equation with a symmetric operator is called a **symmetric integral equation**. For a symmetric operator, the main identity takes the form

$$(K\varphi, \psi) = (\varphi, K\psi), \quad K = K^*. \tag{12.47}$$

Obviously, for the iterations, we have

$$(K^*)^n = (K^n)^* = K^n, \quad \text{i.e., } K_n(x, s) = K_n^*(x, s). \tag{12.48}$$

Theorem 219 *The eigenvalues of a symmetric integral equation are real.*

Proof Let λ_0 be an eigenvalue and φ_0 a corresponding eigenfunction. Then

$$\varphi_0 = \lambda_0 K\varphi_0. \tag{12.49}$$

By the definition of the eigenfunction, φ_0 is not identically zero and hence $\|\varphi_0\| > 0$. For the same reason, $\lambda_0 \neq 0$ because otherwise we would have that $\varphi_0 \equiv 0$.

Consider a scalar product of Eq. (12.49) by φ_0:

$$(\varphi_0, \varphi_0) = \lambda_0 (K\varphi_0, \varphi_0).$$

Then

$$\frac{1}{\lambda_0} = \frac{(K\varphi_0, \varphi_0)}{(\varphi_0, \varphi_0)} = \frac{(K\varphi_0, \varphi_0)}{\|\varphi_0\|^2},$$

and it remains to show that the scalar product $(K\varphi_0, \varphi_0)$ is real. On the one side, since K is symmetric, we have $(K\varphi_0, \varphi_0) = (\varphi_0, K\varphi_0)$, and on the other, due to the property of the scalar product, $(\varphi_0, K\varphi_0) = \overline{(K\varphi_0, \varphi_0)}$. Thus, $(K\varphi_0, \varphi_0) = \overline{(K\varphi_0, \varphi_0)}$, and $(K\varphi_0, \varphi_0)$ is a real number that finishes the proof. ∎

Theorem 220 *The eigenfunctions of a symmetric integral equation corresponding to different eigenvalues are mutually orthogonal.*

Proof Let $\lambda_1 \neq \lambda_2$ be two eigenvalues of a symmetric kernel $K(x, s)$ and φ_1 and φ_2 the corresponding eigenfunctions, i.e., $\varphi_1 = \lambda_1 K \varphi_1$ and $\varphi_2 = \lambda_2 K \varphi_2$. We have

$$(\varphi_1, \varphi_2) = (\lambda_1 K \varphi_1, \varphi_2) = \lambda_1 (\varphi_1, K \varphi_2) = \lambda_1 \left(\varphi_1, \frac{1}{\lambda_2}\varphi_2\right) = \frac{\lambda_1}{\lambda_2}(\varphi_1, \varphi_2).$$

That is,

$$\left(1 - \frac{\lambda_1}{\lambda_2}\right)(\varphi_1, \varphi_2) = 0,$$

which implies $(\varphi_1, \varphi_2) = 0$.   ∎

Theorem 221 *The sequence of eigenfunctions can be orthonormalized.*

Proof Clearly, it is easy to normalize the eigenfunctions by dividing them over their respective norms. If there are several eigenfunctions belonging to the same eigenvalue, they can be orthogonalized by applying the orthogonalization procedure and then normalized. Due to the previous theorem, the eigenfunctions belonging to different eigenvalues are orthogonal.   ∎

Let us agree to write the sequence of the eigenvalues repeating each of them as many times as the number of linearly independent eigenfunctions it has and ordering them by their absolute value:

$$\lambda_1, \lambda_2, \lambda_3, \ldots \qquad |\lambda_1| \leq |\lambda_2| \leq |\lambda_3| \leq \ldots.$$

The corresponding sequence of the eigenfunctions has the form

$$\varphi_1(x), \varphi_2(x), \varphi_3(x), \varphi_4(x), \ldots,$$

with $\varphi_n(x) - \lambda_n K \varphi_n(x) = 0$ and $(\varphi_j, \varphi_k) = \delta_{jk}$.

Thus, we can now assume that to every eigenvalue from the sequence there corresponds one eigenfunction, and among the eigenvalues, there can be equal ones.

Theorem 222 *The set of the eigenvalues of the second iterated kernel coincides with the set of the squares of the eigenvalues of the initial kernel (notice that the theorem is valid also for nonsymmetric kernels).*

Proof Let us show first that if λ_0 is an eigenvalue of the operator K, then λ_0^2 is an eigenvalue of the operator K^2. We have

$$\varphi_0 = \lambda_0 K \varphi_0.$$

Replacing on the right hand side φ_0 by $\lambda_0 K \varphi_0$ gives

$$\varphi_0 = \lambda_0 K \left(\lambda_0 K \varphi_0 \right) = \lambda_0^2 K^2 \varphi_0$$

or more explicitly

$$\varphi_0(x) - \lambda_0^2 \int_a^b K_2(x, s) \varphi_0(s) ds = 0.$$

The opposite is also true. Let μ_0 be an eigenvalue of the second iterated kernel $K_2(x, s)$ and $\varphi_0(x)$ the corresponding eigenfunction:

$$\varphi_0(x) - \mu_0 \int_a^b K_2(x, s) \varphi_0(s) ds = 0 \quad \Longleftrightarrow \quad \varphi_0 - \mu_0 K^2 \varphi_0 = 0.$$

This can be written as

$$\left(I + \sqrt{\mu_0} K \right) \left(I - \sqrt{\mu_0} K \right) \varphi_0 = 0.$$

Let $\psi = \left(I - \sqrt{\mu_0} K \right) \varphi_0$. If $\psi \equiv 0$, then $\sqrt{\mu_0}$ is an eigenvalue of the kernel $K(x, s)$ with the corresponding eigenfunction $\varphi_0(x)$, and in this case the statement is proved. If $\psi \not\equiv 0$, then $\psi(x)$ is an eigenfunction of the kernel $K(x, s)$ belonging to the eigenvalue $-\sqrt{\mu_0}$. This finishes the proof.   ∎

Remark 223 This theorem can be easily extended onto the case of the iterated kernels of any order n.

Corollary 224 *If the kernel $K(x, s)$ is symmetric, the eigenvalues of the second iterated kernel are positive (this is a direct corollary of Theorems 219, 222).*

Theorem 225 (On the Existence of an Eigenvalue) *Any symmetric kernel which is not identically zero possesses at least one eigenvalue.*

Proof Let us show that the kernel $K_2(x, s)$ possesses at least one eigenvalue. This, due to Theorem 222, guarantees the existence of an eigenvalue of the kernel $K(x, s)$. Suppose that in the disk $|\lambda| \leq \mu$ ($\mu > 0$ is arbitrary) there exists no eigenvalue of the kernel $K_2(x, s)$. Take an arbitrary function $\varphi \not\equiv 0$ and consider the function

$$f(x) = \varphi(x) - \mu K^2 \varphi(x) = \varphi(x) - \mu \int_a^b K_2(x, s) \varphi(s) ds. \qquad (12.50)$$

Obviously $f \not\equiv 0$ because otherwise μ would be an eigenvalue of the kernel $K_2(x, s)$.

Considering (12.50) as an integral equation with a given function $f(x)$, we can solve it to find $\varphi(x)$:

$$\varphi(x) = f(x) + \mu \int_a^b \Gamma(x, s, \mu) f(s) ds,$$

where $\Gamma(x, s, \mu)$ is the resolvent of the kernel $K_2(x, s)$. Taking the scalar product of this equality with the function $f(x)$, we obtain

$$(\varphi, f) = \|f\|^2 + \mu \int_a^b \int_a^b \Gamma(x, s, \mu) \overline{f(x)} f(s) dx ds. \tag{12.51}$$

Let us show that the integral here is non-negative. For this purpose, let us note that since there is no eigenvalue of the kernel $K_2(x, s)$ in the disk $|\lambda| \leq \mu$, the resolvent $\Gamma(x, s, \lambda)$ and the double integral admit power series expansions in terms of λ (see Sect. 12.4). We have

$$\int_a^b \int_a^b \Gamma(x, s, \lambda) \overline{f(x)} f(s) dx ds = \sum_{n=1}^{\infty} \alpha_n \lambda^{n-1}.$$

In particular, for $\lambda = \mu$, we obtain

$$\int_a^b \int_a^b \Gamma(x, s, \mu) \overline{f(x)} f(s) dx ds = \sum_{n=1}^{\infty} \alpha_n \mu^{n-1}.$$

An expression for the coefficients α_n can be found from the following considerations. The resolvent admits the series expansion in terms of iterated kernels, equality (12.10). Since $(K^2)^n = K^{2n}$, the n-th iterated kernel for the kernel $K_2(x, s)$ is $K_{2n}(x, s)$. Hence

$$\Gamma(x, s, \lambda) = \sum_{n=1}^{\infty} \lambda^{n-1} K_{2n}(x, s).$$

Multiplication of this series by an $L_2(a, b)$- function f and integration give us the expression for the coefficients α_n:

$$\alpha_n = \int_a^b \int_a^b K_{2n}(x, s) \overline{f(x)} f(s) dx ds.$$

Let us simplify this expression. We have

$$\int_a^b K_{2n}(x, s) f(s) ds = K^{2n} f(x).$$

Thus,

$$\alpha_n = \left(K^{2n} f, f \right) = \left(K^n \left(K^n f \right), f \right).$$

Hence, due to (12.47) and (12.48),

$$\alpha_n = \left(K^n f, K^n f \right) = \left\| K^n f \right\|^2 \geq 0.$$

Now the equality (12.51) takes the form

$$(\varphi, f) = \| f \|^2 + \sum_{n=1}^{\infty} \mu^n \left\| K^n f \right\|^2$$

from which it follows that $(\varphi, f) > 0$.

Returning to equality (12.50), let us take its scalar product with φ:

$$(\varphi, f) = (\varphi, \varphi) - \mu \left(\varphi, K^2 \varphi \right).$$

Since

$$\left(\varphi, K^2 \varphi \right) = (\varphi, K (K\varphi)) = (K\varphi, K\varphi) = \| K\varphi \|^2,$$

we obtain

$$(\varphi, f) = \| \varphi \|^2 - \mu \| K\varphi \|^2.$$

Because of the inequality $(\varphi, f) > 0$, we have that $\| \varphi \|^2 - \mu \| K\varphi \|^2 > 0$. From this inequality, we obtain the inequality

$$\| K\varphi \| \leq \frac{1}{\sqrt{\mu}} \| \varphi \|, \qquad (12.52)$$

which is valid for all $\varphi \in L_2(a, b)$ and for all $\mu > 0$, such that there is no eigenvalue of $K_2(x, s)$ in the disk $|\lambda| \leq \mu$.

Now suppose that $K_2(x, s)$ does not possess any eigenvalue. In this case the number μ can be chosen arbitrarily large. Moreover, we can let $\mu \to \infty$, and hence from (12.52), we obtain $\| K\varphi \| = 0$. Hence

$$\int_a^b K(x, s)\varphi(s)ds \equiv 0$$

for all $\varphi \in L_2(a, b)$. It is not difficult to show that this implies that $K(x, s) \equiv 0$. Indeed, let us fix x, for which the integral $\int_a^b |K(x, s)|^2 \, ds$ exists, and choose $\varphi(s) = \overline{K}(x, s)$. Then for that x (a.e. in (a, b)) we obtain

$$\int_a^b |K(x,s)|^2 \, ds \equiv 0$$

and hence $K(x,s) \equiv 0$.

Thus, if the kernel $K_2(x,s)$ has no eigenvalue, the kernel $K(x,s)$ is necessarily equal to zero identically. Hence if $K(x,s)$ is not zero identically, the kernel $K_2(x,s)$ possesses at least one eigenvalue. This, due to Theorem 222, implies that $K(x,s)$ itself possesses at least one eigenvalue.   ∎

Let us turn to the inequality (12.52). It is valid in any disk $|\lambda| \le \mu$, where there is no eigenvalue of the kernel $K_2(x,s)$. Let λ_1 be the smallest by its absolute value eigenvalue of the kernel $K(x,s)$. Then λ_1^2 is the smallest by its absolute value eigenvalue of the kernel $K_2(x,s)$. The inequality (12.52) holds for all $\mu < \lambda_1^2$. Letting $\mu \to \lambda_1^2$, we obtain the inequality

$$\|K\varphi\| \le \frac{1}{|\lambda_1|} \|\varphi\|.$$

Here the equality is attained for $\varphi(x) = \varphi_1(x)$, where $\varphi_1(x)$ is an eigenfunction of the kernel $K(x,s)$ corresponding to the eigenvalue λ_1. In this case $\varphi_1 = \lambda_1 K\varphi_1$ and $K\varphi_1 = \frac{1}{\lambda_1}\varphi_1 \implies \|K\varphi_1\| = \frac{1}{|\lambda_1|}\|\varphi_1\|$. Hence the norm of a symmetric Fredholm operator is

$$\|K\| = \frac{1}{|\lambda_1|},$$

where λ_1 is the smallest by its absolute value eigenvalue of this operator.

Consider a symmetric kernel $K(x,s)$, and let

$$\lambda_1, \lambda_2, \lambda_3, \ldots \qquad |\lambda_1| \le |\lambda_2| \le |\lambda_3| \le \ldots$$

be the sequence of its eigenvalues, and

$$\varphi_1(x), \varphi_2(x), \varphi_3(x), \varphi_4(x), \ldots$$

the sequence of corresponding eigenfunctions. Let us introduce a new kernel

$$K^{(n)}(x,s) = K(x,s) - \sum_{j=1}^n \frac{\varphi_j(x)\overline{\varphi_j(s)}}{\lambda_j}.$$

This kernel is symmetric:

$$\overline{K^{(n)}(s,x)} = K(x,s) - \sum_{j=1}^n \frac{\overline{\varphi_j(s)}\varphi_j(x)}{\lambda_j} = K^{(n)}(x,s).$$

Lemma 226 *The sequences* $\lambda_{n+1}, \lambda_{n+2}, \ldots$ *and* $\varphi_{n+1}(x), \varphi_{n+2}(x), \ldots$ *represent the sequences of the eigenvalues and corresponding eigenfunctions of the kernel* $K^{(n)}(x, s)$.

Proof Let $m > n$. Consider the expression

$$\varphi_m(x) - \lambda_m \int_a^b K^{(n)}(x, s)\varphi_m(s)ds$$

$$= \varphi_m(x) - \lambda_m \int_a^b K(x, s)\varphi_m(s)ds + \lambda_m \sum_{j=1}^n \frac{\varphi_j(x)}{\lambda_j} \int_a^b \varphi_m(s)\overline{\varphi_j(s)}ds$$

$$= 0$$

because $\varphi_m(x)$ is an eigenfunction belonging to the eigenvalue λ_m of the kernel $K(x, s)$, while $\int_a^b \varphi_m(s)\overline{\varphi_j(s)}ds = (\varphi_m, \varphi_j) = 0$ due to Theorem 220.

Hence λ_m and $\varphi_m(x)$ for $m > n$ are the eigenvalue and the corresponding eigenfunction of the kernel $K^{(n)}(x, s)$. Let us prove the opposite that if μ is an eigenvalue of the kernel $K^{(n)}(x, s)$, and $\psi(x)$ is the corresponding eigenfunction, then they are necessarily among the elements of the sequence for $m > n$. We have

$$\psi(x) - \mu \int_a^b K^{(n)}(x, s)\psi(s)ds = 0.$$

Replacing $K^{(n)}(x, s)$ by its expression gives

$$\psi(x) - \mu \int_a^b K(x, s)\psi(s)ds + \mu \sum_{j=1}^n \frac{\varphi_j(x)}{\lambda_j} \int_a^b \psi(s)\overline{\varphi_j(s)}ds = 0. \qquad (12.53)$$

The scalar product of this equality with $\varphi_l(x)$, $1 \leq l \leq n$ leads to the equation

$$(\psi, \varphi_l) - \mu (K\psi, \varphi_l) + \mu \sum_{j=1}^n (\varphi_j, \varphi_l) \frac{(\psi, \varphi_j)}{\lambda_j} = 0. \qquad (12.54)$$

Due to the symmetry of K, we have that

$$(K\psi, \varphi_l) = (\psi, K\varphi_l) = \left(\psi, \frac{1}{\lambda_l}\varphi_l\right) = \frac{1}{\lambda_l}(\psi, \varphi_l).$$

Furthermore, due to the orthonormality of the eigenfunctions, we have

$$\sum_{j=1}^n (\varphi_j, \varphi_l) \frac{(\psi, \varphi_j)}{\lambda_j} = \frac{(\psi, \varphi_l)}{\lambda_l}.$$

Thus, the second and the third terms in (12.54) cancel each other, and we obtain that

$$(\psi, \varphi_l) = 0, \quad l = 1, 2, \ldots, n,$$

and Eq. (12.53) takes the form

$$\psi(x) - \mu K \psi(x) = 0,$$

i.e., μ is an eigenvalue of K and $\psi(x)$ its corresponding eigenfunction. This means that μ and $\psi(x)$ are elements of the sequence from the condition of the lemma. Moreover, since, as we proved, $\psi(x)$ is orthogonal to the first n eigenfunctions, it can coincide only with an eigenfunction whose index is greater than n. ∎

Corollary 227 *The smallest (by absolute value) eigenvalue of the kernel $K^{(n)}(x, s)$ is λ_{n+1}, if only the kernel $K(x, s)$ has more than n eigenvalues.*

Consider a special case when the kernel $K(x, s)$ has a finite number of the eigenvalues $\lambda_1, \lambda_2, \ldots, \lambda_m$. Then the kernel $K^{(m)}(x, s)$ does not have any eigenvalue. Then due to Theorem 225 on the existence of an eigenvalue, we obtain that $K^{(m)}(x, s) \equiv 0$, which implies that

$$K(x, s) = \sum_{j=1}^{m} \frac{\varphi_j(x)\overline{\varphi_j(s)}}{\lambda_j}. \tag{12.55}$$

That is, the kernel $K(x, s)$ is necessarily degenerate. Recalling that any degenerate kernel has at most a finite number of eigenvalues, we obtain the following statement.

Corollary 228 *The system of eigenvalues and eigenfunctions of a symmetric square integrable kernel is finite if and only if the kernel is degenerate, and in this case the kernel can be written in the form (12.55).*

Corollary 229 *For any $\varphi \in L_2(a, b)$, the equality holds*

$$\lim_{n \to \infty} \left\| K^{(n)} \varphi \right\| = 0. \tag{12.56}$$

Proof This equality is obvious for a degenerate $K(x, s)$, because in this case for a large enough n, we obtain $K^{(n)}\varphi \equiv 0$. If $K(x, s)$ is not degenerate, then λ_{n+1} is the smallest (by absolute value) eigenvalue of the kernel $K^{(n)}(x, s)$, and

$$\left\| K^{(n)} \varphi \right\| \leq \frac{1}{|\lambda_{n+1}|} \|\varphi\|.$$

Since $|\lambda_{n+1}| \to \infty$ when $n \to \infty$, we obtain (12.56). ∎

12.16   Hilbert–Schmidt Theorem

Theorem 230 *Let $K(x, s)$ be a symmetric Fredholm kernel and $h(x)$ an arbitrary $L_2(a, b)$—function. Then the function*

$$f(x) = Kh(x) = \int_a^b K(x, s)h(s)ds$$

admits the following Fourier series representation in terms of the eigenfunctions of the kernel $K(x, s)$:

$$f(x) = \sum_{n=1}^{\infty} \frac{(h, \varphi_n)}{\lambda_n} \varphi_n(x).$$

The series converges in the $L_2(a, b)$-norm.

Proof Consider the Fourier series expansion of the function $h(s)$ in terms of the eigenfunctions of the kernel $K(x, s)$:

$$h(s) \sim \sum_{n=1}^{\infty} h_n \varphi_n(s), \quad h_n = (h, \varphi_n).$$

Due to the Bessel inequality, the series $\sum_{n=1}^{\infty} |h_n|^2$ is convergent. Furthermore, consider the Fourier series expansion of the function $f(x)$:

$$f(x) \sim \sum_{n=1}^{\infty} f_n \varphi_n(x), \quad f_n = (f, \varphi_n).$$

Since $f = Kh$, for the Fourier coefficients, we obtain

$$f_n = (Kh, \varphi_n) = (h, K\varphi_n) = \left(h, \frac{1}{\lambda_n} \varphi_n\right) = \frac{1}{\lambda_n}(h, \varphi_n) = \frac{h_n}{\lambda_n},$$

where we used the fact that φ_n are eigenfunctions and all eigenvalues λ_n are real. Hence the Fourier series for the function $f(x)$ takes the form

$$\sum_{n=1}^{\infty} \frac{h_n}{\lambda_n} \varphi_n(x). \tag{12.57}$$

Since $h \in L_2(a, b)$, the function $f(x)$ is also square integrable on (a, b), and its Fourier series (12.57) converges in the $L_2(a, b)$-norm. Let us prove that the sum of the series is indeed the function $f(x)$. Denote

$$\omega_n(x) = \sum_{k=1}^{n} \frac{h_k}{\lambda_k} \varphi_k(x).$$

Then

$$f(x) - \omega_n(x) = \int_a^b K(x,s)h(s)ds - \sum_{k=1}^{n} \frac{\varphi_k(x)}{\lambda_k} \int_a^b h(s)\overline{\varphi_k(s)}ds.$$

Hence

$$f(x) - \omega_n(x) = \int_a^b K^{(n)}(x,s)h(s)ds = K^{(n)}h(x).$$

Due to Corollary 229,

$$\|f - \omega_n\| = \left\| K^{(n)}h \right\| \to 0 \quad \text{when } n \to \infty.$$

Thus,

$$f(x) = \sum_{n=1}^{\infty} \frac{h_n}{\lambda_n} \varphi_n(x).$$

∎

Remark 231 In the Hilbert–Schmidt theorem, the completeness of the system of the eigenfunctions $\{\varphi_n(x)\}$ is not assumed. In fact, on the contrary, in many cases, the Hilbert–Schmidt theorem allows one to establish the completeness of different orthogonal systems.

As we see from the Hilbert–Schmidt theorem any function which is the image of some function $h \in L_2$ under the action of a symmetric Fredholm operator can be expanded into an L_2-convergent Fourier series in terms of the eigenfunctions of the operator. It is of interest to clarify under which conditions this series converges also absolutely and uniformly (regularly). It can be proved that a sufficient condition for it is the condition (12.44) (see the proof in [71, pp. 152-154]).

12.17   Solution of Symmetric Integral Equations

A symmetric integral equation is a special case of a general Fredholm equation, and for its solution general methods that we studied in previous sections can be used. Here we formulate another problem. Given a system of its eigenvalues and eigenfunctions, how this information can be used to solve the symmetric integral equation? As we will see in this section, in this case solution of the equation is especially simple and elegant.

Consider a symmetric Fredholm integral equation

$$\varphi(x) - \lambda \int_a^b K(x,s)\varphi(s)ds = f(x), \quad f \in L_2(a,b). \tag{12.58}$$

As before, let

$$\lambda_1, \lambda_2, \lambda_3, \ldots \qquad |\lambda_1| \le |\lambda_2| \le |\lambda_3| \le \ldots.$$

be the sequence of its eigenvalues, and

$$\varphi_1(x), \varphi_2(x), \varphi_3(x), \varphi_4(x), \ldots$$

the sequence of corresponding eigenfunctions. We recall that the solution is sought in the space $L_2(a, b)$.

First, let us study the case when λ is regular. Then Eq. (12.58) has a solution $\varphi \in L_2(a, b)$, and for the function $\int_a^b K(x,s)\varphi(s)ds$ the Hilbert–Schmidt theorem is valid. Hence, if $\sum_{n=1}^\infty c_n\varphi_n(x)$ is the Fourier series corresponding to the function $\varphi(x)$,

$$\varphi(x) \sim \sum_{n=1}^\infty c_n\varphi_n(x), \quad c_n = (\varphi, \varphi_n),$$

then by the Hilbert–Schmidt theorem,

$$\int_a^b K(x,s)\varphi(s)ds = \sum_{n=1}^\infty \frac{c_n}{\lambda_n}\varphi_n(x).$$

Let us substitute this series into Eq. (12.58). We obtain

$$\varphi(x) - \lambda \sum_{n=1}^\infty \frac{c_n}{\lambda_n}\varphi_n(x) = f(x). \tag{12.59}$$

This equality will give us the solution $\varphi(x)$ if we find the way to calculate the coefficients c_n. Consider the scalar product of (12.59) with an eigenfunction $\varphi_m(x)$, taking into account the orthonormality of the eigenfunctions. We obtain

$$c_m \left(1 - \frac{\lambda}{\lambda_m}\right) = f_m, \quad f_m = (f, \varphi_m). \tag{12.60}$$

Since λ is regular, we have that $1 - \frac{\lambda}{\lambda_m} \ne 0$ and

$$c_m = \frac{\lambda_m f_m}{\lambda_m - \lambda}, \quad m = 1, 2, \ldots.$$

Substituting these expressions into (12.59), we obtain the solution of Eq. (12.58) in the form

$$\varphi(x) = f(x) + \lambda \sum_{n=1}^{\infty} \frac{f_n}{\lambda_n - \lambda} \varphi_n(x). \tag{12.61}$$

Next, consider the case when λ is an eigenvalue, say, $\lambda = \lambda_p = \lambda_{p+1} = \ldots = \lambda_q$ (here we take into account the multiplicity of the eigenvalue). Assuming that Eq. (12.58) is however solvable, we arrive at the relation (12.60). If m does not coincide with any of the numbers $p, p+1, \ldots, q$, then, as before, $1 - \frac{\lambda}{\lambda_m} \neq 0$, and $c_m = \frac{\lambda_m f_m}{\lambda_m - \lambda}$. If, on the contrary, m equals some of the numbers $p, p+1, \ldots, q$, then from (12.60) we obtain $f_m = 0$ or, which is the same,

$$(f, \varphi_m) = 0, \quad m = p, p+1, \ldots, q. \tag{12.62}$$

Thus, if λ is an eigenvalue, then for the solvability of the integral equation it is necessary that the right hand side f be orthogonal to all the eigenfunctions corresponding to λ. If the conditions (12.62) are fulfilled, Eq. (12.60) becomes an identity for the values $m = p, p+1, \ldots, q$. In this case Eq. (12.58) has an infinite set of solutions which have the form

$$\varphi(x) = f(x) + \lambda \sum_{n=1}^{\infty} {}' \frac{f_n}{\lambda_n - \lambda} \varphi_n(x) + \sum_{n=p}^{q} c_n \varphi_n(x),$$

where the prime in the first sum means that the terms with the numbers $p, p+1, \ldots, q$, for which both the numerator f_n and the denominator $\lambda_n - \lambda$ become zero, are discarded. The coefficients c_n in the second sum are arbitrary constants.

Thus, if λ is an eigenvalue to which there correspond the linearly independent eigenfunctions

$$\varphi_p(x), \varphi_{p+1}(x), \ldots, \varphi_q(x), \tag{12.63}$$

then the necessary and sufficient condition for the solvability of Eq. (12.58) is the condition (12.62) of orthogonality of $f(x)$ to each eigenfunction from the set (12.63). This result is in a full agreement with the Fourth Fredholm theorem (Theorem 199), because in this case the homogeneous adjoint integral equations coincide.

12.18   Bilinear Series

The bilinear series

$$\sum_{n=1}^{\infty} \frac{\varphi_n(x)\overline{\varphi_n(s)}}{\lambda_n} \tag{12.64}$$

in fact appeared in the Hilbert–Schmidt theorem. Here $\{\lambda_n\}$ and $\{\varphi_n(x)\}$ are the system of eigenvalues and eigenfunctions of a symmetric kernel $K(x, s)$. Notice that the bilinear series was obtained as a Fourier series of the kernel $K(x, s)$ in terms of the system of functions $\left\{\overline{\varphi_n(s)}\right\}$.

Theorem 232 *The bilinear series (12.64) converges in the L_2-norm in the main square, and its sum equals the kernel $K(x, s)$ in the sense of $L_2((a, b) \times (a, b))$.*

Proof The system of functions

$$\varphi_k(x)\overline{\varphi_k(s)}, \quad k = 1, 2, \ldots \tag{12.65}$$

is orthonormal in the main square. Indeed,

$$\int_a^b \int_a^b \varphi_k(x)\overline{\varphi_k(s)}\overline{\varphi_m(x)}\varphi_m(s)dxds = \int_a^b \varphi_k(x)\overline{\varphi_m(x)}dx \cdot \int_a^b \varphi_m(s)\overline{\varphi_k(s)}ds = \begin{cases} 0, & k \neq m \\ 1, & k = m. \end{cases}$$

Consider a Fourier series in terms of the functions (12.65) associated with kernel $K(x, s)$:

$$K(x, s) \sim \sum_{n=1}^{\infty} A_n \varphi_n(x)\overline{\varphi_n(s)}.$$

For the coefficients A_n, we obtain

$$A_n = \left(K(x, s), \varphi_n(x)\overline{\varphi_n(s)} \right) = \int_a^b \int_a^b K(x, s)\overline{\varphi_n(x)}\varphi_n(s)dxds$$

$$= \int_a^b \overline{\varphi_n(x)} \int_a^b K(x, s)\varphi_n(s)dsdx.$$

The interior integral equals $\varphi_n(x)/\lambda_n$. Hence

$$A_n = \frac{1}{\lambda_n} \int_a^b |\varphi_n(x)|^2 \, dx = \frac{1}{\lambda_n} \|\varphi_n\|^2 = \frac{1}{\lambda_n}.$$

Thus,

$$K(x, s) \sim \sum_{n=1}^{\infty} \frac{\varphi_n(x)\overline{\varphi_n(s)}}{\lambda_n},$$

and the bilinear series is nothing but the Fourier series of the kernel $K(x, s)$ in terms of the functions (12.65). As every Fourier series it converges in the sense of L_2 in the corresponding domain, which is the main square in the case under consideration, and its sum is square integrable. Our next aim is to prove that the sum of the bilinear series is precisely the kernel $K(x, s)$.

Consider the kernel

$$L(x, s) = K(x, s) - \sum_{n=1}^{\infty} \frac{\varphi_n(x)\overline{\varphi_n(s)}}{\lambda_n},$$

which is obviously square integrable in the main square. Let us prove that it has no eigenvalue. For this purpose, let us consider the homogeneous equation

$$\psi(x) - \lambda \int_a^b L(x, s)\psi(s)ds = 0, \tag{12.66}$$

or

$$\psi(x) - \lambda K\psi(x) + \lambda \int_a^b \sum_{n=1}^{\infty} \frac{\varphi_n(x)\overline{\varphi_n(s)}}{\lambda_n} \psi(s)ds = 0.$$

Let us verify that the series here can be integrated termwise. The integral

$$\int_a^b |K(x, s)|^2 \, ds$$

exists a.e. The bilinear series is a Fourier series of the kernel $K(x, s)$ considered as a function of s with respect to the system of functions $\{\overline{\varphi_n(s)}\}$. Such series converges in the $L_2(a, b)$-norm with respect to s, and it was shown that it can be integrated termwise pre-multiplying it by an arbitrary square integrable function. In this case that function is $\psi(s)$.

Thus, integrating termwise, we obtain the equation

$$\psi(x) - \lambda K\psi(x) + \lambda \sum_{n=1}^{\infty} \frac{(\psi, \varphi_n)}{\lambda_n} \varphi_n(x)ds = 0.$$

Due to the Hilbert–Schmidt theorem,

$$K\psi(x) = \sum_{n=1}^{\infty} \frac{(\psi, \varphi_n)}{\lambda_n} \varphi_n(x),$$

and the equation gives us $\psi \equiv 0$. Thus, Eq. (12.66) has a trivial solution only for all values of λ, and hence the kernel $L(x, s)$ has no eigenvalue. Consequently, due to the theorem on the existence of an eigenvalue (Theorem 225), $L(x, s) \equiv 0$, and hence

$$K(x, s) = \sum_{n=1}^{\infty} \frac{\varphi_n(x)\overline{\varphi_n(s)}}{\lambda_n}. \tag{12.67}$$

∎

Multiplying (12.67) by $\overline{K(x, s)}$ and integrating over the main square, we obtain the formula

$$\sum_{n=1}^{\infty} \frac{1}{\lambda_n^2} = \int_a^b \int_a^b |K(x, s)|^2 \, dx ds = B^2. \tag{12.68}$$

Theorem 233 *Let $\{\varphi_n(x)\}$ be an orthonormal system of functions on (a, b) and $\{\lambda_n\}$ a sequence of real numbers such that the series $\sum_{n=1}^{\infty} \frac{1}{\lambda_n^2}$ converges. Then the series (12.64) converges with respect to the L_2-norm in the main square, and the sum of this series is a Fredholm symmetric kernel for which the numbers $\{\lambda_n\}$ are the eigenvalues and the functions $\{\varphi_n(x)\}$ are the corresponding eigenfunctions.*

Proof The convergence of the series (12.64) follows directly from the Riesz–Fischer theorem. Denote its sum by $K(x, s)$:

$$\sum_{n=1}^{\infty} \frac{\varphi_n(x)\overline{\varphi_n(s)}}{\lambda_n} = K(x, s). \tag{12.69}$$

The square integrability and symmetry of this kernel are evident. Let us multiply this equality by $\lambda_m \varphi_m(s)$ and integrate with respect to s. As was explained above, the series can be integrated termwise. We obtain

$$\varphi_m(x) = \lambda_m \int_a^b K(x, s)\varphi_m(s)ds.$$

This shows that λ_m is an eigenvalue and $\varphi_m(x)$ is a corresponding eigenfunction of the kernel $K(x, s)$.

Let us suppose now that the kernel $K(x, s)$ has another eigenfunction $\varphi(x)$ which is orthogonal to all $\varphi_n(x)$, and let λ be its corresponding eigenvalue. Thus,

$$\varphi(x) = \lambda \int_a^b K(x, s)\varphi(s)ds.$$

Replacing the kernel $K(x, s)$ with its expression (12.69) and integrating termwise give us the equality

$$\varphi(x) = \lambda \sum_{n=1}^{\infty} \frac{(\varphi, \varphi_n)}{\lambda_n} \varphi_n(x).$$

Since each scalar product (φ, φ_n) equals zero, we obtain that $\varphi \equiv 0$. Thus, the functions $\{\varphi_n(x)\}$ represent the system of the eigenfunctions of the kernel (12.69). ∎

12.19   Bilinear Series for Iterated Kernels

Let us write the Hilbert–Schmidt formula in the form

$$f(x) = \int_a^b K(x, t)h(t)dt = \sum_{n=1}^{\infty} \frac{h_n}{\lambda_n} \varphi_n(x) \qquad (12.70)$$

and substitute here $h(t) = K(t, s)$. Then taking into account the symmetry of the kernel, we obtain

$$h_n = \int_a^b K(t, s)\overline{\varphi_n(t)}dt = \int_a^b \overline{K(t, s)\varphi_n(t)}dt = \int_a^b \overline{K(s, t)\varphi_n(t)}dt = \frac{1}{\lambda_n}\overline{\varphi_n(s)}.$$

The last equality is due to the fact that $\varphi_n(x)$ is an eigenfunction of $K(x, t)$. Thus, the Hilbert–Schmidt formula leads to the equality

$$K_2(x, s) = \sum_{n=1}^{\infty} \frac{\varphi_n(x)\overline{\varphi_n(s)}}{\lambda_n^2},$$

where due to the Hilbert–Schmidt theorem the series converges in the $L_2(a, b)$-norm with respect to x. Due to the symmetry of the kernel, the series also converges in the $L_2(a, b)$-norm with respect to s. Moreover, the series converges in the L_2-norm in the main square, because due to (12.68), the series $\sum_{n=1}^{\infty} \frac{1}{\lambda_n^2}$ converges.

In a similar way, considering in the formula (12.70) $h(t) = K_2(t, s)$, we obtain

$$K_3(x, s) = \sum_{n=1}^{\infty} \frac{\varphi_n(x)\overline{\varphi_n(s)}}{\lambda_n^3}.$$

In general, by induction, the following formula is obtained:

$$K_m(x, s) = \sum_{n=1}^{\infty} \frac{\varphi_n(x)\overline{\varphi_n(s)}}{\lambda_n^m}. \tag{12.71}$$

This formula shows that the kernel $K_m(x, s)$ is expanded into a Fourier series in terms of $\{\overline{\varphi_n(s)}\}$. Hence the Parseval identity gives us the equality

$$\sum_{n=1}^{\infty} \frac{|\varphi_n(x)|^2}{\lambda_n^{2m}} = \int_a^b |K_m(x, s)|^2 \, ds.$$

The series can be integrated termwise. Taking into account that $\|\varphi_n\| = 1$, we obtain

$$\sum_{n=1}^{\infty} \frac{1}{\lambda_n^{2m}} = \int_a^b \int_a^b |K_m(x, s)|^2 \, dx ds.$$

Due to Theorem 233, the formula (12.71) gives the expansion of the iterated kernels into bilinear series. Hence the following statement is valid.

Theorem 234 *Let $K(x, s)$ be a Fredholm symmetric kernel and $\{\lambda_n\}$ and $\{\varphi_n(x)\}$ the system of its eigenvalues and eigenfunctions. Then, for $m \geq 2$, the numbers $\{\lambda_n^m\}$ and the functions $\{\varphi_n(x)\}$ represent the system of the eigenvalues and eigenfunctions of the iterated kernel $K_m(x, s)$.*

12.20   Resolvent of a Symmetric Kernel

In the formula (12.61) for the solution of a symmetric integral equation for regular values of the spectral parameter, let us replace the coefficient f_n by its integral expression

$$f_n = (f, \varphi_n) = \int_a^b f(s)\overline{\varphi_n(s)} ds.$$

Then (12.61) can be written as

$$\varphi(x) = f(x) + \lambda \sum_{n=1}^{\infty} \int_a^b \frac{\varphi_n(x)\overline{\varphi_n(s)}}{\lambda_n - \lambda} f(s) ds. \tag{12.72}$$

Let us show that in (12.72) it is possible to change the order of summation and integration. Consider the series

$$\sum_{n=1}^{\infty} \frac{\varphi_n(x)\overline{\varphi_n(s)}}{\lambda_n - \lambda}. \tag{12.73}$$

The functions $\varphi_n(x)\overline{\varphi_n(s)}$, $n = 1, 2, \ldots$, are orthonormal in the main square. Furthermore, the series

$$\sum_{n=1}^{\infty} \frac{1}{|\lambda_n - \lambda|^2} \tag{12.74}$$

converges. Indeed, $\lambda_n \to \infty$, and hence for large enough n the inequality holds $|\lambda_n| > 2|\lambda|$. Then $|\lambda| < |\lambda_n|/2$, $|\lambda_n - \lambda| > |\lambda_n|/2$, and $\frac{1}{|\lambda_n - \lambda|} < \frac{2}{|\lambda_n|}$. Now the convergence of the series (12.74) follows from the shown earlier convergence of the series $\sum_{n=1}^{\infty} \frac{1}{\lambda_n^2}$.

From the Riesz–Fischer theorem, it follows that for any regular value of λ the series (12.73) converges in the main square with respect to the L_2-norm. Repeating the reasonings used in the proof of Theorem 232, we find that the series (12.73) can be integrated termwise with respect to s, pre-multiplying it by an arbitrary square integrable function. This means precisely that in (12.72) the change of the order of summation and integration is permitted. This gives us a new form for the solution

$$\varphi(x) = f(x) + \lambda \int_a^b f(s) \sum_{n=1}^{\infty} \frac{\varphi_n(x)\overline{\varphi_n(s)}}{\lambda_n - \lambda} ds.$$

Comparing it with the representation (12.11), we conclude that the sum of the series (12.73) is the resolvent of the symmetric kernel $K(x, s)$,

$$\Gamma(x, s, \lambda) = \sum_{n=1}^{\infty} \frac{\varphi_n(x)\overline{\varphi_n(s)}}{\lambda_n - \lambda}.$$

From this formula, we observe that the resolvent of a symmetric kernel has only simple poles.

12.21   Extremal Properties of Eigenvalues and Eigenfunctions

By the Hilbert–Schmidt theorem, for any square integrable function $h(s)$, the equality holds

$$\int_a^b K(x,s)h(s)ds = \sum_{n=1}^{\infty} \frac{(h,\varphi_n)}{\lambda_n}\varphi_n(x).$$

Consider the scalar product of this equality with $h(x)$. The series here converges in L_2, and hence the scalar multiplication can be performed termwise:

$$(Kh,h) = \sum_{n=1}^{\infty}\left(\frac{(h,\varphi_n)}{\lambda_n}\varphi_n,h\right) = \sum_{n=1}^{\infty}\frac{(h,\varphi_n)(\varphi_n,h)}{\lambda_n}$$

$$= \sum_{n=1}^{\infty}\frac{|(h,\varphi_n)|^2}{\lambda_n}.$$

Let, as always, the numbers λ_n be ordered by their absolute value, and thus, λ_1 has the smallest absolute value. Then, from the last equality, we obtain

$$|(Kh,h)| \le \frac{1}{|\lambda_1|}\sum_{n=1}^{\infty}|(h,\varphi_n)|^2 ,$$

and using additionally the Bessel inequality,

$$|(Kh,h)| \le \frac{\|h\|^2}{|\lambda_1|}.$$

For the normalized functions $h(x)$, this inequality takes the form

$$|(Kh,h)| \le \frac{1}{|\lambda_1|}, \quad \|h\|=1.$$

Here, the equality is attained for $h(x) = \varphi_1(x)$. Indeed, since $\varphi_1(x)$ is an eigenfunction corresponding to λ_1, $\varphi_1 = \lambda_1 K\varphi_1$, and hence

$$(K\varphi_1,\varphi_1) = \frac{\|\varphi_1\|^2}{\lambda_1} = \frac{1}{\lambda_1}.$$

Thus, the following theorem is valid.

Theorem 235 *The absolute value $|(Kh,h)|$ considered on the set of normalized functions attains its maximum equal to $\frac{1}{\lambda_1}$, and the maximum is attained for $h(x) = \varphi_1(x)$.*

Now, consider the set of all normalized functions which are orthogonal to the first $m-1$ eigenfunctions:

$$\|h\|=1, \quad (h,\varphi_1)=(h,\varphi_2)=\ldots=(h,\varphi_{m-1})=0.$$

Then

$$Kh(x) = \sum_{n=m}^{\infty} \frac{(h, \varphi_n)}{\lambda_n} \varphi_n(x).$$

Repeating the above reasonings, we obtain the following statement.

Theorem 236 *The magnitude $|(Kh, h)|$ considered on the set of normalized functions orthogonal to the first $m - 1$ eigenfunctions of the kernel $K(x, s)$ attains its maximum equal to $\frac{1}{\lambda_m}$, and the maximum is attained for $h(x) = \varphi_m(x)$.*

The last two theorems serve as a base for the variational methods for computing eigenvalues and eigenfunctions of symmetric kernels.

Chapter 13
Solution of Boundary Value Problems for the Laplace Equation

For simplicity of the exposition in this chapter, we restrict our consideration to the case $n = 3$. In Sect. 11.8, we obtained boundary integral equations equivalent to corresponding boundary value problems for the Laplace equation. When the boundary Σ is a Lyapunov surface, the integral equations are equations with weak singularities. Indeed, due to Lemma 150, we have the bound $|\cos(r, \nu)| \leq br^\alpha$, where b is a positive constant and $0 < \alpha \leq 1$. Hence for the kernels in the boundary integral equations obtained in Sect. 11.8, we have the estimate

$$\left| \frac{\cos(r, \nu)}{r^2} \right| \leq \frac{b}{r^{2-\alpha}},$$

which shows that the kernels on Lyapunov surfaces (which are two-dimensional) in $\mathbb{R}^3$ possess only weak singularities (see Sect. 12.13). Thus, the Fredholm theory is applicable to them, which gives us a convenient tool for complementing the uniqueness theorems for boundary value problems (from Sect. 10.2) with the results on the existence of solutions. This is the aim of this chapter. We begin with the interior Dirichlet and exterior Neumann problems.

13.1 Problems D_+ and N_-

The boundary Σ is assumed to be a closed Lyapunov surface. Notice that the integral equations for the interior Dirichlet problem (Problem D_+)

$$q(P) + \frac{1}{2\pi} \int_\Sigma \frac{\cos(r, \nu_Q)}{r^2} q(Q) d\sigma_Q = -\frac{1}{2\pi} f(P), \quad P \in \Sigma \tag{13.1}$$

and for the exterior Neumann problem (Problem N_-)

© The Author(s), under exclusive license to Springer Nature Switzerland AG 2022
A. N. Karapetyants, V. V. Kravchenko, *Methods of Mathematical Physics*,
https://doi.org/10.1007/978-3-031-17845-0_13

$$q(P) - \frac{1}{2\pi} \int_\Sigma \frac{\cos(r, \nu_P)}{r^2} q(Q) d\sigma_Q = -\frac{1}{2\pi} f(P), \quad P \in \Sigma \qquad (13.2)$$

are mutually adjoint, as well as the other two equations, that for the exterior Dirichlet problem (Problem D_-)

$$q(P) - \frac{1}{2\pi} \int_\Sigma \frac{\cos(r, \nu_Q)}{r^2} q(Q) d\sigma_Q = \frac{1}{2\pi} f(P), \quad P \in \Sigma \qquad (13.3)$$

and that for the interior Neumann problem (Problem N_+)

$$q(P) + \frac{1}{2\pi} \int_\Sigma \frac{\cos(r, \nu_P)}{r^2} q(Q) d\sigma_Q = \frac{1}{2\pi} f(P), \quad P \in \Sigma. \qquad (13.4)$$

Indeed, consider the kernel in (13.1):

$$K(P, Q) = \frac{\cos(r, \nu_Q)}{2\pi r^2}.$$

Since it is real valued, its adjoint has the form

$$K^*(P, Q) = K(Q, P),$$

that is, in the adjoint kernel, the points P and Q are permuted, and correspondingly, the normal is taken at the point P and the direction r from the point Q toward the point P (see Fig. 13.1). Then the angle (r, ν_Q) changes to $\pi - (r, \nu_P)$, and we obtain

$$K^*(P, Q) = \frac{\cos(\pi - (r, \nu_P))}{2\pi r^2} = -\frac{\cos(r, \nu_P)}{2\pi r^2}. \qquad (13.5)$$

Thus, $K^*(P, Q)$ is precisely the kernel in (13.2), and the integral equations for the interior Dirichlet and exterior Neumann problems are mutually adjoint.

Fig. 13.1 Illustration for the equality (13.5)

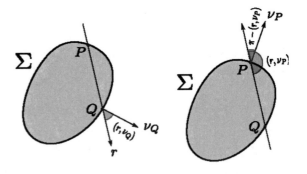

Analogously, the integral equations for the exterior Dirichlet and interior Neumann problems are adjoint.

Let us prove that Eqs. (13.1) and (13.2) are solvable for all $f \in C(\Sigma)$. Due to their mutual adjointness, it suffices to conduct the proof for one of them. Let us do it for the exterior Neumann problem (Eq. (13.2)). The corresponding homogeneous equation has the form

$$q_0(P) - \frac{1}{2\pi} \int_\Sigma \frac{\cos(r, \nu_P)}{r^2} q_0(Q) d\sigma_Q = 0, \quad P \in \Sigma. \qquad (13.6)$$

Let q_0 be any of its solution. Then it is continuous, due to Corollary 216.

Consider the single layer potential

$$v_0(P) = \int_\Sigma q_0(Q) \frac{d\sigma_Q}{r_{PQ}}.$$

According to Theorem 160, the limit value of its normal derivative has the form

$$\left(\frac{\partial v_0}{\partial \nu}\right)_- (P) = -2\pi q_0(P) + \int_\Sigma \frac{\cos(r, \nu_P)}{r^2} q_0(Q) d\sigma_Q,$$

which is zero due to (13.6). Thus, the function $v_0(P)$ is harmonic in the exterior domain V_- and equals zero on the boundary Σ. Hence, due to Theorem 133 on the uniqueness of the solution of the exterior Neumann problem, v_0 is identically zero in V_-. Due to the continuity of the single layer potential on the surface Σ (Theorem 154), we have that v_0 is harmonic in the interior domain V_+ and equals zero on its boundary Σ. Hence, due to the uniqueness theorem for the interior Dirichlet problem (Theorem 129), we obtain that $v_0 \equiv 0$ in V_+. Then $\left(\frac{\partial v_0}{\partial \nu}\right)_+ \equiv 0$ on Σ, and due to Corollary 161,

$$q_0 = \frac{1}{4\pi}\left(\left(\frac{\partial v_0}{\partial \nu}\right)_+ - \left(\frac{\partial v_0}{\partial \nu}\right)_-\right) \equiv 0 \quad \text{on } \Sigma.$$

Thus, the homogeneous equation (13.6) possesses a trivial solution only. Hence, the value of the parameter $\lambda = -1$ is regular, and Eq. (13.2) is always solvable. This means that when Σ is a Lyapunov surface, and the given normal derivative of the solution on Σ is continuous, the exterior Neumann problem is solvable, and its solution can be represented as a single layer potential.

Equation (13.1) is the adjoint of (13.2), and hence, it is also solvable for any continuous right hand side. Hence, if Σ is a Lyapunov surface, and the given value of the solution on Σ is continuous, the interior Dirichlet problem is solvable, and its solution can be represented in the form of a double layer potential.

Taking into account Theorems 129 and 133, we obtain that **the interior Dirichlet and exterior Neumann problems for the Laplace equation are uniquely solvable.**

13.2 Problem N_+

Now let us consider the mutually adjoint integral equations for the exterior Dirichlet problem

$$q(P) - \frac{1}{2\pi} \int_\Sigma \frac{\cos(r, v_Q)}{r^2} q(Q) d\sigma_Q = \frac{1}{2\pi} f(P), \quad P \in \Sigma \tag{13.7}$$

and interior Neumann problem

$$q(P) + \frac{1}{2\pi} \int_\Sigma \frac{\cos(r, v_P)}{r^2} q(Q) d\sigma_Q = \frac{1}{2\pi} f(P), \quad P \in \Sigma. \tag{13.8}$$

Let us show that for these equations the value of the parameter $\lambda = 1$ is an eigenvalue.

Consider the homogeneous equation corresponding to (13.7)

$$q_0(P) - \frac{1}{2\pi} \int_\Sigma \frac{\cos(r, v_Q)}{r^2} q_0(Q) d\sigma_Q = 0, \quad P \in \Sigma. \tag{13.9}$$

It is easy to see that this equation admits the solution

$$q_0(P) \equiv 1,$$

because with the aid of the formula (11.17) for the value of the Gauss integral from Sect. 11.5, we have

$$\int_\Sigma \frac{\cos(r, v_Q)}{r^2} d\sigma_Q = - \int_\Sigma \frac{\partial}{\partial v_Q} \left(\frac{1}{r} \right) d\sigma_Q = 2\pi, \quad P \in \Sigma.$$

Thus, Eq. (13.9) admits a nontrivial solution, and hence, $\lambda = 1$ is an eigenvalue of the kernel

$$K(P, Q) = \frac{1}{2\pi} \frac{\cos(r, v_Q)}{r^2},$$

and Eqs. (13.7) and (13.8) may be unsolvable. Due to the third Fredholm theorem (Sect. 12.13), the adjoint to (13.9) equation

$$\mu(P) + \frac{1}{2\pi} \int_{\Sigma} \frac{\cos(r, \nu_P)}{r^2} \mu(Q) d\sigma_Q = 0, \quad P \in \Sigma \qquad (13.10)$$

has an eigenfunction that we denote by $\mu_0(P)$. Let us prove that Eq. (13.10) does not have other linearly independent eigenfunctions. Let $\mu_1(P)$ be a solution of (13.10). Then

$$\mu_k(P) + \frac{1}{2\pi} \int_{\Sigma} \frac{\cos(r, \nu_P)}{r^2} \mu_k(Q) d\sigma_Q = 0, \quad k = 0, 1, \quad P \in \Sigma. \qquad (13.11)$$

Consider the single layer potentials

$$v_k(P) = \int_{\Sigma} \mu_k(Q) \frac{d\sigma_Q}{r}, \quad k = 0, 1, \quad P \in V_+.$$

The limit values of their normal derivatives have the form (Theorem 160)

$$\left(\frac{\partial v_k}{\partial \nu} \right)_+ (P) = 2\pi \mu_k(P) + \int_{\Sigma} \frac{\cos(r, \nu_P)}{r^2} \mu_k(Q) d\sigma_Q,$$

which is zero due to (13.11). Both functions $v_k(P)$ are harmonic in V_+, and their normal derivatives equal zero on Σ; hence by Theorem 130, the functions $v_k(P)$ are constant:

$$v_k(P) \equiv C_k, \quad k = 0, 1, \quad P \in V_+.$$

Let us show that $C_k \neq 0$. Suppose the opposite. Then $v_k(P) \equiv 0$ in V_+. Due to the continuity of the single layer potential on the surface, $v_k(P) \equiv 0$ on Σ as well. In the exterior domain V_-, the function $v_k(P)$ is harmonic, and since $v_k(P) \equiv 0$ on Σ, due to the uniqueness theorem for the exterior Dirichlet problem (Theorem 132), $v_k(P) \equiv 0$ in V_- as well. Hence,

$$\left(\frac{\partial v_k}{\partial \nu} \right)_- (P) \equiv 0,$$

and due to Corollary 161,

$$\mu_k = \frac{1}{4\pi} \left(\left(\frac{\partial v_k}{\partial \nu} \right)_+ - \left(\frac{\partial v_k}{\partial \nu} \right)_- \right) \equiv 0,$$

which contradicts the assumption that μ_k is an eigenfunction. Thus, $C_k \neq 0$, $k = 0, 1$.

Consider the function

$$\mu_2(P) = C_1 \mu_0(P) - C_0 \mu_1(P)$$

together with the corresponding single layer potential

$$v_2(P) = \int_\Sigma \mu_2(Q) \frac{d\sigma_Q}{r} = C_1 v_0(P) - C_0 v_1(P).$$

Obviously, $v_2(P) \equiv 0$ in V_+. Then repeating the reasoning above, we obtain that $v_2(P) \equiv 0$ in the whole space, and hence, $\mu_2(P) \equiv 0$, which means that

$$\mu_1(P) = \frac{C_1}{C_0} \mu_0(P).$$

This finishes the proof of the fact that (13.10) possesses a unique linearly independent solution.

By the fourth Fredholm theorem (Sect. 12.13) for the solvability of the integral equation (13.8), it is necessary and sufficient that the right hand side $\frac{1}{2\pi} f(P)$ be orthogonal to all solutions of the adjoint homogeneous equation (13.9). We showed that (13.9) has one linearly independent solution $q_0(P) \equiv 1$. Thus the necessary and sufficient conditions reduce to the single equality

$$\int_\Sigma f(P) d\sigma_P = 0. \tag{13.12}$$

Thus, we proved the sufficiency of the equality from Property 1 of harmonic functions, which is exactly (13.12) (see Sect. 9.3) for the solvability of the interior Neumann problem. **Hence, if the condition (13.12) is fulfilled, the interior Neumann problem is (uniquely) solvable, and its solution is representable in the form of a single layer potential.**

13.3   Problem D_-

In the preceding section, we showed that in general the integral equation (13.7) may be unsolvable. This means that in general the solution of the exterior Dirichlet problem is not representable as a double layer potential. This is quite natural because the double layer potential decays at infinity as R^{-2} (R is the distance of the point P with respect to the origin), while an arbitrary harmonic function (when $n = 3$) decays at infinity as R^{-1}. For simplicity, let us put the origin inside the surface Σ and look for the solution of the exterior Dirichlet problem in the form of a sum of the double layer potential with an unknown density $\mu(Q)$ and the harmonic function

$$\frac{1}{R} \int_\Sigma \mu(Q) d\sigma_Q,$$

which decays at infinity as R^{-1}.

Thus, the solution of the exterior Dirichlet problem is sought in the form

$$u(P) = \int_\Sigma \mu(Q) \frac{\partial}{\partial v} \left(\frac{1}{r} \right) d\sigma_Q + \frac{1}{R} \int_\Sigma \mu(Q) d\sigma_Q, \quad P \in V_-. \tag{13.13}$$

On the surface Σ, the Dirichlet boundary condition must be fulfilled

$$u(P) = f(P), \quad P \in \Sigma.$$

The limit formula (11.20) from Theorem 157 gives us the following equation for the unknown function μ:

$$\mu(P) - \frac{1}{2\pi} \int_\Sigma \left(\frac{\cos(r, v_Q)}{r^2} - \frac{1}{R} \right) \mu(Q) d\sigma_Q = \frac{1}{2\pi} f(P), \quad P \in \Sigma. \tag{13.14}$$

The kernel of this integral equation consists of two terms, the first one having a weak singularity, the second one being a continuous function. Thus, this is an integral equation with a weak singularity. According to the Fredholm theory, it is solvable if and only if the corresponding homogeneous equation has a trivial solution only.

Let us study the homogeneous equation

$$\mu_0(P) - \frac{1}{2\pi} \int_\Sigma \left(\frac{\cos(r, v_Q)}{r^2} - \frac{1}{R} \right) \mu_0(Q) d\sigma_Q = 0, \quad P \in \Sigma. \tag{13.15}$$

Let μ_0 be its solution. The left hand side of (13.15) can be regarded as the limit value divided by 2π of the harmonic function

$$u_0(P) = \int_\Sigma \mu_0(Q) \frac{\partial}{\partial v} \left(\frac{1}{r} \right) d\sigma_Q + \frac{1}{R} \int_\Sigma \mu_0(Q) d\sigma_Q$$

on the surface Σ. Thus, $u_0 \equiv 0$ in $\overline{V}_-$:

$$\int_\Sigma \mu_0(Q) \frac{\partial}{\partial v} \left(\frac{1}{r} \right) d\sigma_Q + \frac{1}{R} \int_\Sigma \mu_0(Q) d\sigma_Q \equiv 0, \quad P \in \overline{V}_-.$$

Multiplication of this equality by R gives

$$R \int_\Sigma \mu_0(Q) \frac{\partial}{\partial v} \left(\frac{1}{r} \right) d\sigma_Q + \int_\Sigma \mu_0(Q) d\sigma_Q \equiv 0, \quad P \in \overline{V}_-.$$

Letting $R \to \infty$, we see that the first term here tends to zero and hence

$$\int_\Sigma \mu_0(Q) d\sigma_Q = 0. \tag{13.16}$$

Thus, Eq. (13.15) reduces to Eq. (13.9). Since μ_0 is its solution, we obtain that $\mu_0(P) \equiv$ Const. From (13.16), we find that this constant is zero. Thus, the homogeneous equation (13.15) admits a trivial solution only. Hence, Eq. (13.14) is always solvable.

Thus, the exterior Dirichlet problem is solvable, and its solution can be represented in the form (13.13). Combining this conclusion with Theorem 132 on the uniqueness of the solution, we obtain that when Σ is a Lyapunov surface, and the given value of the solution on Σ is continuous, **the exterior Dirichlet problem for the Laplace equation is uniquely solvable.**

Chapter 14
Helmholtz Equation

14.1 Definition and Relations with Time-Dependent Models

An elliptic equation of the form

$$\Delta u + cu = -f, \tag{14.1}$$

where $c = \text{Const} \neq 0$ is known as the Helmholtz equation. The right hand side f describes physical sources or external forces and their density distribution in the space. When $f \equiv 0$, we obtain the homogeneous Helmholtz equation

$$\Delta u + cu = 0.$$

First, let us determine what is the possible physical meaning of the solutions of the Helmholtz equation and, consequently, why it is one of the most important equations of mathematical physics.

Consider the equation

$$\Delta w = a_0 \frac{\partial^2 w}{\partial t^2} + 2a_1 \frac{\partial w}{\partial t} + a_2 w, \tag{14.2}$$

where a_0, a_1, and a_2 are some constants. When $a_0 > 0$, $a_1 > 0$, and $a_2 > 0$, this is the telegraph equation. We encountered it in Sect. 3.11. When $a_1 = a_2 = 0$, and $a_0 > 0$, this is the wave equation; when $a_0 = a_2 = 0$, this is the heat equation. Finally, when $a_0 = 0$, $a_1 > 0$, and $a_2 \neq 0$, the equation becomes the diffusion equation for a medium in which chemical or chain reactions take place.

Following the method of separation of variables, we can look for solutions of Eq. (14.2) having the form

$$w(x, t) = u(x)v(t),$$

A. N. Karapetyants, V. V. Kravchenko, *Methods of Mathematical Physics*,
https://doi.org/10.1007/978-3-031-17845-0_14

where $u(x)$ is a function only of space coordinates and $v(t)$ is a function only of time. Substitution of this expression into Eq. (14.2) leads to the equation

$$\frac{\Delta u}{u} = \frac{1}{v}\left(a_0\frac{\partial^2 v}{\partial t^2} + 2a_1\frac{\partial v}{\partial t} + a_2 v\right).$$

Since the left side of this equation is independent of t, while the right side depends only on t, we obtain two separate equations

$$\Delta u + cu = 0 \tag{14.3}$$

and

$$a_0\frac{d^2 v}{dt^2} + 2a_1\frac{dv}{dt} + (a_2 + c)\,v = 0,$$

where c is some constant.

Hence, we see that the Helmholtz equation describes the intensity of physical phenomena taking place at all points of a region under consideration and obeying a single evolution in time law.

Of special importance is the case when $v(t) = e^{-i\omega t}$ or $v(t) = e^{i\omega t}$. Then

$$w(x, t) = u(x)e^{-i\omega t} \tag{14.4}$$

or

$$w(x, t) = u(x)e^{i\omega t} \tag{14.5}$$

describes the so-called time-harmonic or simply harmonic oscillations. The importance of this special case of the evolution in time law is based on the observation that essentially arbitrary time-dependent phenomena can be represented as superpositions of time-harmonic components.

The choice between the substitutions (14.4) or (14.5) is a matter of convention. They are equivalent. Therefore, we may use only one of them. We shall use the substitution (14.4).

14.2   Fundamental Solutions

In our study of the Helmholtz equation, we shall mainly consider the three-dimensional situation

$$\Delta u(M) + cu(M) = -f(M), \quad M \in \mathbb{R}^3$$

and occasionally mention the corresponding results for the two-dimensional case.

Let us obtain spherically symmetric solutions of the homogeneous Helmholtz equation. Let M_0 be a fixed point of $\mathbb{R}^3$. Equation (14.3) in a spherical coordinate system (r, θ, ϕ) centered at the point M_0 has the form

$$\frac{1}{r^2}\frac{\partial}{\partial r}\left(r^2\frac{\partial u}{\partial r}\right) + \frac{1}{r^2 \sin\theta}\frac{\partial}{\partial \theta}\left(\sin\theta\frac{\partial u}{\partial \theta}\right) + \frac{1}{r^2 \sin^2\theta}\frac{\partial^2 u}{\partial \phi^2} + cu = 0.$$

Since we are interested in the solutions of this equation that depend only on the distance $r = r_{M_0 M}$ between the points M_0 and M, i.e., $u(M) = u(r)$, the equation simplifies to the form

$$\frac{1}{r^2}\frac{d}{dr}\left(r^2\frac{du}{dr}\right) + cu = 0 \tag{14.6}$$

or

$$\frac{d^2 u}{dr^2} + \frac{2}{r}\frac{du}{dr} + cu = 0.$$

Multiplying this equation by r and performing the obvious transformation, we obtain

$$\frac{d^2(ru)}{dr^2} + cru = 0.$$

Thus, the function $U = ru$ satisfies the equation

$$\frac{d^2 U}{dr^2} + cU = 0,$$

which has two linearly independent solutions

$$U_1(r) = e^{-ikr} \quad \text{and} \quad U_2(r) = e^{ikr},$$

where $k^2 = c$. Hence, Eq. (14.6) has two linearly independent solutions

$$u_1(r) = \frac{e^{-ikr}}{r} \quad \text{and} \quad u_2(r) = \frac{e^{ikr}}{r}.$$

Notice that when $r \to 0$ $(M \to M_0)$, both solutions have a singularity of the same order as the fundamental solution of the Laplacian (see Sect. 9.1). However, only one solution of (14.6) can represent a physical wave produced by a point source.

When $\operatorname{Im} k \neq 0$, it is easy to see that only one of these solutions is bounded at infinity. Indeed, suppose, e.g., that $k'' = \operatorname{Im} k > 0$. Then $(k' = \operatorname{Re} k)$

$$u_1(r) = \frac{e^{-i(k'+ik'')r}}{r} = e^{-ik'r}\frac{e^{k''r}}{r}$$

is unbounded at infinity, while

$$u_2(r) = e^{ik'r}\frac{e^{-k''r}}{r}$$

is bounded, and hence, no linear combination of $u_1(r)$ and $u_2(r)$ can represent a physically meaningful solution but only $au_2(r)$, where a is an arbitrary constant.

When $\operatorname{Im} k = 0$, both solutions $u_1(r)$ and $u_2(r)$ decay at infinity as $O(1/r)$, and in order to choose the physically meaningful one, we need to consider them in the corresponding physical context. For definiteness, let us consider a special case of Eq. (14.2), the wave equation

$$w_{tt} = a^2 \Delta w, \quad a \neq 0, a \in \mathbb{R}. \tag{14.7}$$

Its time-harmonic solutions have the form (substitution (14.4))

$$w(x, t) = u(x)e^{-i\omega t},$$

where the factor $u(x)$ satisfies the Helmholtz equation $\Delta u + cu = 0$ with $c = \omega^2/a^2$. Suppose that $\omega \in \mathbb{R}, \omega \neq 0$. Then $c > 0$. Take the solution $u_2(r)$. The corresponding solution of (14.7) has the form

$$w_2(r, t) = e^{-i\omega t}u_2(r) = \frac{1}{r}e^{-i\omega\left(t-\frac{r}{a}\right)} = \frac{1}{r}e^{ik(r-at)}.$$

Considering the surfaces on which the expression $k(r - at)$ remains constant in time, we see that it is a system of concentric spheres, the radius r of which grows in time with the speed $r = a$. Hence, the solution $u_2(r)$ corresponds to the system of outgoing spherical waves with the phase velocity a. Similarly, it can be established that $u_1(r)$ corresponds to the system of spherical waves incoming from infinity, and the concentric spheres converge into the point M_0. This solution is not physically meaningful for the study of wave propagation phenomena generated by sources located in a bounded region of $\mathbb{R}^3$. Notice that our analysis of the spherically symmetric solutions depended on the choice between the substitutions (14.4) and (14.5). If instead of (14.4) we would prefer (14.5), then $u_1(r)$ and not $u_2(r)$ would correspond to a physically meaningful solution.

14.3 Integral Representation for Solutions

In Sect. 9.2, the Green's identity for the Laplace operator was obtained. A similar formula is valid for the Helmholtz operator. Denote by $Lu = \Delta u + cu$ the Helmholtz operator. Then the second Green's identity leads to the equality

$$\int_V (uLv - vLu)\, dV = \int_\Sigma \left(u\frac{\partial v}{\partial \nu} - v\frac{\partial u}{\partial \nu} \right) d\sigma. \tag{14.8}$$

Substituting the fundamental solution $v = \frac{e^{ikr}}{r}$ and repeating the reasoning from Sect. 9.2 lead to the formula

$$u(M_0) = -\frac{1}{4\pi} \int_\Sigma \left(u\frac{\partial}{\partial \nu}\left(\frac{e^{ikr}}{r} \right) - \frac{e^{ikr}}{r}\frac{\partial u}{\partial \nu} \right) d\sigma$$

$$+ \frac{1}{4\pi} \int_V f(M)\frac{e^{ikr}}{r}\, dV, \quad M_0 \in V. \tag{14.9}$$

In particular, in a sourceless situation ($f \equiv 0$), we have

$$u(M_0) = -\frac{1}{4\pi} \int_\Sigma \left(u\frac{\partial}{\partial \nu}\left(\frac{e^{ikr}}{r} \right) - \frac{e^{ikr}}{r}\frac{\partial u}{\partial \nu} \right) d\sigma.$$

This formula implies that any solution of the homogeneous Helmholtz equation at interior points of the domain V possesses derivatives of any order, that is, $u \in C^\infty(V)$.

Let in (14.8) $Lv = 0$ and $Lu = -f$. Then (14.8) takes the form

$$0 = -\int_\Sigma \left(u\frac{\partial v}{\partial \nu} - v\frac{\partial u}{\partial \nu} \right) d\sigma + \int_V fv\, dV.$$

Multiplying this equality by $1/(4\pi)$ and adding it to (14.9) give

$$u(M_0) = \int_\Sigma \left(u\frac{\partial}{\partial \nu}G(M, M_0) - G(M, M_0)\frac{\partial u}{\partial \nu} \right) d\sigma$$

$$+ \int_V G(M, M_0)f(M)\, dV, \quad M_0 \in V,$$

where

$$G(M, M_0) = \frac{e^{ikr}}{r} + v(M, M_0).$$

Let us require that

$$G(M, M_0)|_\Sigma = 0 \quad \text{for } M \in \Sigma \text{ and } M_0 \in V. \tag{14.10}$$

Definition 237 A function $G(M, M_0)$ is called the Green's function of the Dirichlet problem for the equation $Lu = \Delta u + cu = 0$ in the domain $V \subset \mathbb{R}^3$, if it satisfies the following conditions:

(1) $LG(M, M_0) = 0$ in $V\backslash\{M_0\}$, i.e., $G(M, M_0)$ as a function of the coordinates $M = (x, y, z)$ is a solution of the homogeneous Helmholtz equation in V except for the point M_0, where it has a singularity as a fundamental solution.
(2) $G(M, M_0)|_\Sigma = 0$ for $M \in \Sigma$ and $M_0 \in V$.
(3) In the domain V, the function $G(M, M_0)$ admits the representation

$$G(M, M_0) = q(M, M_0) + v(M, M_0),$$

where $q(M, M_0)$ is a fundamental solution of the equation $Lu = 0$, e.g.,

$$q(M, M_0) = \frac{e^{ikr}}{4\pi r}, \quad r = r_{MM_0},$$

and $v(M, M_0)$ is a solution of $Lu = 0$, continuous in $\overline{V}$.

The existence of a Green's function of the Dirichlet problem is equivalent to the existence of a continuous function v satisfying the equation $Lu = 0$ in V and the boundary condition

$$v(M, M_0)|_\Sigma = -q(M, M_0) \quad \text{for } M \in \Sigma \text{ and } M_0 \in V.$$

In a similar way, the Green's function for the Neumann problem is defined. In this case, it is frequently called the **Neumann function** and denoted by $N(M, M_0)$. Instead of the boundary condition (14.10), the Neumann function satisfies the boundary condition

$$\frac{\partial}{\partial \nu_M} N(M, M_0)\bigg|_\Sigma = 0 \quad \text{for } M \in \Sigma \text{ and } M_0 \in V.$$

14.4   Interior Boundary Value Problems

Let $V \subset \mathbb{R}^3$ be a bounded domain with a Lyapunov boundary Σ. The following three main boundary value problems will be considered.

First Boundary Value Problem (Dirichlet Problem) To find a function $u(M)$ satisfying the following conditions $u \in C(\overline{V}) \cap C^2(V)$:

$$Lu(M) = \Delta u(M) + cu(M) = -f(M), \quad M \in V;$$

$$u(M) = \varphi(M), \quad M \in \Sigma.$$

Second Boundary Value Problem (Neumann Problem) To find a function $u(M)$ satisfying the following conditions $u \in C^1(\overline{V}) \cap C^2(V)$:

$$Lu(M) = -f(M), \quad M \in V;$$

$$\frac{\partial u(M)}{\partial \nu} = \psi(M), \quad M \in \Sigma.$$

Third Boundary Value Problem (Robin Problem) To find a function $u(M)$ satisfying the following conditions $u \in C^1(\overline{V}) \cap C^2(V)$:

$$Lu(M) = -f(M), \quad M \in V;$$

$$\frac{\partial u(M)}{\partial \nu} + h(M)u(M) = \chi(M), \quad M \in \Sigma.$$

In all three problems, the functions f, φ, ψ, χ, and h are supposed to be known and at least continuous functions.

The exterior boundary value problems are formulated analogously for the exterior domain $\Omega = \mathbb{R}^3 \backslash \overline{V}$ with an additional condition at infinity.

14.5   Maximum and Minimum Principles

The study of the formulated boundary value problems in the case $c < 0$ is based on the maximum and minimum principle.

Theorem 238 (Interior Extremum Principle) *For $c < 0$ and $f \leq 0\ (\geq 0)$, a solution of (14.1) cannot attain its local positive (negative) maximum (minimum) at an interior point of the domain V.*

Proof Suppose that a solution u of (14.1) attains its local positive maximum at a point $M_0 \in V$. Then at this point, due to the necessary condition of extremum, we have that

$$\Delta u(M_0) = u_{xx}(M_0) + u_{yy}(M_0) + u_{zz}(M_0) \leq 0,$$

and hence,

$$Lu(M_0) = \Delta u(M_0) + cu(M_0) < 0,$$

which contradicts the equality

$$Lu(M_0) = -f(M_0) \geq 0.$$

The impossibility of a local negative minimum in V is proved analogously.   ∎

Corollary 239 (Principle of Maximum and Minimum) *When $c < 0$, any continuous in $\overline{V}$ solution of the homogeneous Helmholtz equation $Lu = 0$ attains its largest positive value and smallest negative value only on the boundary Σ, and for all $M_0 \in \overline{V}$, the following bound is valid*

$$|u(M_0)| \leq \max_{M \in \Sigma} |u(M)|.$$

With the aid of these statements, in the case $c < 0$, the uniqueness and stability of the solutions of the Dirichlet problem for the equation $Lu = 0$ can be proved, as well as the uniqueness of the solution of the first, second, and third (with $h > 0$) boundary value problems for the equation $Lu = -f$.

For $c > 0$, the statement of Theorem 238 in general is not true, and the uniqueness of the boundary value problems stated above can be broken. Let us study some corresponding examples.

Example 240 Let V be the following cube in $\mathbb{R}^3$:

$$V = \left\{ (x, y, z) \in \mathbb{R}^3 \mid 0 < x, y, z < \frac{\pi}{k},\ k > 0 \right\}.$$

Consider the Helmholtz equation

$$\Delta u(M) + cu(M) = 0, \quad M \in V$$

with $c = 3k^2 > 0$.

The function

$$u(x, y, z) = \sin kx \sin ky \sin kz$$

is a solution of this equation. It attains its maximum (equal to 1) at an interior point $\left(\frac{\pi}{2k}, \frac{\pi}{2k}, \frac{\pi}{2k}\right)$ of V. Hence, in this case ($c > 0$), the statements of Theorem 238 and Corollary 239 are not true.

Example 241 Let $V = \{(x, y) \in \mathbb{R}^2 \mid 0 < x < a, 0 < y < b; a, b > 0\}$ be a rectangle. Let us find the values of λ, for which the homogeneous Dirichlet problem

$$\begin{cases} u_{xx} + u_{yy} + \lambda u = 0, & (x, y) \in V, \\ u|_{\Sigma = \partial V} = 0 \end{cases}$$

admits nontrivial solutions.

Applying the method of separation of variables $u(x, y) = X(x)Y(y)$, we obtain two Sturm–Liouville problems and eventually find out that for

$$\lambda = \lambda_{mn} = \left(\frac{m\pi}{a}\right)^2 + \left(\frac{n\pi}{b}\right)^2$$

the problem admits the nontrivial solutions

$$u_{mn}(x, y) = \sin\left(\frac{m\pi}{a}x\right)\sin\left(\frac{n\pi}{b}y\right).$$

Notice that all such values of λ (the Dirichlet eigenvalues) are positive that confirms that, in general, when $c > 0$ the uniqueness of the solution of the Dirichlet problem for the Helmholtz equation can be broken.

Example 242 Let $V = \{(x, y) \in \mathbb{R}^2 \mid x^2 + y^2 < R^2\}$ be a disk of radius R centered at the origin. Let us find the values of λ, for which the homogeneous Dirichlet problem

$$\begin{cases} u_{xx} + u_{yy} + \lambda u = 0, & (x, y) \in V, \\ u|_{r=R} = 0 \end{cases}$$

admits nontrivial solutions.

It is convenient to use polar coordinates and look for the solutions depending only on $r = \sqrt{x^2 + y^2}$. All such solutions have the form

$$u(r) = C_1 J_0(kr) + C_2 Y_0(kr),$$

where $J_0(kr)$ and $Y_0(kr)$ are Bessel functions of order zero, of the first and second kinds, respectively. Since we are interested in bounded solutions, C_2 must equal zero because $Y_0(kr) \to \infty$ when $r \to 0$. Hence, $u(r)$ is a solution of the homogeneous Dirichlet problem if the equality holds

$$J_0(kR) = 0. \tag{14.11}$$

The Bessel function $J_0(kR)$ has a countable set of positive zeros. Let α_n be the n-th zero: $J_0(\alpha_n) = 0$. Then $k_n = \alpha_n/R$ satisfies the equality (14.11), and hence, for all

$$\lambda_n = \left(\frac{\alpha_n}{R}\right)^2$$

the homogeneous Dirichlet problem admits the nontrivial solutions

$$u_n(r) = J_0(\sqrt{\lambda_n}r) = J_0(\frac{\alpha_n r}{R}).$$

Example 243 Let $B = \{(x, y, z) \in \mathbb{R}^3 \mid x^2 + y^2 + z^2 < R^2\}$ be a ball of radius R centered at the origin. Let us find the values of λ, for which the homogeneous Dirichlet problem

$$\begin{cases} \Delta u + \lambda u = 0, & (x, y, z) \in B, \\ u|_{r=R} = 0 \end{cases}$$

admits nontrivial solutions.

Let us look for spherically symmetric solutions (depending only on $r = \sqrt{x^2 + y^2 + z^2}$). Then we come to Eq. (14.6) with $c = \lambda$, the general solution of which can be written in the form

$$u(r) = C_1 \frac{\sin kr}{r} + C_2 \frac{\cos kr}{r}.$$

Since we are interested in bounded solutions, C_2 must equal zero because $\frac{\cos kr}{r}$ has a singularity at $r = 0$. Hence, $u(r)$ is a nontrivial solution of the homogeneous Dirichlet problem if $\sin kR = 0$, i.e., $k = n\pi/R$, $n = 1, 2, \ldots$. Thus, for

$$\lambda = \lambda_n = \left(\frac{n\pi}{R}\right)^2$$

the homogeneous Dirichlet problem admits the nontrivial solutions

$$u_n(r) = \frac{1}{r} \sin\left(\frac{n\pi r}{R}\right).$$

In a similar way, the eigenvalues and eigenfunctions can be found for the second and third boundary value problems in some simple domains.

14.6   Exterior Boundary Value Problems: Uniqueness

Let us consider Eq. (14.1) in an unbounded domain Ω exterior with respect to V, i.e., $\Omega = \mathbb{R}^3 \backslash \overline{V}$. In order to identify a unique solution of a boundary value problem in Ω with boundary conditions on $\Sigma = \partial \Omega$, it is necessary to impose additional conditions at infinity. When $c = -\varkappa^2 < 0$ (or more generally, $c = k^2$, $\mathrm{Im}\, k \neq 0$), such additional condition can be the requirement of the uniform convergence of the solution to zero at infinity.

Theorem 244 *Let $c = -\varkappa^2 < 0$. Then the first, second, and third (with $h > 0$) boundary value problems for Eq. (14.1) in the unbounded domain Ω can have at most one solution from the class of functions that at infinity tend to zero uniformly.*

Proof Suppose any of these problems possesses two different solutions $u_1(M)$ and $u_2(M)$ tending to zero uniformly, when $r \to +\infty$. Then their difference $u = u_1 - u_2$ is a solution of the homogeneous Helmholtz equation

$$\Delta u - \varkappa^2 u = 0 \quad \text{in } \Omega, \tag{14.12}$$

it satisfies a corresponding homogeneous boundary condition on Σ, and $u(M) \to 0$, when $r \to +\infty$. Let us prove that $u(M) \equiv 0$ in Ω.

Suppose that there exists such point $M_0 \in \Omega$ that $u(M_0) \neq 0$. Without loss of generality, let us suppose that $u(M_0) > 0$. Since $u(M) \to 0$, when $r \to +\infty$, for any $\varepsilon > 0$, there exists a ball B_R of a radius R large enough, centered at the origin, such that $M_0 \in B_R$, and $|u(M)| < \varepsilon$ for all $M \in \Omega \backslash B_R$. Let $G = \Omega \cap B_R$. Then the function $u(M)$ is a solution of (14.12) in G. It is continuous on $\overline{G}$. Hence, due to Corollary 239, the positive global $\max_{M \in \overline{G}} u(M)$ can be attained only on the spherical surface ∂B_R, where $|u(M)| < \varepsilon$. Thus, $|u(M)| < \varepsilon$ everywhere in $\overline{G}$. The number ε can be chosen being less than $u(M_0)$, and hence, we obtain a contradiction. Thus, $u(M) \equiv 0$ in Ω. ∎

Let us study the case $c = k^2 > 0$. To explain the difference with the previous case, we begin with an example. Let Ω be the complement of the ball B from Example 243. Then the homogeneous Helmholtz equation $\Delta u + k^2 u = 0$ admits the linearly independent solutions

$$u_1(r) = \frac{\sin kr}{r} \quad \text{and} \quad u_2(r) = \frac{\cos kr}{r}.$$

Both decay at infinity as $O(\frac{1}{r})$. The first one is identically zero on the sphere $r = R$ for $k_n = \pi n / R$, $n \in \mathbb{N}$, and the second one is identically zero on the sphere $r = R$

for $k_m = \pi (1 + 2m)/(2R)$, $m \in \mathbb{N}$. This example shows that the requirement of a uniform convergence to zero at infinity is insufficient for the uniqueness of the solution. Indeed, considering the Dirichlet problem in Ω for the homogeneous Helmholtz equation $\Delta u + k^2 u = 0$ with k from the set $\{k_n = \pi n/R\}_{n=1}^\infty$, we find out that besides the trivial solution $u \equiv 0$ in Ω, there exists an infinite number of nontrivial solutions $u(r) = a \frac{\sin kr}{r}$, where a is an arbitrary constant. All of them tend to zero uniformly at infinity.

To simplify the study that follows, let us consider Eq. (14.1) in the whole space $\mathbb{R}^3$ and suppose that the right side $f(M)$ is a compactly supported function, i.e., it equals zero outside some bounded domain V. As was discussed above, the solution of (14.1) for $c = k^2 > 0$ can be regarded as the amplitude of a time-harmonic wave generated by periodic in time sources (or external forces) $f(M)e^{-i\omega t}$ with the amplitude $f(M)$. Since $f(M)$ is compactly supported, all the sources are located in a bounded region V. Far away from this region, the waves behave as outgoing spherical waves. Hence, a physical condition for identifying a unique physically meaningful solution of the equation $\Delta u + k^2 u = -f$ in the whole space $\mathbb{R}^3$ is to find a solution corresponding at infinity to outgoing spherical waves.

Now let us formalize this physical requirement mathematically. Consider the spherical waves, i.e., the spherically symmetric solutions of the homogeneous wave equation in $\mathbb{R}^3$:

$$\frac{1}{a^2} \frac{\partial^2 U}{\partial t^2} = \frac{1}{r^2} \frac{\partial}{\partial r} \left(r^2 \frac{\partial U}{\partial r} \right), \tag{14.13}$$

where r is the distance with the respect to the origin. This equation can be written as a one-dimensional wave equation for the function rU:

$$\frac{1}{a^2} \frac{\partial^2 (rU)}{\partial t^2} = \frac{\partial^2}{\partial r^2} (rU).$$

Its general solution has the form

$$V = rU = f_1(r - at) + f_2(r + at),$$

where $f_1, f_2 \in C^2(\mathbb{R})$. The functions

$$V_1(r, t) = f_1(r - at) \quad \text{and} \quad V_2(r, t) = f_2(r + at)$$

represent general solutions of the partial differential equations of first order:

$$\frac{\partial V_1}{\partial r} + \frac{1}{a} \frac{\partial V_1}{\partial t} = 0 \quad \text{and} \quad \frac{\partial V_2}{\partial r} - \frac{1}{a} \frac{\partial V_2}{\partial t} = 0. \tag{14.14}$$

Thus, Eq. (14.13) has the solutions

$$U_1(r, t) = \frac{f_1(r - at)}{r} \quad \text{and} \quad U_2(r, t) = \frac{f_2(r + at)}{r}.$$

The first one represents an outgoing (to infinity) spherical wave, while the second one represents an incoming (from infinity) spherical wave.

Let us apply the partial differential expressions from (14.14) to the functions $U_1(r, t)$ and $U_2(r, t)$. We have

$$\frac{\partial U_1}{\partial r} + \frac{1}{a} \frac{\partial U_1}{\partial t} = -\frac{f_1(r - at)}{r^2}$$

and

$$\frac{\partial U_2}{\partial r} - \frac{1}{a} \frac{\partial U_2}{\partial t} = -\frac{f_2(r + at)}{r^2}.$$

From here and under the assumption of boundedness of the functions f_1 and f_2, we obtain

$$\frac{\partial U_1}{\partial r} + \frac{1}{a} \frac{\partial U_1}{\partial t} = O\left(\frac{1}{r^2}\right) = o\left(\frac{1}{r}\right), \quad r \to +\infty$$

and

$$\frac{\partial U_2}{\partial r} - \frac{1}{a} \frac{\partial U_2}{\partial t} = O\left(\frac{1}{r^2}\right) = o\left(\frac{1}{r}\right), \quad r \to +\infty.$$

If we consider the time-harmonic waves with the dependence on time as $e^{-i\omega t}$:

$$U_{1,2}(r, t) = u_{1,2}(r)e^{-i\omega t}, \quad k = \omega/a,$$

then we obtain

$$\frac{\partial u_1}{\partial r} - iku_1 = o\left(\frac{1}{r}\right), \quad \text{when } r \to +\infty$$

for the outgoing waves, and

$$\frac{\partial u_2}{\partial r} + iku_2 = o\left(\frac{1}{r}\right), \quad \text{when } r \to +\infty$$

for the incoming waves.

Since waves generated by sources located in a bounded region V at infinity are similar to outgoing spherical waves, it is natural to propose the mathematical condition identifying the outgoing waves in the form

Fig. 14.1 Illustration to
(14.16)

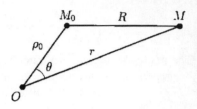

$$\frac{\partial u}{\partial r} - iku = o\left(\frac{1}{r}\right), \qquad \text{when } r \to +\infty, \tag{14.15}$$

which is called the **Sommerfeld radiation condition**.

Among the fundamental solutions of the equation $\Delta u + k^2 u = 0$, only the solution

$$q(M, M_0) = u_1(R) = \frac{e^{ikR}}{R}$$

satisfies (14.15). Here $R = r_{MM_0} = |M - M_0|$. Let us verify this fact.

We consider a point source located at some point $M_0 \in V$. The amplitude of the generated spherical wave is given by $u_1(R) = \frac{e^{ikR}}{R}$, where the distance R can be written as follows:

$$R = \sqrt{r^2 + \rho_0^2 - 2r\rho_0 \cos\theta}, \tag{14.16}$$

where ρ_0 is the distance of M_0 from the origin, r is the distance of M from the origin, and θ is the angle between the rays OM_0 and OM (see Fig. 14.1).

Hence,

$$R = r + O(1), \quad \frac{\partial R}{\partial r} = \frac{r - \rho_0 \cos\theta}{R} = 1 + O\left(\frac{1}{r}\right).$$

Taking into account these asymptotic relations, we obtain

$$\left(\frac{\partial}{\partial r} - ik\right)\frac{e^{ikR}}{R} = \frac{\partial}{\partial R}\left(\frac{e^{ikR}}{R}\right)\frac{\partial R}{\partial r} - ik\frac{e^{ikR}}{R} \tag{14.17}$$

$$= \left(ik - \frac{1}{R}\right)\frac{e^{ikR}}{R}\frac{\partial R}{\partial r} - ik\frac{e^{ikR}}{R}$$

$$= \left(ik - \frac{1}{R}\right)\frac{e^{ikR}}{R}\left(1 + O\left(\frac{1}{r}\right)\right) - ik\frac{e^{ikR}}{R}$$

$$= -\frac{e^{ikR}}{R^2} + O\left(\frac{1}{r^2}\right) = O\left(\frac{1}{r^2}\right) = o\left(\frac{1}{r}\right).$$

Theorem 245 *Let the support of the function* f *be contained in a bounded domain* V, *and* $f \in C^1(\overline{V})$. *Then there exists a unique solution of the inhomogeneous Helmholtz equation*

$$\Delta u + k^2 u = -f \quad in \ \mathbb{R}^3 \tag{14.18}$$

satisfying the Sommerfeld radiation condition at infinity (14.15). It is defined by the volume potential

$$u(M) = \frac{1}{4\pi} \int_V f(P) \frac{e^{ikR}}{R} dV_P, \quad R = r_{MP}. \tag{14.19}$$

Proof The fact that the function (14.19) is a solution of (14.18) is proved similarly to Theorem 139 from Sect. 11.1. Moreover, it tends to zero at infinity as $1/r$. Let us prove that it satisfies the Sommerfeld radiation condition. With the aid of (14.17), we obtain

$$\left(\frac{\partial}{\partial r} - ik\right) u(M) = \frac{1}{4\pi} \int_V f(P) \left(\frac{\partial}{\partial r} - ik\right) \frac{e^{ikR}}{R} dV_P$$

$$= \frac{1}{4\pi} \int_V f(P) o\left(\frac{1}{r}\right) dV_P = o\left(\frac{1}{r}\right).$$

Thus, the function ((14.19) is a solution of (14.18) and satisfies the Sommerfeld radiation condition (14.15).

It remains to prove its uniqueness. Suppose the opposite that there exist two different solutions of the problems $u_1(M)$ and $u_2(M)$. Then their difference $u = u_1 - u_2$ is a solution of the homogeneous Helmholtz equation $\Delta u + k^2 u = 0$ in $\mathbb{R}^3$ and satisfies the Sommerfeld radiation condition (14.15). Let M_0 be an arbitrary fixed point of $\mathbb{R}^3$. Consider the ball B_ρ of radius $\rho > 0$ centered at the origin and containing the point M_0. Then from (14.9), we obtain

$$u(M_0) = -\frac{1}{4\pi} \int_{\Sigma_\rho} \left(u \frac{\partial}{\partial \nu} \left(\frac{e^{ikR}}{R}\right) - \frac{e^{ikR}}{R} \frac{\partial u}{\partial \nu}\right) d\sigma,$$

where $\Sigma_\rho = \partial B_\rho$ and $R = r_{MM_0}$, $M \in \Sigma_\rho$. Let us estimate the integrand when $\rho \to +\infty$. Taking into account the Sommerfeld radiation condition (14.15) and (14.17), we have

$$\left(u \frac{\partial}{\partial \nu} \left(\frac{e^{ikR}}{R}\right) - \frac{e^{ikR}}{R} \frac{\partial u}{\partial \nu}\right)\bigg|_{\Sigma_\rho} = \left(u \frac{\partial}{\partial r} \left(\frac{e^{ikR}}{R}\right) - \frac{e^{ikR}}{R} \frac{\partial u}{\partial r}\right)\bigg|_{\Sigma_\rho}$$

$$= \left(u \left(ik - \frac{1}{R}\right) \frac{e^{ikR}}{R} \left(1 + o\left(\frac{1}{r}\right)\right)\right)$$

$$-\frac{e^{ikR}}{R}\left(iku + O\left(\frac{1}{r^2}\right)\right)\Big|_{r=\rho}$$

$$= \left(u\frac{e^{ikR}}{R}\left(O\left(\frac{1}{r}\right)\left(ik - \frac{1}{R}\right) - \frac{1}{R}\right) - \frac{e^{ikR}}{R}O\left(\frac{1}{r^2}\right)\right)\Big|_{r=\rho}$$

$$= O\left(\frac{1}{r^3}\right)\Big|_{r=\rho} = O\left(\frac{1}{\rho^3}\right), \quad \text{when } \rho \to +\infty.$$

Hence,

$$|u(M_0)| \le \frac{1}{4\pi}\int_{\Sigma_\rho}\frac{C}{r^3}d\sigma = \frac{C \cdot 4\pi\rho^2}{4\pi\rho^3} = \frac{C}{\rho} \to 0, \quad \text{when } \rho \to +\infty.$$

Here, C is a positive constant.

Thus, $u(M_0) = 0$, i.e., $u_1(M_0) = u_2(M_0)$. Due to the arbitrariness of the point M_0, we obtain that $u_1(M_0) = u_2(M_0)$ for all $M_0 \in \mathbb{R}^3$. ∎

Remark 246 The Sommerfeld radiation condition (14.15) corresponds to the chosen dependence on time $e^{-i\omega t}$. In case of the choice $e^{i\omega t}$, the corresponding radiation condition identifying the outgoing waves has the form

$$\frac{\partial u}{\partial r} + iku = o\left(\frac{1}{r}\right), \quad \text{when } r \to +\infty.$$

Remark 247 In the two-dimensional case, the radiation conditions have the form

$$\frac{\partial u}{\partial r} - iku = o\left(\frac{1}{\sqrt{r}}\right), \quad \text{when } r \to +\infty$$

for the dependence on time $e^{-i\omega t}$, and

$$\frac{\partial u}{\partial r} + iku = o\left(\frac{1}{\sqrt{r}}\right), \quad \text{when } r \to +\infty$$

for the dependence on time $e^{i\omega t}$.

Remark 248 The Sommerfeld radiation conditions are applicable when the Helmholtz equation is studied in the whole space or in a complement of a bounded domain, i.e., when the boundary Σ is located in a bounded region. If the boundary goes to infinity, in general, other types of radiation conditions associated with a problem under consideration should be formulated.

We finish this chapter with a theorem establishing the uniqueness of a general exterior boundary value problem for the Helmholtz equation in a dissipative medium. The wave propagation in a dissipative medium is modeled by the equation

$$\Delta U = \frac{1}{a^2} U_{tt} + \beta U_t - F,$$

where the coefficient $\beta > 0$ characterizes the dissipation in the medium. Supposing that the source function $F(M, t)$ is time-harmonic, $F(M, t) = f(M)e^{-i\omega t}$, we look for the solutions in the form

$$U(M, t) = u(M)e^{-i\omega t}.$$

Then the function $u(M)$ satisfies the Helmholtz equation

$$\Delta u(M) + s^2 u(M) = -f(M)$$

with the complex coefficient $s^2 = k^2 + i\omega\beta$, $k = \omega/a$.

Theorem 249 *The exterior boundary value problem*

$$\Delta u(M) + s^2 u(M) = -f(M), \quad M \in \Omega, \ s^2 = k^2 + i\omega\beta, \ \beta > 0,$$

$$\left(\alpha \frac{\partial u}{\partial \nu} + hu\right)\Big|_\Sigma = \varphi(M), \quad M \in \Sigma,$$

where α and h are the real valued functions defined on Σ, and $|\alpha| + |h| > 0$ can have at most one solution in the class of functions tending to zero uniformly at infinity.

Proof Let $u(M)$ be a solution of the corresponding homogeneous problem, tending to zero uniformly at infinity. Let B_ρ be a ball of a large enough radius $\rho > 0$, centered at the origin and containing the surface Σ. Let $\Omega_\rho = B_\rho \cap \Omega$. The boundary of Ω_ρ is $\partial\Omega_\rho = \Sigma \cup \Sigma_\rho$, where $\Sigma_\rho = \partial B_\rho$. Let us consider the following identity in Ω_ρ:

$$0 = \bar{u}\left(\Delta u + s^2 u\right) = \bar{u}\left(u_{xx} + u_{yy} + u_{zz}\right) + s^2 |u|^2$$

$$= (\bar{u}u_x)_x + (\bar{u}u_y)_y + (\bar{u}u_z)_z - \bar{u}_x u_x - \bar{u}_y u_y - \bar{u}_z u_z + s^2 |u|^2$$

$$= (\bar{u}u_x)_x + (\bar{u}u_y)_y + (\bar{u}u_z)_z - |\text{grad } u|^2 + s^2 |u|^2,$$

where $\bar{u}$ is the complex conjugate of u. Integrating this equality and applying the Gauss–Ostrogradsky formula, we obtain

$$\int_{\Sigma \cup \Sigma_\rho} \bar{u}\frac{\partial u}{\partial \nu} d\sigma - \int_{\Omega_\rho}\left(|\text{grad } u|^2 - s^2 |u|^2\right) d\Omega_\rho = 0.$$

Consider the imaginary part of this equality

$$\text{Im} \int_\Sigma \bar{u} \frac{\partial u}{\partial v} d\sigma - \text{Im} \int_{\Sigma_\rho} \bar{u} \frac{\partial u}{\partial v} d\sigma + \omega\beta \int_{\Omega_\rho} |u|^2 d\Omega_\rho = J_1 + J_2 + J_3 = 0. \quad (14.20)$$

In the case of the first and second boundary value problems, the surface integral

$$J_1 = \text{Im} \int_\Sigma \bar{u} \frac{\partial u}{\partial v} d\sigma = 0,$$

while in the case of the third problem, due to the boundary condition (in which we suppose that $\alpha \neq 0$ because otherwise this would be the first problem), we have

$$J_1 = \text{Im} \int_\Sigma \bar{u} \left(-\frac{h}{\alpha} u \right) d\sigma = 0.$$

The integral

$$J_2 = -\text{Im} \int_{\Sigma_\rho} \bar{u} \frac{\partial u}{\partial v} d\sigma$$

tends to zero when $\rho \to +\infty$ because

$$\left| \bar{u} \frac{\partial u}{\partial v} \right|_{\Sigma_\rho} = \left| \bar{u} \frac{\partial u}{\partial \rho} \right|_{\Sigma_\rho} \leq C \frac{e^{-2\,\text{Re}\,s\rho}}{\rho^2}.$$

Thus, letting $\rho \to +\infty$ in (14.20), we obtain

$$J_3 = \omega\beta \int_{\Omega_\rho} |u|^2 d\Omega_\rho = 0,$$

which implies that $u \equiv 0$ in Ω. ∎

14.6.1 Existence

Let us return to the integral representation (14.9) for any solution of the inhomogeneous Helmholtz equation in a bounded domain V:

$$u(M_0) = \int_\Sigma q(M, M_0) \frac{\partial u(M)}{\partial v_M} d\sigma_M - \int_\Sigma \frac{\partial q(M, M_0)}{\partial v_M} u(M) d\sigma_M$$

$$+ \int_V f(M) q(M, M_0) dV_M. \quad (14.21)$$

Since the fundamental solution

$$q(M, M_0) = \frac{e^{ikr_{MM_0}}}{4\pi r_{MM_0}} = \frac{1}{4\pi r_{MM_0}} + \frac{ik}{4\pi} + \frac{(ik)^2 r_{MM_0}}{8\pi} + \dots$$

possesses the same singularity as the fundamental solution for the Laplace equation, the properties of the integral operators in (14.21) are similar to the properties of the single layer, double layer, and volume potentials that were studied in detail in Chap. 11. Theorems proved there on the boundedness, continuity and jump discontinuity of the corresponding potentials remain valid in the case of their generalizations corresponding to the Helmholtz equation. Hence, we will use these facts here without proofs.

Note that the volume potential

$$w(M_0) = \int_V f(M)q(M, M_0)dV_M \tag{14.22}$$

gives us a particular solution of the inhomogeneous Helmholtz equation, satisfying the Sommerfeld condition (see Theorem 245). Thus, with its aid, the study of a problem for the inhomogeneous equation reduces to the study of a corresponding problem for a homogeneous equation. For example, if we are interested in solving an interior Dirichlet problem for the inhomogeneous equation

$$\begin{cases} (\Delta + c)u = -f & \text{in } V \\ u|_\Sigma = \varphi, \end{cases}$$

then considering

$$u = v + w$$

(where w is the volume potential (14.22)), for the function v, we obtain the following problem for the homogeneous Helmholtz equation:

$$\begin{cases} (\Delta + c)v = 0 & \text{in } V \\ v|_\Sigma = (\varphi - w)|_\Sigma. \end{cases}$$

Thus, in what follows, we restrict our consideration to problems for the homogeneous Helmholtz equation.

The following analogue of the integral representation (14.21) is valid for the exterior domains.

Theorem 250 *Let $u \in C^2(\Omega) \cap C^1(\overline{\Omega})$ be a solution of the Helmholtz equation*

$$(\Delta + c)u = 0 \quad \text{in } \Omega \tag{14.23}$$

satisfying the Sommerfeld condition at infinity. Then

$$u(M_0) = -\int_{\Sigma} q(M, M_0) \frac{\partial u(M)}{\partial \nu_M} d\sigma_M + \int_{\Sigma} \frac{\partial q(M, M_0)}{\partial \nu_M} u(M) d\sigma_M$$

for all $M_0 \in \Omega$.

An immediate corollary of this theorem is the following Atkinson's theorem.

Theorem 251 *Let $u \in C^2(\Omega)$ be a solution of (14.23) satisfying the Sommerfeld condition at infinity. Let $R_0 > 0$ be such that $\Sigma_{R_0} := \{ M \in \mathbb{R}^3 : |M| = R_0 \} \subset \Omega$, and r, θ, ϕ spherical coordinates of the point M. Then $u(M)$ admits the series representation*

$$u(M) = \frac{e^{ikr}}{r} \sum_{n=0}^{\infty} \frac{F_n(\theta, \phi)}{r^n}, \qquad (14.24)$$

which is valid for $r \geq R_0$, and the series converges absolutely and uniformly with respect to r, θ, ϕ. The coefficients $F_n(\theta, \phi)$ are defined in Corollary 253 below. The series can be differentiated termwise with respect to r, θ, ϕ any number of times, and the resulting series converges absolutely and uniformly as well.

For the detailed proof, we refer to [24].

Corollary 252 *Any solution u of the Helmholtz equation in Ω satisfying the Sommerfeld condition at infinity has the asymptotic behavior*

$$u(M) = \frac{e^{ikr}}{r} F_0(\theta, \phi) + O\left(\frac{1}{r^2}\right), \qquad r \to \infty.$$

The function $F_0(\theta, \phi)$ is called the **radiation pattern** or **far-field pattern** of u.

Corollary 253 *The coefficients $F_n(\theta, \phi)$ in (14.24) are recursively determined in terms of $F_0(\theta, \phi)$ by the formula*

$$2ikn F_n = n(n-1) F_{n-1} + B F_{n-1}, \qquad n = 1, 2, \ldots,$$

where

$$B := \frac{1}{\sin\theta} \frac{\partial}{\partial \theta} \left(\sin\theta \frac{\partial}{\partial \theta} \right) + \frac{1}{\sin^2\theta} \frac{\partial^2}{\partial \phi^2}$$

*is **Beltrami's operator** for the sphere.*

The proof is obtained by substituting the series (14.24) into the Helmholtz equation, differentiating termwise and equating the coefficients at equal powers of $1/r$.

Corollary 254 *Let $u \in C^2(\Omega)$ be a solution of (14.23) satisfying the Sommerfeld condition at infinity, for which the far-field pattern vanishes identically. Then $u \equiv 0$ in Ω.*

For the proof, we refer to [24, p. 75].

Theorems on the existence of solutions to boundary value problems are proved usually by reducing the problem to a boundary integral equation the solvability of which is then studied. The reduction to the boundary integral equations is given by the following two theorems.

Theorem 255 *The double layer potential*

$$u(M_0) = \int_\Sigma \frac{\partial q(M, M_0)}{\partial \nu_M} \psi(M) d\sigma_M, \quad M_0 \in \mathbb{R}^3 \backslash \Sigma \qquad (14.25)$$

with continuous density ψ is a solution of the interior Dirichlet problem

$$\begin{cases} (\Delta + c) u = 0 & \text{in } V, \\ u|_\Sigma = \varphi, & \varphi \in C(\Sigma) \end{cases}$$

provided ψ is a solution of the integral equation

$$\psi(M_0) - 2 \int_\Sigma \frac{\partial q(M, M_0)}{\partial \nu_M} \psi(M) d\sigma_M = -2\varphi(M_0), \quad M_0 \in \Sigma.$$

It solves the exterior Dirichlet problem

$$\begin{cases} (\Delta + c) u = 0 & \text{in } \Omega = \mathbb{R}^3 \backslash \overline{V}, \\ u|_\Sigma = \varphi, & \varphi \in C(\Sigma), \\ \frac{\partial u}{\partial r} - iku = o\left(\frac{1}{r}\right), & \text{when } r \to +\infty \end{cases}$$

provided ψ is a solution of the integral equation

$$\psi(M_0) + 2 \int_\Sigma \frac{\partial q(M, M_0)}{\partial \nu_M} \psi(M) d\sigma_M = 2\varphi(M_0), \quad M_0 \in \Sigma.$$

Proof The function u defined by (14.25) is a solution of the Helmholtz equation in $\mathbb{R}^3 \backslash \Sigma$ and satisfies the Sommerfeld condition at infinity. Moreover, the following formulas (analogous to those from Theorem 157) for the limit values on the boundary of the double layer potential are valid

$$u_\pm(M_0) = \int_\Sigma \frac{\partial q(M, M_0)}{\partial \nu_M} \psi(M) d\sigma_M \mp \frac{1}{2} \psi(M_0), \quad M_0 \in \Sigma,$$

which show that the double layer potential u is a solution of the respective Dirichlet problem if and only if ψ satisfies the respective boundary integral equation.   ∎

Analogously, the following theorem reducing the Neumann problems to the corresponding boundary integral equations is obtained.

Theorem 256 *The single layer potential*

$$u(M_0) = \int_\Sigma q(M, M_0)\psi(M)d\sigma_M, \quad M_0 \in \mathbb{R}^3 \backslash \Sigma$$

with continuous density ψ is a solution of the interior Neumann problem

$$\begin{cases} (\Delta + c)u = 0 & in\ V, \\ \frac{\partial u}{\partial v}\big|_\Sigma = \varphi, & \varphi \in C(\Sigma) \end{cases}$$

provided ψ is a solution of the integral equation

$$\psi(M_0) + 2\int_\Sigma \frac{\partial q(M, M_0)}{\partial v_{M_0}}\psi(M)d\sigma_M = 2\varphi(M_0), \quad M_0 \in \Sigma.$$

It solves the exterior Neumann problem

$$\begin{cases} (\Delta + c)u = 0 & in\ \Omega = \mathbb{R}^3 \backslash \overline{V}, \\ \frac{\partial u}{\partial v}\big|_\Sigma = \varphi, & \varphi \in C(\Sigma), \\ \frac{\partial u}{\partial r} - iku = o\left(\frac{1}{r}\right), & when\ r \to +\infty \end{cases}$$

provided ψ is a solution of the integral equation

$$\psi(M_0) - 2\int_\Sigma \frac{\partial q(M, M_0)}{\partial v_{M_0}}\psi(M)d\sigma_M = -2\varphi(M_0), \quad M_0 \in \Sigma.$$

For the detailed proof, we refer, e.g., to [24].

The theory of the boundary integral equations from Theorems 255 and 256 is developed in detail (see, e.g., [24, Sect. 3.4]). Based on this theory, the following theorem on the solvability and uniqueness of the boundary value problems is proved:

Theorem 257

(1) The exterior Dirichlet and Neumann problems are uniquely solvable.
(2) The interior Dirichlet problem is solvable if and only if

$$\int_\Sigma \varphi \frac{\partial w}{\partial v}d\Sigma = 0$$

for all solutions w to the homogeneous interior Dirichlet problem.
(3) The interior Neumann problem is solvable if and only if

$$\int_{\Sigma} \varphi w \, d\Sigma = 0$$

for all solutions w to the homogeneous interior Neumann problem.

For the detailed proof, we refer to [24, Sect. 3.4].

Remark 258 Methods of the potential theory are well developed for a much more general class of elliptic equations with variable coefficients. Although the fundamental solutions that appear in the construction of the potential operators are not available in explicit form, their basic properties required for the proof of the main theorems are well studied. Thus, theorems analogous to Theorem 257 are known for general elliptic equations (especially those of second order), see, e.g., [8, Chapter 5]. In particular, the result that we will use below establishes that for a strictly positive function q that is continuous in $\overline{V}$, the interior Dirichlet and Neumann problems for the equation $(\Delta - q) u = 0$ are uniquely solvable.

Additionally to the boundary value problems stated for an interior or exterior domain separately, an important role in practical applications belongs to the **transmission problem** involving both the interior and exterior domains. It describes the wave propagation in a space containing two different material media.

Transmission Problem Find $u_1 \in C^2(V) \cap C(\overline{V})$ and $u_2 \in C^2(\Omega) \cap C(\overline{\Omega})$, $\Omega = \mathbb{R}^3 \setminus \overline{V}$ such that

$$\left(\Delta + k_1^2\right) u_1 = 0 \quad \text{in } V,$$

$$\left(\Delta + k_2^2\right) u_2 = 0 \quad \text{in } \Omega,$$

u_2 satisfies the Sommerfeld condition (14.15) at infinity, and u_1 and u_2 satisfy the

transmission conditions on the boundary

$$\begin{cases} \mu_2 u_2 - \mu_1 u_1 = f, \\ \frac{\partial u_2}{\partial \nu} - \frac{\partial u_1}{\partial \nu} = g \end{cases} \quad \text{on } \Sigma,$$

where k_1, k_2, μ_1, μ_2 are some given complex numbers; $f, g \in C(\Sigma)$.

With the aid of the techniques based on the surface potentials, the following theorem is proved.

Theorem 259 *Let $k_1, k_2, \mu_1, \mu_2 > 0$. Then the transmission problem is uniquely solvable.*

For the proof, we refer to [24, Sect. 3.8]. For an extension of this theorem onto the case of complex values of the numerical parameters involved, we refer to [61].

Chapter 15
Method of Non-orthogonal Series

In this chapter, we explain the idea of one of the simplest, direct, and efficient numerical approaches for solving boundary value problems of mathematical physics. The approach is often referred to as the method of discrete sources (see, e.g., [28]) or the method of non-orthogonal series (see, e.g., [35]). The idea of the method will be explained on the example of the Dirichlet problem for the Helmholtz equation.

Suppose that an infinite system of solutions u_k of the Helmholtz equation is known, such that any solution u of the Helmholtz equation in a domain V can be approximated arbitrarily closely by a finite linear combination of the functions u_k with respect to some appropriate norm. That is, for any $\varepsilon > 0$, there exists such number $N \in \mathbb{N}$ and constants a_k that

$$\left\| u - \sum_{k=1}^{N} a_k u_k \right\|_V < \varepsilon.$$

In this case, we say that the system of functions $\{u_k\}_{k=1}^{\infty}$ is a **complete system of solutions** of the Helmholtz equation in V with respect to the chosen norm.

Then it is natural to look for an approximate solution of the Dirichlet problem in the form

$$u_N = \sum_{k=1}^{N} a_k u_k.$$

Since each u_k is a solution of the Helmholtz equation and the equation is linear, any such linear combination u_N is an exact solution of the Helmholtz equation. Thus, in order to obtain an approximate solution of the Dirichlet problem, one needs to find such constants a_k, $k = 1, 2, \ldots, N$, that the boundary condition $u = \varphi$ on Σ be fulfilled approximately: $u_N \cong \varphi$. For this purpose, there are developed several numerical methods. Perhaps, the simplest one is the collocation method. Take N

A. N. Karapetyants, V. V. Kravchenko, *Methods of Mathematical Physics*,
https://doi.org/10.1007/978-3-031-17845-0_15

different points M_j, $j = 1, 2, \ldots, N$, on the boundary Σ and require the fulfillment of the boundary condition at these points:

$$\sum_{k=1}^{N} a_k u_k(M_j) = \varphi(M_j), \quad j = 1, 2, \ldots, N. \tag{15.1}$$

This gives us an $N \times N$ system of linear algebraic equations for computing the numbers a_k, $k = 1, 2, \ldots, N$, which in turn leads to an approximate solution u_N of the problem. Obviously, with natural modifications, the same approach can be applied to all boundary value problems considered in the previous chapter. In the case of exterior problems, it is convenient to dispose of a complete system of solutions satisfying the Sommerfeld condition at infinity.

Thus, the necessary ingredient of the method of non-orthogonal series is a complete system of solutions that should be complete on the boundary Σ in the sense that any given function $\varphi \in L_2(\Sigma)$ can be approximated arbitrarily closely by a linear combination of the functions u_k:

$$\left\| \varphi - \sum_{k=1}^{N} a_k u_k \right\|_{L_2(\Sigma)} < \varepsilon.$$

Then using the stability of the problem with respect to the boundary data, one can show that

$$\| u - u_N \| < C\varepsilon, \quad C = \text{Const}$$

in the domain under consideration. As an example, let us prove this fact in the case of the exterior Dirichlet problem

$$\begin{cases} \left(\Delta + k^2 \right) u = 0 & \text{in } \Omega = \mathbb{R}^3 \backslash \overline{V}, \\ u|_\Sigma = \varphi, & \varphi \in C\left(\Sigma \right), \\ \frac{\partial u}{\partial r} - iku = o\left(\frac{1}{r} \right), & \text{when } r \to +\infty. \end{cases} \tag{15.2}$$

Let $\{u_n\}_{n=1}^{\infty}$ be a system of solutions of the Helmholtz equation satisfying the Sommerfeld condition at infinity, and the system $\{\varphi_n = u_n|_\Sigma\}_{n=1}^{\infty}$ of their boundary traces be complete in $L_2(\Sigma)$. Consider a function $u_N = \sum_{n=1}^{N} a_n u_n$, where the number N and the coefficients $a_1, \ldots, a_N$ are chosen such that the boundary data φ is approximated with a prescribed accuracy $\varepsilon > 0$:

$$\left\| \varphi - \sum_{n=1}^{N} a_n \varphi_n \right\|_{L_2(\Sigma)} < \varepsilon.$$

Let us show that on any closed subdomain $\overline{\Omega}_s$ of the domain Ω, the function u_N approximates uniformly the exact solution of the problem.

Indeed, the solution of (15.2) can be written in the form of a double layer potential

$$u(M) = -\int_\Sigma \frac{\partial G(M, P)}{\partial \nu_P} \varphi(P) d\Sigma_P,$$

where $G(M, P)$ is a Green's function of the Dirichlet problem in the domain Ω. Analogously,

$$u_N(M) = -\int_\Sigma \frac{\partial G(M, P)}{\partial \nu_P} u_N(P) d\Sigma_P.$$

Hence, with the aid of Cauchy–Bunyakovsky–Schwarz inequality, we obtain

$$|u(M) - u_N(M)| \leq \int_\Sigma \left| \frac{\partial G(M, P)}{\partial \nu_P} \right| |\varphi(P) - u_N(P)| d\Sigma_P$$

$$\leq \left(\int_\Sigma \left| \frac{\partial G(M, P)}{\partial \nu_P} \right|^2 d\Sigma_P \right)^{1/2} \left(\int_\Sigma \left| \varphi(P) - \sum_{n=1}^N a_n \varphi_n(P) \right|^2 d\Sigma_P \right)^{1/2}.$$

In $\overline{\Omega}_s$, we have

$$\left(\int_\Sigma \left| \frac{\partial G(M, P)}{\partial \nu_P} \right|^2 d\Sigma_P \right)^{1/2} < C.$$

Thus,

$$|u(M) - u_N(M)| < C\varepsilon, \quad M \in \overline{\Omega}_s.$$

That is, the sequence $u_N(M)$ converges uniformly in $\overline{\Omega}_s$ to the solution $u(M)$.

In order to obtain a good system of linear algebraic equations (15.1), it is necessary that the functions u_k be linearly independent. Let us give some examples of complete systems of linearly independent solutions for the Helmholtz equation that are frequently used in practice.

15.1 Complete Systems of Solutions

15.1.1 Spherical Wave Functions

Denote

$$u_{mn} = j_n(kr) P_n^{|m|}(\cos\theta) e^{im\phi},$$

$$v_{mn} = h_n^{(1)}(kr) P_n^{|m|}(\cos\theta) e^{im\phi},$$

$$n = 0, 1, \dots, \quad m = -n, \dots, n,$$

where $j_n(z)$ stands for the spherical Bessel function of order n (see Sect. 7.6) and $h_n^{(1)}(z) = \sqrt{\frac{\pi}{2z}} H_{n+\frac{1}{2}}^{(1)}(z) = j_n(z) + i n_n(z)$ is the Hankel spherical function. Here $H_{n+\frac{1}{2}}^{(1)}(z)$ is the Hankel function of the first kind, $n_n(z)$ is the spherical Bessel function of the second kind, $n_n(z) = (-1)^{n+1} \sqrt{\frac{\pi}{2z}} J_{-n-\frac{1}{2}}(z)$, and $P_n^{|m|}$ stands for the associated Legendre polynomials.

All u_{mn} are solutions of the Helmholtz equation $(\Delta + k^2) u = 0$ in $\mathbb{R}^3$, and all v_{mn} are solutions of the Helmholtz equation in $\mathbb{R}^3 \setminus \{0\}$ satisfying the Sommerfeld condition (14.15) at infinity. The system $\{u_{mn}\}$ is used for solving interior boundary value problems, while the system $\{v_{mn}\}$ is better suited for exterior problems.

The following series expansion of the fundamental solution $g(M, P)$ in terms of the spherical wave functions is used for studying properties of these two systems of solutions,

$$q(M, P) = \frac{ik}{\pi} \sum_{n=0}^{\infty} \sum_{m=-n}^{n} d_{mn} \begin{cases} v_{-mn}(P) u_{mn}(M), & |P| > |M| \\ u_{-mn}(P) v_{mn}(M), & |P| < |M| \end{cases}, \tag{15.3}$$

where the normalization constants are given by

$$d_{mn} = \frac{2n+1}{4} \frac{(n-|m|)!}{(n+|m|)!}.$$

Theorem 260 *Let $\Sigma = \partial V$ be a closed surface of class C^2. Then each of the systems of functions is complete in $L_2(\Sigma)$:*

$$\{v_{mn}\}, \quad \left\{ \frac{\partial v_{mn}}{\partial v} \right\}, \quad \{u_{mn}\}, \quad \left\{ \frac{\partial u_{mn}}{\partial v} \right\}$$

(for the completeness of the last two systems, we suppose additionally that k^2 is not a Dirichlet eigenvalue of the domain V and k^2 is not a Neumann eigenvalue of the domain V, respectively).

Proof Let us show that the system $\{v_{mn}\}$ is closed in $L_2(\Sigma)$, which is equivalent to its completeness (see, e.g., [3, Sect. 9]). Let $a \in L_2(\Sigma)$, and suppose that

$$\int_{\Sigma} \overline{a}(P) v_{mn}(P) d\Sigma_P = 0 \quad \text{for all } n = 0, 1, \dots \quad \text{and } m = -n, \dots, n.$$

That is, we suppose that there exists a nontrivial function $a \in L_2(\Sigma)$ orthogonal to all the functions v_{mn}. Consider a ball of radius r: $B_r \subset V$ and

$$u(M) = \int_{\Sigma} q(M, P)\overline{a}(P) d\Sigma_P, \quad M \in B_r.$$

Substituting the series expansion (15.3) corresponding to $|P| > |M|$, we obtain that $u \equiv 0$ in B_r. Since u is a solution of the Helmholtz equation, and due to the analyticity of the solutions of the Helmholtz equation, this implies that $u \equiv 0$ in V, which in turn implies that $u \equiv 0$ on Σ (due to the continuity of u). This implies that $a = 0$ on Σ a.e. The proof of this step can be found, e.g., in [28, p. 27].

In a similar way, the rest of the theorem is proved. ∎

15.1.2 Fundamental Solutions

Let Σ be a closed surface in $\mathbb{R}^3$ of class C^2, and let $\Sigma^+ \in C^2$ be a closed surface enclosing Σ, while $\Sigma^- \in C^2$ a closed surface enclosed by Σ. Figure 15.1 depicts a schematic representation of the surface Σ and both auxiliary surfaces.

Let $\{P_n^-\}_{n=1}^{\infty}$ be a dense set of points on Σ^- and $\{P_n^+\}_{n=1}^{\infty}$ a dense set of points on Σ^+. Denote

$$\varphi_n^-(M) = q(M, P_n^-) \quad \text{and} \quad \varphi_n^+(M) = q(M, P_n^+).$$

By V, we denote the interior domain with the boundary Σ and by V_i the interior domain bounded by Σ^-. Thus, $V_i \subset V$, $\Sigma = \partial V$, $\Sigma^- = \partial V_i$.

Theorem 261 ([63]) *Let k^2 be not a Dirichlet eigenvalue of the domain V_i. Then each of the systems of functions is complete in $L_2(\Sigma)$:*

$$\{\varphi_n^-\}, \quad \left\{\frac{\partial \varphi_n^-}{\partial \nu}\right\}, \quad \{\varphi_n^+\}, \quad \left\{\frac{\partial \varphi_n^+}{\partial \nu}\right\}.$$

Proof Consider $a \in L_2(\Sigma)$, and suppose that $\int_{\Sigma} \overline{a}(M)\varphi_n^-(M) d\Sigma_M = 0$ for all $n = 1, 2, \dots$. Hence, we have that a single layer potential with the density $\overline{a}$ vanishes at a dense set of points on Σ^-: $u_{\overline{a}}(P_n^-) = 0$, $n = 1, 2, \dots$. This implies that $u_{\overline{a}} \equiv 0$ on Σ^-. Moreover, since k^2 is not a Dirichlet eigenvalue of the domain V_i, we obtain that necessarily $u_{\overline{a}} \equiv 0$ in V_i. The rest of the proof of the completeness of the system $\{\varphi_n^-\}$ repeats the reasonings from the previous theorem.

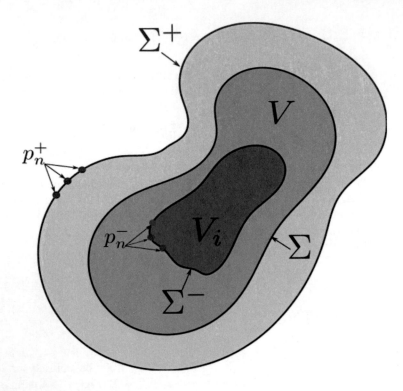

Fig. 15.1 Schematic representation of the auxiliary surfaces and points

Repeating the above arguments for the double layer potential instead of the single layer one, we obtain the completeness of the system $\left\{ \frac{\partial \varphi_n^-}{\partial \nu} \right\}$. The proof of the second part of the theorem is similar.   ∎

Remark 262 In [38], it was shown that the systems of functions $\left\{ \varphi_n^\pm \right\}$ are complete in any Sobolev space $H^s\,(\Sigma),\, s \in \mathbb{R}$.

Each function φ_n^- represents a wave generated by a point source located inside the domain V. In the exterior domain $\Omega = \mathbb{R}^3 \backslash \overline{V}$, it is a regular solution of the Helmholtz equation, satisfying the Sommerfeld condition at infinity. Thus, the system of functions $\left\{ \varphi_n^- \right\}$ is well suited for solving exterior boundary value problems for the Helmholtz equation. Any solution of the Helmholtz equation satisfying the Sommerfeld condition can be approximated arbitrarily closely by a linear combination of the functions φ_n^-. The coefficients of the linear combination can be found from the boundary condition on Σ. Similarly, the system of functions $\left\{ \varphi_n^+ \right\}$ is convenient for solving interior boundary value problems for the Helmholtz equation.

15.1.3   Plane Waves

Another important complete system of solutions of the Helmholtz equation is known as plane waves. Denote

$$w(\overrightarrow{P}, \overrightarrow{k}) = e^{i\left(\overrightarrow{P}, \overrightarrow{k}\right)},$$

where $\overrightarrow{P}$ is the coordinate vector corresponding to the point $P = (x, y, z)$, and $\overrightarrow{k}$ is the wave vector, such that $\left(\overrightarrow{k}, \overrightarrow{k}\right) = k^2$ is the coefficient in the Helmholtz equation. Any such vector $\overrightarrow{k}$ admits the representation

$$\overrightarrow{k} = \overrightarrow{k}(\alpha, \beta) = k \begin{pmatrix} \cos\alpha \sin\beta \\ \sin\alpha \sin\beta \\ \cos\beta \end{pmatrix},$$

where α and β can be interpreted as angles identifying the direction in the space $\mathbb{R}^3$. Let us choose points $\omega_n = (\alpha_n, \beta_n)$ on a unitary sphere S_1 centered at the origin and define the **plane waves** as the functions

$$w_n(P) = w(\overrightarrow{P}, \overrightarrow{k}_n) = e^{i\left(\overrightarrow{P}, \overrightarrow{k}_n\right)}, \quad n = 1, 2, \dots,$$

where $\overrightarrow{k}_n = \overrightarrow{k}(\alpha_n, \beta_n)$. Each plane wave $w_n(P)$ is a solution of the Helmholtz equation in $\mathbb{R}^3$. The following theorem on the completeness of the plane waves is valid.

Theorem 263 *Let* $\{\omega_n\}_{n=1}^{\infty}$ *be a dense set of points of the unit sphere* S_1. *Let* V *be a bounded domain in* $\mathbb{R}^3$ *with the boundary* Σ *of class* C^2. *Then the set of functions* $\{w_n\}_{n=1}^{\infty}$ *is complete in* $L_2(\Sigma)$ *if* k^2 *is not a Dirichlet eigenvalue of the domain* V. *Also the set* $\left\{\frac{\partial w_n}{\partial v}\right\}_{n=1}^{\infty}$ *is complete in* $L_2(\Sigma)$.

For the proof, a series expansion of the plane wave in terms of the spherical waves is used after which the reasonings are similar to those from the proof of the previous theorem. For details, we refer to [28, p. 58].

Thus, the system of the plane waves can also be used for solving boundary value problems for the Helmholtz equation, though since they do not satisfy the Sommerfeld condition at infinity, they are not convenient for solving exterior problems.

15.1.4   Linear Independence

All three systems of solutions introduced in this section are linearly independent. Let us prove this fact in the case of the fundamental solutions.

Theorem 264 *Let k^2 be not a Dirichlet eigenvalue of the domain V. Then the functions $\{\varphi_n^+\}_{n=1}^{\infty}$ are linearly independent on $\Sigma = \partial V$.*

Proof Suppose the opposite that there exist such constants c_i that

$$u(P) = \sum_{i=1}^{N} c_i \varphi_{k_i}^+(P) \equiv 0 \quad \text{on } \Sigma,$$

where k_i are distinct natural numbers, and at least $c_r \neq 0$, $r \leq N$. Due to the uniqueness of the solution of the Dirichlet problem, we obtain that $u \equiv 0$ in $\overline{V}$. Due to the analyticity property of the solutions of the Helmholtz equation, $u \equiv 0$ in V_e, where V_e is the domain bounded by the auxiliary surface Σ^+ (see Sect. 15.1.2). Now let $P \to P_{k_r}^+$. Then $\left| c_r \varphi_{k_r}^+ \right| \to \infty$ due to the singularity of the function $\varphi_{k_r}^+$ at the point $P_{k_r}^+$, but the rest of the terms in the linear combination $\sum_{i=1}^{N} c_i \varphi_{k_i}^+$ remain bounded, which contradicts the equality $\sum_{i=1}^{N} c_i \varphi_{k_i}^+ \equiv 0$ in V_e. Thus, $c_r = 0$. $\blacksquare$

Numerical approach based on the use of complete systems of solutions is a frequently used tool in practice. We refer the reader to [5, 28, 29, 35, 38, 62, 63], and numerous references therein. In the next chapter, we explain how the complete systems of solutions can be used for constructing the so-called reproducing (or Bergman) kernels that allow one to solve boundary value and spectral problems.

Chapter 16
Bergman Kernel Approach

The aim of this chapter is to introduce and explain an extremely useful tool for practical solution of boundary value and spectral problems for elliptic equations of mathematical physics. Originally developed in [12], it received relatively little attention in bibliography, though its range of applicability is not inferior to that of Green's function, and moreover, the concept of the complete system of solutions acquires even more profound meaning when viewed through the prism of the Bergman kernel approach that we shall study in this chapter.

In order to simplify the exposition, we will restrict it to the case of the two-dimensional Schrödinger equation, which is also often referred to as the two-dimensional Helmholtz equation with variable coefficient

$$\Delta u(x, y) - q(x, y)u(x, y) = 0, \quad \Delta = \frac{\partial^2}{\partial x^2} + \frac{\partial^2}{\partial y^2}. \tag{16.1}$$

First, we shall consider the case $q > 0$ and later on the case of a general complex valued potential q.

We consider Eq. (16.1) in a finite domain $V \subset \mathbb{R}^2$ bounded by a finite number of closed Lyapunov curves C_k. We denote by C the boundary of V.

16.1 Fundamental Solutions

Let q be a continuously differentiable positive function in $\overline{V}$ and Q a point of V. Equation (16.1) possesses fundamental solutions that are regular solutions at any point P of V except for the point Q where they have a logarithmic singularity coinciding with the singularity of the fundamental solution of the Laplacian $\sim \frac{1}{2\pi} \ln \frac{1}{r_{PQ}}$ (see, e.g., [31]). Any fundamental solution of (16.1) has the form

© The Author(s), under exclusive license to Springer Nature Switzerland AG 2022
A. N. Karapetyants, V. V. Kravchenko, *Methods of Mathematical Physics*,
https://doi.org/10.1007/978-3-031-17845-0_16

$$S(P, Q) = \frac{1}{2\pi} \ln \frac{1}{r_{PQ}} + s(P, Q),$$

where $s(P, Q)$ is twice continuously differentiable in $\overline{V}$ except for the point Q where it is still continuously differentiable (see, e.g., [12, p. 268]). The fundamental solution is not of course unique, since adding to it any regular solution of (16.1) we again obtain a fundamental solution.

Similarly to Eq. (9.3) from Sect. 9.1, the following representation for any function $v \in C^2(V) \cap C^1(\overline{V})$ is obtained

$$v(Q) = -\int_C v(P) \frac{\partial S(P, Q)}{\partial v_P} ds_P + \int_C S(P, Q) \frac{\partial v(P)}{\partial v_P} ds_P$$

$$- \int_V S(P, Q)\left(\Delta v(P) - q(P)v(P)\right) dV. \tag{16.2}$$

From here, we obtain an integral representation for solutions of (16.1):

$$u(Q) = \int_C \left(S(P, Q) \frac{\partial u(P)}{\partial v_P} - u(P) \frac{\partial S(P, Q)}{\partial v_P} \right) ds_P. \tag{16.3}$$

16.2   Green's and Neumann's Functions

In what follows, we understand by $S(P, Q)$ an arbitrarily chosen but fixed fundamental solution of (16.1). The general theory of boundary value problems of Eq. (16.1) guarantees the existence of a regular solution $g(P, Q)$ of (16.1) that has the boundary values $-S(P, Q)$ for fixed $Q \in V$ and $P \in C$. Thus, the function

$$G(P, Q) = S(P, Q) + g(P, Q)$$

is a solution of (16.1) and is twice continuously differentiable in $\overline{V}$ except for the point Q where it behaves like $S(P, Q)$. Moreover,

$$G(P, Q) = 0, \quad \text{for all } P \in C.$$

$G(P, Q)$ is called **Green's function** of Eq. (16.1) in the domain V.

Green's function is, of course, a particular fundamental solution. Hence, the integral representation (16.3) may be applied with $S(P, Q) = G(P, Q)$. In view of the vanishing of Green's function on the boundary $C = \partial V$, (16.3) takes the form

$$u(Q) = -\int_C u(P) \frac{\partial G(P, Q)}{\partial v_P} ds_P. \tag{16.4}$$

Thus, (16.4) provides a representation of any regular solution of (16.1) in V in terms of its boundary values. Hence, the knowledge of Green's function allows one to obtain a solution of the Dirichlet boundary value problem for an arbitrary boundary data.

Next, let us solve the following Neumann boundary value problem. Find a solution $n(P, Q)$ of (16.1) that on C has the normal derivative

$$\frac{\partial n(P, Q)}{\partial \nu_P} = -\frac{\partial S(P, Q)}{\partial \nu_P}$$

for fixed $Q \in V$.

Then the function

$$N(P, Q) = S(P, Q) + n(P, Q)$$

is a solution of (16.1) that is twice continuously differentiable in $\overline{V}$ except for the point Q where it behaves like $S(P, Q)$. Moreover,

$$\frac{\partial N(P, Q)}{\partial \nu_P} = 0, \quad \text{for all } P \in C.$$

$N(P, Q)$ is called **Neumann's function** of Eq. (16.1) in the domain V. Applying (16.3) with $S(P, Q) = N(P, Q)$, we obtain

$$u(Q) = \int_C N(P, Q) \frac{\partial u(P)}{\partial \nu_P} ds_P, \quad Q \in V$$

for any regular solution u of (16.1) in V. Thus, Neumann's function provides a representation of every regular solution in terms of its normal derivative on C and solves the general Neumann boundary value problem in V.

Let $\lambda(s)$ be a continuous non-negative function of the arc length s along C. It is well known that the boundary value problem of the third kind always has a solution. This problem is that of finding a regular solution u of (16.1) that on C satisfies the condition

$$\frac{\partial u}{\partial \nu} - \lambda u = f$$

for any given continuous function f. We may solve a particular problem of this kind with

$$f(s) = -\frac{\partial S(P, Q)}{\partial \nu_P} + \lambda(s) S(P, Q)$$

for $Q \in V$. Thus, we obtain a solution $r_\lambda(P, Q)$ of (16.1) such that

$$R_\lambda(P, Q) = S(P, Q) + r_\lambda(P, Q)$$

is a solution of (16.1) that is twice continuously differentiable in V, except for the point Q where it behaves like $S(P, Q)$ and which on C satisfies the condition

$$\frac{\partial R_\lambda(P, Q)}{\partial \nu_P} - \lambda(s) R_\lambda(P, Q) = 0, \quad \text{for all } P \in C.$$

$R_\lambda(P, Q)$ is called **Robin's function** of Eq. (16.1) in the domain V, corresponding to the weight function λ. Again, using (16.3) with $S(P, Q) = R_\lambda(P, Q)$, we obtain

$$u(Q) = \int_C R_\lambda(P, Q) \left(\frac{\partial u(P)}{\partial \nu_P} - \lambda u(P) \right) ds_P$$

$$= \int_C f(s) R_\lambda(P, Q) ds_P.$$

Thus, Robin's function enables us to solve the general boundary value problem of the third kind.

It is easy to prove that all three special fundamental solutions: the Green's, Neumann's, and Robin's functions are symmetric in P and Q,

$$G(P, Q) = G(Q, P), \quad N(P, Q) = N(Q, P), \quad R_\lambda(P, Q) = R_\lambda(Q, P).$$

Furthermore, it is convenient to assume from the beginning that $S(P, Q)$ is symmetric in P and Q, since we know that such fundamental solutions exist. Under this assumption, the functions $g(P, Q), n(P, Q), r_\lambda(P, Q)$ also become symmetric in their argument points.

16.3   Bergman's Kernel

Let us consider the function

$$K(P, Q) = N(P, Q) - G(P, Q).$$

Note that for $Q \in V$ and $P \in C$, we have

$$K(P, Q) = N(P, Q) \quad \text{and} \quad \frac{\partial K(P, Q)}{\partial \nu_P} = -\frac{\partial G(P, Q)}{\partial \nu_P}, \quad P \in C.$$

Hence, for any regular solution u of (16.1) in V, we have the equalities

$$u(Q) = \int_C K(P, Q) \frac{\partial u(P)}{\partial \nu_P} ds_P, \quad Q \in V \tag{16.5}$$

and

$$u(Q) = \int_C u(P) \frac{\partial K(P, Q)}{\partial v_P} ds_P, \quad Q \in V. \tag{16.6}$$

Thus, the kernel K enables us to solve both the Dirichlet and Neumann problems, but differently from the Green's and Neumann's functions the kernel K is a regular solution of (16.1). Indeed, we have that

$$K(P, Q) = n(P, Q) - g(P, Q),$$

where both n and g are regular solutions of (16.1) in V.

The function K is called **Bergman's kernel** of Eq. (16.1) in V.

16.4   Energy Integral and Scalar Product

Let $v, w \in C^2(\overline{V})$ be real valued functions. Consider the following operation:

$$
\{v, w\} = \int_V \left(v_x w_x + v_y w_y + q v w \right) dV
$$

$$
= \int_V \left((\nabla v, \nabla w) + q v w \right) dV,
$$

which satisfies all the properties of a scalar product. The norm generated by this scalar product has the form

$$
\|v\|_E = \left(\int_V \left(v_x^2 + w_y^2 + q v^2 \right) dV \right)^{\frac{1}{2}},
$$

where the integral $\int_V \left(v_x^2 + w_y^2 + q v^2 \right) dV$ is frequently referred to as the **energy integral** associated with Eq. (16.1). Of course, when introducing the norm and the scalar product in this way, we use the fact that q is a positive function.

One can easily verify the fulfillment of the Cauchy–Bunyakovsky–Schwarz inequality

$$
|\{v, w\}| \leq \|v\|_E \|w\|_E
$$

and the triangle inequality

$$
\|v + w\|_E \leq \|v\|_E + \|w\|_E.
$$

Due to Green's identity, we have the equality

$$\{v, w\} = \int_C v \frac{\partial w}{\partial v} ds - \int_V v (\Delta w - qw) \, dV \qquad (16.7)$$

for all $v, w \in C^2(\overline{V})$.

The class of all functions from $C^2(\overline{V})$ with the metric based on the scalar product $\{\cdot, \cdot\}$ is a linear space that we denote by Ω. Let us introduce two subspaces of Ω: (a) the linear space Ω^0 of all functions $v \in \Omega$ that vanish on C and (b) the linear space Σ of all functions $u \in \Omega$ that satisfy equation (16.1). In view of (16.7), we conclude that

$$\{u, v\} = 0 \quad \text{if } u \in \Sigma \text{ and } v \in \Omega^0.$$

Thus the spaces Σ and Ω^0 are orthogonal to each other with respect to the scalar product $\{\cdot, \cdot\}$.

Let $w \in \Omega$. From the existence theorem for the Dirichlet boundary value problem for Eq. (16.1) in V (see Remark 258), we know that there exists a function $u \in \Sigma$ with the same values as w on C. Hence, $w - u = v \in \Omega^0$. Thus, we have

$$\Omega = \Sigma + \Omega^0, \qquad (16.8)$$

i.e., the linear space Ω admits this decomposition into two orthogonal subspaces Σ and Ω^0. The only common element of both spaces is the zero function.

Note that due to (16.7), the integral representation (16.2) for any function $v \in C^2(V) \cap C^1(\overline{V})$ can be written in the form

$$v(Q) = - \int_C v(P) \frac{\partial S(P, Q)}{\partial v_P} ds_P + \{S(P, Q), v(P)\}, \qquad (16.9)$$

where $S(P, Q)$ is an arbitrary fundamental solution of (16.1). Choosing $S(P, Q) = N(P, Q)$ and taking into account that the normal derivative of $N(P, Q)$ on C vanishes, we obtain

$$v(Q) = \{N(P, Q), v(P)\} \quad \text{for any } v \in \Omega. \qquad (16.10)$$

Thus, the scalar product of any function v from Ω with Neumann's function reproduces v. Neumann's function appears as the unit multiplier or the **reproducing kernel** in the algebra induced by the energy integral.

Similarly, for every $v \in \Omega^0$, we have the identity

$$v(Q) = \{G(P, Q), v(P)\},$$

i.e., Green's function is a reproducing kernel in the subspace Ω^0.

For all $u \in \Sigma$, we have on the other hand from (16.9), in view of (16.4):

$$\{G(P, Q), u(P)\} = 0, \tag{16.11}$$

i.e., Green's function is orthogonal to the space Σ of all regular solutions of (16.1) in V.

Let $u \in \Sigma$. In view of (16.10) and (16.11), we have

$$\{K(P, Q), u(P)\} = u(Q). \tag{16.12}$$

Thus, the Bergman kernel $K(P, Q)$ is a reproducing kernel of the space Σ with respect to the scalar product $\{\cdot, \cdot\}$. We notice that the geometric relation (16.8) between spaces is translated into the same relation between their reproducing kernels

$$N = K + G.$$

Let $w \in \Omega$. According to (16.8), we may split it up into functions $u \in \Sigma$ and $v \in \Omega^0$, i.e.,

$$w = u + v$$

in a unique way. From (16.10), we have

$$
\begin{aligned}
w(Q) = \{N(P, Q), w(P)\} &= \{K(P, Q) + G(P, Q), w(P)\} \\
&= \{K(P, Q), w(P)\} + \{G(P, Q), u(P) + v(P)\} \\
&= \{K(P, Q), w(P)\} + \{G(P, Q), v(P)\} \\
&= \{K(P, Q), w(P)\} + v(Q).
\end{aligned}
$$

Hence,

$$
\begin{aligned}
\{K(P, Q), w(P)\} &= w(Q) - v(Q) \\
&= u(Q).
\end{aligned}
$$

Thus, the kernel $K(P, Q)$ maps an arbitrary function $w \in \Omega$ into a solution u of (16.1). Moreover, the solution u has the same boundary values on C as the function w.

We finish this section by noticing that a reproducing kernel for the third boundary value problem can also be constructed in a similar way (see [12, p. 280]).

16.5   Complete Systems of Solutions and Construction of Bergman's Kernel

Let us assume that an infinite system of solutions $\{u_n\}_{n=1}^{\infty}$ of (16.1) is given, which is complete with respect to the norm $\|\cdot\|_E$. We may assume without loss of generality that the functions u_n are orthonormal with respect to the scalar product $\{\cdot, \cdot\}$, i.e.,

$$\{u_n, u_m\} = \delta_{nm} = \begin{cases} 1, & m = n, \\ 0, & m \neq n. \end{cases}$$

Each system $\{u_n\}_{n=1}^{\infty}$ may be brought into this form by the Gram–Schmidt process of orthonormalization.

Notice that as it follows from (16.7), for any regular solutions u and v of (16.1), the scalar product takes the form

$$\{u, v\} = \int_C u \frac{\partial v}{\partial v} ds$$

and thus reduces to the computation of a contour integral.

Any function $u \in \Sigma$ can be represented in the form of the series

$$u(P) = \sum_{n=1}^{\infty} a_n u_n(P)$$

with the coefficients

$$a_n = \{u, u_n\}.$$

The series converges with respect to the norm $\|\cdot\|_E$. Additionally, the uniform convergence of the series on every compact subset of V can be proved [12, p. 281].

Let us apply this result, in particular, to the function $K(P, Q)$, which is a regular solution of (16.1). Thus,

$$K(P, Q) = \sum_{n=1}^{\infty} a_n(Q) u_n(P)$$

with

$$a_n(Q) = \{K(P, Q), u_n(P)\}.$$

Due to the reproducing property of the kernel $K(P, Q)$ (equality (16.12)), we obtain $a_n(Q) = u_n(Q)$, and thus,

$$K(P, Q) = \sum_{n=1}^{\infty} u_n(P)u_n(Q). \tag{16.13}$$

This formula provides a simple and feasible way for constructing the kernel $K(P, Q)$ in terms of an arbitrary complete orthonormal set of solutions of (16.1) in V or at least for approximating it, if only a finite set of exact solutions are available.

Thus, the Bergman kernel $K(P, Q)$ preserves important features of Green's and Neumann's functions that allow one to solve boundary value problems, but additionally, it can be efficiently constructed from any complete system of solutions.

16.6   Reproducing Kernels for Arbitrary q

The condition $q > 0$ in V is obviously quite restrictive and is imposed to make it possible to work with the scalar product introduced above. When this condition is not fulfilled, other scalar products can be used associated with other reproducing kernels. Thus, let q be a continuous real valued function defined on $\overline{V}$. Consider the space Σ equipped with one of the following scalar products:

$$\langle u, v \rangle_1 = \int_C uv \, ds \quad \text{or} \quad \langle u, v \rangle_2 = \int_C \frac{\partial u}{\partial v} \frac{\partial v}{\partial v} \, ds. \tag{16.14}$$

Let us look for the existence of kernels $F(P, Q)$ and $M(P, Q)$ enjoying reproducing properties with respect to the scalar product $\langle \cdot, \cdot \rangle_1$ and $\langle \cdot, \cdot \rangle_2$, respectively. Note that the kernel F would be suitable for solving the Dirichlet problem; meanwhile, for the Neumann problem, one could use the kernel M. The reproducing property of the kernel F has the form

$$\langle F(P, Q), u(P) \rangle_1 = u(Q) \tag{16.15}$$

and guarantees that if F exists, then the homogeneous Dirichlet problem possesses only a trivial solution that implies the existence of Green's function G. From the equality

$$u(Q) = -\int_C u(P) \frac{\partial G(P, Q)}{\partial v_P} \, ds_P, \tag{16.16}$$

we see that if for $P \in C$ we choose $F(P, Q) := -\frac{\partial G(P, Q)}{\partial v_P}$, then (16.15) holds.

On the other hand, $F(P, Q)$ must be a regular solution of (16.1), and its values in V are determined by its values on the boundary C due to (16.16). Thus we obtain

$$F(P, Q) = \int_C \frac{\partial G(R, Q)}{\partial \nu_R} \frac{\partial G(R, P)}{\partial \nu_R} ds_R. \tag{16.17}$$

Hence, we have proved the following statement.

Proposition 265 *The reproducing kernel F exists if and only if the Green's function exists, and equality (16.17) holds.*

Let $\{u_n\}_{n=1}^{\infty}$ be a complete system of functions in Σ, orthonormal with respect to the scalar product $\langle \cdot, \cdot \rangle_1$; then similar to (16.13), one can prove the equality $F(P, Q) = \sum_{n=1}^{\infty} u_n(P)u_n(Q)$, where the series converges in the norm generated by the scalar product $\langle \cdot, \cdot \rangle_1$ and uniformly on any compact subset of V.

Analogous results are valid for the reproducing kernel $M(P, Q)$. Namely, the following equality holds

$$M(P, Q) = \int_C N(R, Q)N(R, P)ds_R,$$

and $M(P, Q) = \sum_{n=1}^{\infty} u_n(P)u_n(Q)$, where $\{u_n\}_{n=1}^{\infty}$ is a complete orthonormal system of functions in Σ with respect to the scalar product $\langle \cdot, \cdot \rangle_2$.

Moreover, similar results are valid in the case when q is a continuous complex valued function in $\overline{V}$. Consider, e.g., the case of the Dirichlet boundary value problem. A corresponding scalar product has the form

$$\langle u, v \rangle_3 = \int_C u\bar{v}ds.$$

Let $F(P, Q)$ be a solution of (16.1) in the variable P satisfying the boundary condition

$$F(P, Q) = -\frac{\partial \overline{G}(P, Q)}{\partial \nu_P}, \quad P \in C \text{ and } Q \in V$$

(we assume the existence of G). Then

$$F(P, Q) = \int_C \frac{\partial \overline{G}(R, Q)}{\partial \nu_R} \frac{\partial G(R, P)}{\partial \nu_R} ds_R.$$

By construction, $F(P, Q)$ is a reproducing kernel, that is, if $u \in \Sigma$, then $\langle F(P, Q), u(P) \rangle_3 = \bar{u}(Q)$ or which is the same $\langle u(P), F(P, Q) \rangle_3 = u(Q)$, $Q \in V$.

If $\{u_n\}_{n=1}^{\infty}$ is now a complete system of functions in Σ, orthonormal with respect to the scalar product $\langle \cdot, \cdot \rangle_3$, then

$$F(P, Q) = \sum_{n=1}^{\infty} u_n(P)\overline{u}_n(Q), \tag{16.18}$$

where the series converges in the norm generated by the scalar product $\langle \cdot, \cdot \rangle_3$ and uniformly on any compact subset of V.

Moreover, the kernel F can be used for solving Dirichlet eigenvalue problems as well. Considering it for the equation $(-\Delta + q - \lambda)u = 0$ as a function of the spectral parameter λ, one can look for zeros of the function $1/F(P_0, Q_0, \lambda)$, where P_0 and Q_0 are arbitrary fixed points of the domain of interest. These zeros coinciding with the singularities of $F(P_0, Q_0, \lambda)$ give the eigenvalues of the Dirichlet problem (see [15]).

In a similar way, using the scalar product $\langle \cdot, \cdot \rangle_2$, one can construct a reproducing kernel for solving the Neumann problem and the corresponding eigenvalue problem. For solving the Robin problem, another scalar product can be introduced. Finally, we notice that the described definitions and properties can be extended onto the slightly more general equation

$$(\operatorname{div} p \operatorname{grad} + q)\, u = 0. \tag{16.19}$$

16.7   Construction of Complete Systems of Solutions

Efficient construction of complete systems of solutions for partial differential equations with variable coefficients is often regarded as a not feasible task, and nevertheless, in many practically interesting situations, such a construction is quite possible and easily realizable numerically. The mathematical tools that help to cope with this task are transmutation operators. The idea to use them for obtaining complete systems of solutions of PDEs was explored, for example, in [9, 11, 23]. The difficulty of this approach always consisted in the necessity of constructing the corresponding integral transmutation kernel (by successive approximation method). Recently, it was shown that often it is not necessary. Surprisingly enough, it is possible to construct the complete systems of solutions directly, by using the known mapping properties of the transmutation operators like that from (7.40) and without constructing the transmutation operator itself (see [16, 17, 57, 58] and [54]). Here we explain this approach on the example of a Schrödinger equation with a potential that has a separable form in Cartesian variables.

Consider equation (16.1) supposing that the complex valued potential $q \in C\left(\overline{V}\right)$ has the form

$$q(x, y) = q_1(x) + q_2(y).$$

We emphasize that even the case when q_1 or q_2 is a constant is already of great interest in practical applications, which include wave propagation models in stratified media.

First, let us suppose that the domain $V \subset \mathbb{R}^2$ in which the equation

$$(\Delta - q_1(x) - q_2(y)) \, u(x, y) = 0 \tag{16.20}$$

is considered represents a rectangle $V = (-a_1, a_1) \times (-a_2, a_2)$, with $a_j > 0$, $j = 1, 2$. Later we will show that this restriction is not essential and can be overcome.

Fix $j \in \{1, 2\}$ and let $f_j \in C^2(-a_j, a_j) \cap C^1[-a_j, a_j]$ be a solution of $-f_j'' + q_j(x)f = 0$, $x \in (-a_j, a_j)$, which does not vanish in $[-a_j, a_j]$ and satisfies the condition $f_j(0) = 1$. Such solution always exists (see Remark 76).

Then, due to Theorem 89, there exists a transmutation operator in the form of a Volterra integral operator of the second kind

$$\mathbf{T}_j v(x) = v(x) + \int_{-x}^{x} K_j(x, t) v(t) dt \tag{16.21}$$

satisfying the relation

$$\left(\frac{\partial^2}{\partial x^2} - q_j(x) \right) \mathbf{T}_j v(x) = \mathbf{T}_j \left(\frac{\partial^2}{\partial t^2} v(t) \right), \quad \text{for } v \in C^2[-a_j, a_j]. \tag{16.22}$$

The kernel $K_j(x, t)$ is continuous in the domain $|t| \leqslant |x| \leqslant a_j$, and the operator $\mathbf{T}_j : C[-a_j, a_j] \to C[-a_j, a_j]$ is bounded and invertible, $\mathbf{T}_j[1] = f_j$.

Take $u \in C(\overline{V})$. The operators $\mathbf{T}_j$ act on u as follows:

$$\mathbf{T}_1 u(x, y) = u(x, y) + \int_{-x}^{x} K_1(x, t) u(t, y) dt$$

and

$$\mathbf{T}_2 u(x, y) = u(x, y) + \int_{-y}^{y} K_2(y, t) u(x, t) dt.$$

$\mathbf{T}_1$ and $\mathbf{T}_2$ commute on $C(\overline{V})$.

Let us define the operator $\mathcal{T} := \mathbf{T}_1 \mathbf{T}_2$. From (16.22), it follows that

$$(\Delta - q_1(x) - q_2(y)) \, \mathcal{T} u = \mathcal{T} \Delta u, \quad \text{for } u \in C^2(V),$$

and due to the boundedness and invertibility of $\mathcal{T}$, we obtain that the space Σ of all regular solutions of (16.20) is the image of the space of harmonic functions in V, denoted by $\text{Har}(V)$, under the action of $\mathcal{T}$:

$$\Sigma(V) = \mathcal{T} \left(\text{Har}(V) \right).$$

An important complete system of harmonic functions is the system of harmonic polynomials that can be defined as follows:

$$p_0(z) = 1, \quad p_m(z) = \mathrm{Re}(z^m) \quad \text{if } m \text{ is odd,}$$

$$p_m(z) = \mathrm{Re}(iz^m) \quad \text{if } m \text{ is even,}$$

where $z = x + iy$. Or, equivalently,

$$p_0(x, y) = 1,$$

$$p_m(x, y) = \sum_{\substack{\text{even } k=0}}^{\frac{m+1}{2}} (-1)^{\frac{k}{2}} \binom{\frac{m+1}{2}}{k} x^{\frac{m+1}{2}-k} y^k \quad \text{for an odd } m,$$

$$p_m(x, y) = \sum_{\substack{\text{odd } k=1}}^{\frac{m}{2}} (-1)^{\frac{k+1}{2}} \binom{\frac{m}{2}}{k} x^{\frac{m}{2}-k} y^k \quad \text{for an even } m.$$

With the aid of (7.40), it is easy to find out what are the images of the harmonic polynomials under the action of the operator $\mathcal{T}$. Let us denote them by

$$u_m = \mathcal{T}[p_m].$$

First, let us notice that due to the boundedness and invertibility of $\mathcal{T}$, the infinite set of the functions u_m is a complete system of solutions of (16.20) in V, and using (7.40), we obtain their explicit form: $u_0(x, y) = f_1(x) f_2(y)$, and for $m > 0$

$$u_m(x, y) = \begin{cases} \displaystyle\sum_{\substack{\text{even } k=0}}^{\frac{m+1}{2}} (-1)^{\frac{k}{2}} \binom{\frac{m+1}{2}}{k} \Phi^1_{\frac{m+1}{2}-k}(x) \Phi^2_k(y), & \text{if } m \text{ is odd,} \\[2em] \displaystyle\sum_{\substack{\text{odd } k=0}}^{\frac{m}{2}} (-1)^{\frac{k+1}{2}} \binom{\frac{m}{2}}{k} \Phi^1_{\frac{m}{2}-k}(x) \Phi^2_k(y), & \text{if } m \text{ is even,} \end{cases}$$

$$\tag{16.23}$$

where Φ^1_n and Φ^2_n are the formal powers from Sect. 7.2 corresponding to q_1 and q_2, respectively.

Finally, using the so-called Runge property of the elliptic equation (16.20), one can extend the completeness property of the constructed system of solutions u_m onto an essentially arbitrary domain and not necessarily a rectangle (see [55]). The Runge property is a general term for the property of the equation guaranteeing that its complete system of solutions in a larger domain remains complete in its subdomains. In particular, the following statement is valid (see [55]).

Proposition 266 *Let $q_j \in C[-a_j, a_j]$, $j = 1, 2$, and $V \subset V$ be a C^2-domain. Then the system $\{u_m\}_{m=0}^\infty$ is a complete system of solutions of (16.20) in V in the sense of the uniform convergence on compact subsets.*

Furthermore, the results on the completeness of $\{u_m\}_{m=0}^\infty$ with respect to other norms are available (see [55]).

It is worth emphasizing that in order to compute the system of solutions $\{u_m\}_{m=0}^\infty$, it is not necessary to dispose of the transmutation operator that in fact was needed for proving its nice properties. The computation of $\{u_m\}_{m=0}^\infty$ is reduced to a simple recursive integration procedure from Sect. 7.2.

The case of the elliptic equation with a coefficient in a separable form in Cartesian coordinates is only an example of the equation admitting such a successful application of the transmutation operators and related formal powers technique for constructing complete systems of solutions. Another interesting example is the spherically symmetric Schrödinger equation (see [54, 55]). Many other equations admitting special symmetries can be solved in a similar way.

Moreover, of course, the harmonic polynomials do not exhaust all complete systems of solutions of equations with constant coefficients that can be quite easily transmuted into complete systems of solutions of equations with variable coefficients. The systems of solutions of the Helmholtz equation from Sect. 15.1 can also be transformed into solutions of more complicated equations with the aid of the transmutation operators technique.

Bibliography

1. M.J. Ablowitz, D.J. Kaup, A.C. Newell, H. Segur, The inverse scattering transform - Fourier analysis for nonlinear problems. Studies Appl. Math. **53**, 249–315 (1974)
2. M. Abramovitz, I.A. Stegun, *Handbook of Mathematical Functions* (Dover, New York, 1972)
3. N.I. Akhiezer, I.M. Glazman, *Theory of Linear Operators in Hilbert Space* (Dover, New York, 1993)
4. T. Aktosun, P. Sacks, Potential splitting and numerical solution of the inverse scattering problem on the line. Math. Methods Appl. Sci. **25**(4), 347–355 (2002)
5. M.A. Alexidze, *Fundamental Functions in Approximate Solutions of Boundary Value Problems* (Nauka, Moscow, 1991) (in Russian)
6. F.V. Atkinson, *Discrete and Continuous Boundary Problems* (Academic, New York, 1964)
7. S. Axler, P. Bourdon, W. Ramey, *Harmonic Function Theory*. GTM, vol. 137 (Springer, New York, 2013)
8. V.M. Babich, M.B. Kapilevich, S.G. Mikhlin, G.I. Natanson, P.M. Riz, L.N. Slobodeckii, M.M. Smirnov, *The Linear Equations of Mathematical Physics*, ed. by S.G. Mikhlin (Nauka, Moscow, 1964) (in Russian)
9. H. Begehr, R.P. Gilbert, *Transformations, Transmutations, and Kernel Functions*, vol. 1 (Longman Scientific & Technical, Harlow, 1992)
10. R. Bellman, *Perturbation Techniques in Mathematics, Engineering and Physics* (Dover Publications, New York, 2003)
11. S. Bergman, *Integral Operators in the Theory of Linear Partial Differential Equations* (Springer, Berlin, 1961, 2. Rev. Print., 1969)
12. S. Bergman, M. Schiffer, *Kernel Functions and Elliptic Differential Equations in Mathematical Physics* (Academic, New York, 1953)
13. W.E. Boyce, R.C. DiPrima, *Elementary Differential Equations and Boundary Value Problems* (Wiley, New York, 2009)
14. R. Camporesi, A.J. Di Scala, A generalization of a theorem of Mammana. Colloq. Math. **122**(2), 215–223 (2011)
15. H. Campos, R. Castillo-Perez, V.V. Kravchenko, Construction and application of Bergman-type reproducing kernels for boundary and eigenvalue problems in the plane. Complex Variables Elliptic Equ. **57**, 787–824 (2012)
16. H. Campos, V.V. Kravchenko, L.M. Mendez, Complete families of solutions for the Dirac equation: an application of bicomplex pseudoanalytic function theory and transmutation operators. Adv. Appl. Clifford Algeb. **22**, 577–594 (2012)

17. H. Campos, V.V. Kravchenko, S.M. Torba, Transmutations, L-bases and complete families of solutions of the stationary Schrödinger equation in the plane. J. Math. Anal. Appl. **389**(2), 1222–1238 (2012)
18. R. Castillo-Perez, V.V. Kravchenko, S.M. Torba, Spectral parameter power series for perturbed Bessel equations. Appl. Math. Comput. **220**(1), 676–694 (2013)
19. R. Castillo-Perez, V.V. Kravchenko, S.M. Torba, Analysis of graded-index optical fibers by the spectral parameter power series method. J. Optics **17**, 025607 (9pp) (2015)
20. Kh. Chadan, P.C. Sabatier, *Inverse Problems in Quantum Scattering Theory* (Springer, New York, 1989)
21. B. Chanane, Eigenvalues of Sturm-Liouville problems using Fliess series. Appl. Analy. **69**, 233–238 (1998)
22. E.A. Coddington, N. Levinson, *Theory of Ordinary Differential Equations* (Tata McGraw-Hill, New Delhi, 1987)
23. D.L. Colton, *Solution of Boundary Value Problem by the Method of Integral Operator* (Pitman Publishing, London, 1976)
24. D. Colton, R. Kress, *Integral Equations Methods in Scattering Theory* (Wiley, New York, 1983)
25. B.B. Delgado, K.V. Khmelnytskaya, V.V. Kravchenko, The transmutation operator method for efficient solution of the inverse Sturm-Liouville problem on a half-line. Math. Methods Appl. Sci. **42**(18), 7359–7366 (2019)
26. B.B. Delgado, K.V. Khmelnytskaya, V.V. Kravchenko, A representation for Jost solutions and an efficient method for solving the spectral problem on the half line. Math. Methods Appl. Sci. **43**, 9304–9319 (2020)
27. J. Delsarte, J.L. Lions, Transmutations d'opérateurs différentiels dans le domaine complexe. Comment. Math. Helv. **32**, 113–128 (1956)
28. A. Doicu, Yu. Eremin, Th. Wriedt, *Acoustic and Electromagnetic Scattering Analysis* (Academic, London, 2000)
29. G. Fairweather, A. Karageorghis, The method of fundamental solutions for elliptic boundary value problems. Adv. Comp. Math. **9**, 69–95 (1998)
30. G. Freiling, V. Yurko, *Inverse Sturm-Liouville Problems and Their Applications* (Nova Science Publishers, Huntington, 2001)
31. P.R. Garabedian, *Partial Differential Equations* (Chelsea Publishing, New York, 1986)
32. I.M. Gelfand, B.M. Levitan, On the determination of a differential equation from its spectral function. Izvestiya AN SSSR, Ser. matem. **15**(4), 309–360 (1951)
33. I. Gradshteyn, I. Ryzhik, *Table of Integrals, Series, and Products* (Academic, New York, 1980)
34. Ph. Hartman, *Ordinary Differential Equations* (Wiley, New York, 1964)
35. A.S. Ilyinski, V.V. Kravtsov, A.G. Sveshnikov, *Mathematical Models of Electrodynamics* (Vysshaya Shkola, Moscow, 1991) (in Russian)
36. S.I. Kabanikhin, *Inverse and Ill-posed Problems: Theory and Applications* (De Gruyter, Berlin, 2012)
37. A.N. Karapetyants, K.V. Khmelnytskaya, V.V. Kravchenko, A practical method for solving the inverse quantum scattering problem on a half line. J. Phys. Confer. Ser. **1540**, 012007, 7 pp. (2020)
38. K.V. Khmelnytskaya, V.V. Kravchenko, V.S. Rabinovich, Quaternionic fundamental solutions for electromagnetic scattering problems and application. Zeitschrift für Analysis und ihre Anwendungen **22**(1), 147–166 (2003)
39. K.V. Khmelnytskaya, V.V. Kravchenko, J.A. Baldenebro-Obeso, Spectral parameter power series for fourth-order Sturm-Liouville problems. Appl. Math. Comput. **219**(8), 3610–3624 (2012)
40. K.V. Khmelnytskaya, V.V. Kravchenko, H.C. Rosu, Eigenvalue problems, spectral parameter power series, and modern applications. Math. Methods Appl. Sci. **38**, 1945–1969 (2015)
41. K.V. Khmelnytskaya, V.V. Kravchenko, S.M. Torba, A representation of the transmutation kernels for the Schrödinger operator in terms of eigenfunctions and applications. Appl. Math. Comput. **353**, 274–281 (2019)

42. A.N. Kolmogorov, S.V. Fomin, *Elements of the Theory of Functions and Functional Analysis* (Dover, New York, 1957)
43. N.S. Koshlyakov, M.M. Smirnov, E.B. Gliner, *Differential Equations of Mathematical Physics* (North-Holland Publishing/Interscience Publishers Wiley, Amsterdam/New York, 1964)
44. A. Kostenko, G. Teschl, On the singular Weyl–Titchmarsh function of perturbed spherical Schrödinger operators. J. Differ. Equ. **250**, 3701–3739 (2011)
45. V.V. Kravchenko, On a method for solving the inverse Sturm–Liouville problem. J. Inverse Ill-posed Problems **27**, 401–407 (2019)
46. V.V. Kravchenko, On a method for solving the inverse scattering problem on the line. Math. Methods Appl. Sci. **42**, 1321–1327 (2019)
47. V.V. Kravchenko, *Direct and Inverse Sturm-Liouville Problems: A Method of Solution* (Birkhäuser, Cham, 2020)
48. V.V. Kravchenko, R.M. Porter, Spectral parameter power series for Sturm-Liouville problems. Math. Methods Appl. Sci. **33**, 459–468 (2010)
49. V.V. Kravchenko, S. Torba, Modified spectral parameter power series representations for solutions of Sturm-Liouville equations and their applications. Appl. Math. Comput. **238**, 82–105 (2014)
50. V.V. Kravchenko, S.M. Torba, Construction of transmutation operators and hyperbolic pseudo-analytic functions. Complex Anal. Oper. Theory, **9**, 389–429 (2015)
51. V.V. Kravchenko, S.M. Torba, Analytic approximation of transmutation operators and applications to highly accurate solution of spectral problems. J. Comput. Appl. Math. **275**, 1–26 (2015)
52. V.V. Kravchenko, S.M. Torba, A direct method for solving inverse Sturm-Liouville problems. Inverse Probl. **37**, 015015 (2021)
53. V.V. Kravchenko, S.M. Torba, A practical method for recovering Sturm-Liouville problems from the Weyl function. Inverse Probl. **37**, 065011 (2021)
54. V.V. Kravchenko, V.A. Vicente-Benitez, Transmutation operators and complete systems of solutions for the radial Schrödinger equation. Math. Methods Appl. Sci. **43**, 9455–9486 (2020)
55. V.V. Kravchenko, V.A. Vicente-Benitez, Runge property and approximation by complete systems of solutions for strongly elliptic equations. Complex Var. Elliptic Equ. **67**, 661–682 (2022)
56. V.V. Kravchenko, L.J. Navarro, S.M. Torba, Representation of solutions to the one-dimensional Schrödinger equation in terms of Neumann series of Bessel functions. Appl. Math. Comput. **314**(1), 173–192 (2017)
57. V.V. Kravchenko, J.A. Otero, S.M. Torba, Analytic approximation of solutions of parabolic partial differential equations with variable coefficients. Adv. Math. Phys. **2017**, Article ID 2947275, 5 pp. (2017)
58. I.V. Kravchenko, V.V. Kravchenko, S.M. Torba, Solution of parabolic free boundary problems using transmuted heat polynomials. Math. Methods Appl. Sci. **42**(15), 5094–5105 (2019)
59. V.V. Kravchenko, R.M. Porter, S.M. Torba, Spectral parameter power series for arbitrary order linear differential equations. Math. Methods Appl. Sci. **42**(15), 4902–4908 (2019)
60. V.V. Kravchenko, E.L. Shishkina, S.M. Torba, A transmutation operator method for solving the inverse quantum scattering problem. Inverse Probl. **36**, 125007 (23pp) (2020)
61. R. Kress, G.F. Roach, Transmission problems for the Helmholtz equation. J. Math. Phys. **19**, 1433–1437 (1978)
62. V.D. Kupradze, On the approximate solution of problems of mathematical physics. Russian Math. Surveys **22**, 59–107 (1967)
63. V.D. Kupradze, M.A. Alexidze, On an approximate method for solving boundary value problems. Soobshcheniya Akademii Nauk GSSR **30**(5), 529–536 (1963) (in Russian)
64. A.F. Leontiev, *Generalizations of Exponential Series* (Nauka, Moscow, 1981) (in Russian)
65. B.Ya. Levin, Fourier and Laplace type transforms by means of solutions to the second order differential equations. Dokl. AN SSSR [Rep. Acad. Sci. USSR] **106**(2), 187–190 (1956) (in Russian)
66. B.M. Levitan, *Inverse Sturm-Liouville Problems* (VSP, Zeist, 1987)

67. B.M. Levitan, I.S. Sargsjan, *Sturm-Liouville and Dirac Operators* (Kluwer Academic Publishers, Dordrecht, 1991)
68. V.A. Marchenko, Some questions on one-dimensional linear second order differential operators. Trans. Moscow Math. Soc. **1**, 327–420 (1952)
69. V.A. Marchenko, *Sturm-Liouville Operators and Applications: Revised Edition* (AMS Chelsea Publishing, Providence, 2011)
70. N.M. Matveev, *Methods of Integration of Ordinary Differential Equations* (Vysshaya Shkola, Moscow, 1967) (in Russian)
71. S.G. Mikhlin, *Lectures on Linear Integral Equations* (Fizmatlit, Moscow, 1959) (in Russian)
72. Y. Pinchover, J. Rubinstein, *An Introduction to Partial Differential Equations* (Cambridge University Press, Cambridge, 2005)
73. J. Poschel, E. Trubowitz, *Inverse Spectral Theory* (Academic, London, 1987)
74. A.P. Prudnikov, Yu.A. Brychkov, O.I. Marichev, *Integrals and Series. Vol. 2. Special Functions* (Gordon & Breach Science Publishers, New York, 1986)
75. J.D. Pryce, *Numerical Solution of Sturm-Liouville Problems* (Clarendon Press, Oxford, 1993)
76. A.G. Ramm, *Inverse Problems: Mathematical and Analytical Techniques with Applications to Engineering* (Springer, Boston, 2005)
77. W. Rudin, *Real and Complex Analysis* (McGraw-Hill, Ljubljana, 1986)
78. K.B. Sabitov, *Equations of Mathematical Physics* (Fizmatlit, Moscow, 2013) (in Russian)
79. J.K. Shaw, *Mathematical Principles of Optical Fiber Communications* (SIAM, Philadelphia, 2004)
80. S.M. Sitnik, E.L. Shishkina, *Transmutations, Singular and Fractional Differential Equations with Applications to Mathematical Physics* (Elsevier, Amsterdam, 2020)
81. P.K. Suetin, *Classical Orthogonal Polynomials*, 3rd edn. (Fizmatlit, Moscow, 2005) (in Russian)
82. A.N. Tikhonov, A.A. Samarskii, *Equations of Mathematical Physics* (Dover Publications, New York, 1990)
83. V.S. Vladimirov, *Equations of Mathematical Physics* (Dekker, New York, 1971)
84. G.N. Watson, *A Treatise on the Theory of Bessel Functions*, 2nd edn. reprinted (Cambridge University Press, Cambridge, 1996)
85. H. Weyl Über gewöhnliche, Differentialgleichungen mit Singularitäten und die zugehörigen Entwicklungen willkürlicher Funktionen. Math. Ann. **68**(2), 220–269 (1910) (in German)
86. V.A. Yurko, *Introduction to the Theory of Inverse Spectral Problems* (Fizmatlit, Moscow, 2007) (in Russian)
87. D. Zwillinger, *Handbook of Differential Equations* (Academic, San Diego, 1997)

Index

© The Author(s), under exclusive license to Springer Nature Switzerland AG 2022
A. N. Karapetyants, V. V. Kravchenko, *Methods of Mathematical Physics*,
https://doi.org/10.1007/978-3-031-17845-0